Methods and Techniques in Clinical Chemistry

STANFORD SERIES ON METHODS AND TECHNIQUES
IN THE CLINICAL LABORATORY

Edited by Paul L. Wolf, M.D.,
Director of Clinical Laboratory,
Clinical Pathology and School of Medical Technology,
Stanford University Medical Center,
Stanford, California

METHODS AND TECHNIQUES IN CLINICAL CHEMISTRY
by Paul L. Wolf, Dorothy Williams,
Tashiko Tsudaka, and Leticia Acosta

CLINICAL HEMATOLOGY PROCEDURES: INTERPRETATIONS AND TECHNIQUES
by Paul L. Wolf, Elisabeth Von der Muehll,
Irma Mills, and Patricia Ferguson

METHODS, TECHNIQUES, AND INTERPRETATIONS OF BLOOD TRANSFUSION
by Paul L. Wolf, Betsy Hafleigh, and
Gabriel Korn

Methods and Techniques in Clinical Chemistry

Paul L. Wolf, M.D.
Director of Clinical Laboratory,
Clinical Pathology and
School of Medical Technology
Associate Professor of Pathology

Dorothy Williams, M.T. (ASCP)
Chief Technologist
Clinical Chemistry Laboratory

Tashiko Tsudaka, M.T. (ASCP)
Supervisor
Clinical Chemistry Laboratory

Leticia Acosta, M.S.
Research Chemist
Clinical Chemistry Laboratory

Stanford University Medical Center
Stanford, California

WILEY-INTERSCIENCE

a division of John Wiley & Sons, Inc.

New York London Sydney Toronto

LIBRARY OF CONGRESS CATALOGING IN PUBLICATION DATA:

Main entry under title:

Methods and techniques in clinical chemistry.

1. Medicine, Clinical. 2. Chemistry, Clinical
I. Wolf, Paul L.

RB40.M37 616.07'5 78-39685
ISBN 0-471-95900-6

Printed in the United States of America

10 9 8 7 6 5 4 3 2

Dedication

To all of our devoted teachers and especially to the residents, medical students, and medical technology students from whom we receive continual inspiration and knowledge.

Preface

This book is intended as a review and reference text book for Clinical Pathologists and Medical Technologists who are concerned with daily problems in diagnostic Clinical Chemistry. We have cited a majority of the past and current technical information relevant to the determination of common Clinical Laboratory Chemistry procedures and their diagnostic interpretation. A great proliferation of knowledge has occurred in the last fifteen years in this field. The emphasis of this book is on the practical aspects of Clinical Chemistry with a tendency to refer to only a minimal amount of esoteric clinically irrelevant information. We have attempted to cover each Clinical Chemistry determination performed in the Stanford Clinical Laboratory completely from a technical and clinical viewpoint. The book is organized such that, the chemical determinations are listed alphabetically to enable the reader to quickly obtain the needed information. All of the important diagnostic tests pertaining to carbohydrate, lipid, protein, enzyme, electrolyte or hormone metabolism are included in this book.

We thus hope that this book will provide the necessary current knowledge required by Clinical Pathologists, Medical Technologists, and Medical Technology Students for optimal care of their patients.

The Authors gratefully acknowledge the cooperation of Anne Davis, M.T. (ASCP), Jean Yanagihara, M.T. (ASCP), Dee Evans, M.T. (ASCP), Janice Gebhardt, M.T. (ASCP), Elsie Sandner, M.T. (ASCP), Carol Glenn, M.T. (ASCP), Suzanne Gibian, M.T. (ASCP), and others of our technical staff for their valuable contribution in compiling information and helping to revise these techniques.

We also gratefully acknowledge the excellent secretarial work of Miss Elisabeth Von der Muehll, M.T. (ASCP), Supervisor School of Medical Technology, who typed and edited the entire manuscript.

THE AUTHORS

Contents

Contents (Continued):

Contents (Continued):

Contents (Continued):

Contents (Continued):

ACID PHOSPHATASE DETERMINATION
(Bessey-Lowry)

PRINCIPLE: Acid phosphatases are found in liver, muscle, spleen, and prostatic tissue as well as in erythrocytes. They may be isolated by hydrolysis at pH 4.8 - 4.9. The enzyme hydrolyzes p-nitrophenyl phosphate to produce p-nitrophenol and phosphate.

$$\text{p-Nitrophenyl phosphate} + H_2O \xrightarrow{\text{(p'tase)}} \text{p-nitrophenol} + \text{phosphate}$$

The citric acid buffered substrate is incubated with serum for 30 minutes. Alkali is added, raising the pH and stopping enzyme action, and diluting up the end product to a convenient concentration. Because p-nitrophenol is an indicator which is yellow in alkaline solutions, blanks are prepared to correct for serum color and turbidity.

The absorbance of p-nitrophenol is measured at 405 nm. The maximum absorbance occurs at 400 nm., but at this wavelength the interference from unhydrolyzed reagent is considerably greater than at 405 nm.

Enzyme activity is expressed in Bessey-Lowry units. One Bessey-Lowry unit is equal to that amount of phosphatase which will liberate 1 millimole of p-nitrophenol per hour per liter of serum at 37°C.

SPECIMEN: 0.4 ml. of unhemolyzed, non-fasting serum. Do not use hemolyzed serum. Serum acid phosphatase is very unstable at room temperature, demonstrating considerable loss of activity within 1 hour at room temperature. Serum or plasma separated rapidly and frozen until analysis will not lose appreciable activity. Specimens should be kept covered at all times due to pH change.

REAGENTS AND EQUIPMENT:

1. Water bath at 37°C. constant temperature.

2. <u>Stock p-Nitrophenol Standard Solution</u>, 10 mmoles/Liter
   Sigma Reagent #104-1 (or):
   In a 1000 ml. volumetric flask dissolve 1.3911 gm. p-nitrophenol in distilled water and bring to volume. Stable for one year at 4°C.

3. <u>Working p-Nitrophenol Solution</u>

Acid Phosphatase Determination (Continued):

Into a 100 ml. volumetric flask pipette 0.5 ml. stock p-nitrophenol standard and bring to volume with 0.02 N. NaOH. Mix thoroughly. Stable for one day.

4. Acid Buffer Reagent
Sigma Reagent #104-4 (or):
In a 1000 ml. volumetric flask, dissolve 18.907 gm. citric acid in 180 ml. 0.1 N. NaOH and 100 ml. 0.1 N. HCl. Bring to volume with distilled water. Add a few drops of chloroform. With the aid of a pH meter, check the pH of the buffer; this should be 4.8. Stable when stored at refrigerator temperatures.

5. p-nitrophenylphosphate, 0.04 gm./ml.
In a 25 ml. volumetric flask, dissolve 0.1 gm. p-nitrophenyl phosphate in distilled water and bring to volume. The dry reagent is also available in pre-weighed capsules (Sigma) containing 0.1 gm. p-nitrophenyl phosphate; these are stable for 1 year in the freezer.

6. NaOH, 0.2 N.

7. NaOH, 0.02 N.

8. NaOH, 0.1 N.

9. Working Buffered Substrate
Just prior to use mix equal parts of the acid buffer and the p-nitrophenyl phosphate substrate sufficient for a day's determinations.

10. Calibration Curve
Pipette the solutions indicated in the following chart into six clean tubes, in duplicate. Mix the contents of each tube thoroughly.

	1	2	3	4	5	6
Ml. Working Standard	1.00	2.00	4.00	6.00	8.00	10.00
Ml. 0.02 N. NaOH	10.10	9.10	7.10	5.10	3.10	1.10
Equivalent Bessey-Lowry Units	0.28	0.56	1.12	1.67	2.23	2.80

Acid Phosphatase Determination (Continued):

Immediately read the absorbance of each tube and its duplicate, and determine the average at 405 nm. with 0.02 N. NaOH as a reference solution. Plot A. against equivalent units on graph paper.

PROCEDURE:

1. Label three clean test tubes as follows:
   Reagent Blank, Control, Specimen.
   Pipette 1.0 ml. Working Buffered Substrate into each.

2. Place in 37°C. water bath. Allow 5 minutes for temperature equilibration.

3. Pipette 0.2 ml. distilled water to tube labelled Reagent Blank. Mix by gentle lateral shaking. Replace in water bath. Start stop watch.

4. At 30 second intervals, add 0.2 ml. sera to control tube and specimen tubes.

5. After exactly 30 minutes incubation, add 4.0 ml. 0.1 N. NaOH. Cap with parafilm, and mix well by inversion.

6. Record the A. of the tests using the reagent blank as reference at 405 nm. on Gilford 300 N flow cell.

7. Since the yellow color of the serum also absorbs at 405 nm., prepare another set of tubes with 0.2 ml. serum plus 5.0 ml. 0.1 N. NaOH. Mix well by inversion. Read these serum blanks against 0.1 N. NaOH as reference. Record absorbance.

8. To calculate, subtract the Absorbance of the serum blank from the Absorbance of the incubated test. Obtain B-L units from the calibration curve.

9. Sera with activity greater than that of the calibration chart must be repeated on dilution.

Acid Phosphatase Determination (Continued):

NORMAL VALUES: 0.13 - 0.64 Bessey-Lowry units.

REFERENCES:

1. Lowry, et. al.: J. Biol. Chem., 20:207, 1954.

2. O'Brien & Ibbott: LABORATORY MANUAL OF PEDIATRIC MICRO CHEMISTRY, Ed. 3, pg. 245.

3. Sigma Chemical Company: Technical Bulletin #104, "Determination of Alkaline and Acid Phosphatases".

4. STANDARD METHODS OF CLINICAL CHEMISTRY, Vol. 5, pg. 2 and 11.

5. Henry, R. B.: CLINICAL CHEMISTRY: PRINCIPLES AND TECHNIQUES, Hoeber, pg. 482, 1964.

CLINICAL INTERPRETATION:

Serum acid phosphatase is a useful procedure which is primarily of importance in the assessment of metastatic carcinoma of the prostate. At times, approximately 40% of patients with metastatic carcinoma of the prostate will not have an elevation of serum acid phosphatase. Slight increase in the serum acid phosphatase after treatment may be of diagnostic value. 90% of patients had this fraction elevated when there was metastases to distinct sites. Recent reports have challenged the specificity of the tartrate-inhibited acid phosphatase for metastatic carcinoma of the prostate. It is important to follow the effectiveness of therapy in patients who have carcinoma of the prostate by serial determinations of the serum acid phosphatase. When orchiectomy is performed and the serum acid phosphatase has been elevated prior to the orchiectomy, there should be a marked decrease in the serum acid phosphatase approximately 4 to 6 weeks after the orchiectomy. Estrogen therapy will also indice a decrease in elevated serum acid phosphatase. If acid phosphatase levels remain elevated following castration and/or estrogen therapy, the therapy has not been successful. If therapy is successful and there is a relapse of the carcinoma, then acid phosphatase may rise again in the serum. When there is hyperplasia of the prostate gland, simple manipulation of the gland may cause an elevation of the serum acid phosphatase. Elevation of the serum acid phosphatase following a rectal examination was more frequent in patients with cancer of the prostate than in controls without malignancies. Rectal examination and palpation of the prostate gland which is non-diseased

Acid Phosphatase Determination (Continued):

may produce a slight increase in the tartrate-inhibited fraction of the acid phosphatase but this is slight. Rarely infarction of the prostate has been associated with a rise in the acid phosphatase. In addition, trauma to the prostate either by surgery or catheterization may produce transient elevation of the acid phosphatase. Other diseases may cause a rise in the serum acid phosphatase. Acid phosphatase may be elevated when there is prominent destruction of blood platelets. Thus, patients with acute thrombocytopenic purpura have elevated phosphatase levels. The determination of acid phosphatase in patients with thrombocytopenia may be a method to differentiate thrombocytopenia. Those patients exhibiting thrombocytopenia with high acid phosphatase have been found to have normal or elevated numbers of megakaryocytes in the bone marrow. In contrast, those patients with thrombocytopenia with diminished numbers of megakaryocytes in the bone marrow will not demonstrate an elevated serum acid phosphatase. Another clinical situation where acid phosphatase is elevated is in Gaucher's disease. In this disease, there is a cerebrosidase deficiency with the presence of mononuclear cells in the reticuloendothelial tissue containing cerebroside. The acid phosphatase which is present in the serum is not inhibited by L-tartrate or formaldehyde. The elevated serum enzyme most likely arises from the Gaucher cells. Another disease in which acid phosphatase is elevated is when there is necrosis of tissue following thromboembolism. The acid phosphatase in this situation may be derived from necrotic tissue such as necrotic myocardial cells in the myocardial infarct. It may also arise from the breakdown of red cells and platelets in the area of infarction. Patients with hemolytic anemia may show elevations in the serum acid phosphatase. Red cells contain acid phosphatase and thus when there is hemolysis, there is liberation from the hemolyzed red cells. Various types of bone disease, such as Paget's disease may be associated with an elevated serum acid phosphatase, especially if the disease is polyostotic. In addition, patients with extensive bone disease associated with multiple myeloma may present with an elevated serum acid phosphatase. Rarely acid phosphatase may be elevated in patients with liver or kidney disease. The acid phosphatase may be liberated from necrotic renal tubules, and if there is oliguria or anuria, there will be inability to excrete the enzyme. In liver disease such as viral hepatitis or cirrhosis, there are a few reports of elevated serum acid phosphatase. In the liver problem, the acid phosphatase is most likely being liberated from the liver cells.

PROSTATIC ACID PHOSPHATASE
(Tartrate Inhibition)

PRINCIPLE: In healthy men and women, the serum manifests only slight acid phosphatase activity, much of which is due to enzyme liberated by platelets in the clotting process. The liver, bone, spleen, kidney, platelets, and erythrocytes all exhibit acid phosphatase activity, but to a lesser extent than the prostate. It is occasionally desirable to distinguish between the amount of serum acid phosphatase contributed by the prostate gland and that which is of non-prostatic origin. Since a substrate which is specific for prostatic acid phosphatase has not been successfully achieved, the approach has been to utilize prostatic enzyme inhibitors.

The inhibitor employed in this methodology is L (+) tartrate. A total serum acid phosphatase determination is performed on a given specimen, as well as a determination which includes tartrate. The difference between the two activities is presumably due to acid phosphatase of prostatic origin.

SPECIMEN: 0.2 ml. of unhemolyzed serum. Serum acid phsophatase is unstable at room temperature; refrigerate quickly or freeze, if test is not to be run that day.

REAGENTS: All reagents used are the same as those for the total "Acid Phosphatase Determination" except for those indicated beneath.

1. Tartrate Acid Buffer, 0.04 M., pH 4.8
   Tartrate is in 0.09 M. Citrate and contains chloroform as a preservative. Store at 0 - 5°C. Available as Sigma #104-12.

2. Working Tartrate Substrate
   Prepare a 1:1 dilution of acid phosphatase substrate plus tartrate acid buffer.

PROCEDURE:

1. Allow two aliquots of serum from the same specimen. On one aliquot determine total acid phosphatase. On the second aliquot determine the activity with tartrate inhibition, as indicated beneath. The two determinations may be run simultaneously.

2. Prepare test tubes for reagent blank, control, and specimens. Pipette 1.0 ml. of working tartrate substrate into each tube and place in a water bath (37°C.) for 5 minutes.

Prostatic Acid Phosphatase (Continued):

3. Pipette 0.2 ml. of water into the test tube for the reagent blank and time with stopwatch. At exactly 30 second intervals deliver 0.2 ml. of control sera and specimens into their respective tubes. Mix each tube after addition, and incubate all for exactly 30 minutes at 37°C.

4. After exactly 30 minutes remove the test tubes from the water bath and add 4.0 ml. of 0.1 N. NaOH. Mix well by inversion against parafilm.

5. Determine the absorbance of each specimen against the reagent blank at 405 nm. in a Gilford 300N flow cell.

6. CALCULATION:

   Substract the prostatic test A from the A of the uncorrected total test. Determine units from the calibration chart.

NORMAL VALUES:

Male:	0.01 - 0.15 Sigma Units
Borderline:	0.15 - 0.20 Sigma Units

REFERENCES:

1. Fishman and Lerner: Journal of Biological Chemistry, 200:89, 1953.

2. Ozar and Issac: Journal of Urology, 74:150, 1955.

3. Andersch and Szezypinski: Am. J. Clinical Path., 17:571, 1947.

4. Batsakis and Brierre: INTERPRETIVE ENZYMOLOGY, C. Thomas, 1967.

5. Murphy, G., et al: Cancer, 23:1309, 1969.

CLINICAL INTERPRETATION:

Acid phosphatase of the prostate also liver and spleen is inhibited by L-tartrate, whereas that of other tissues is not. Measurements Of acid phosphatase activity before and after tartrate inhibition

Prostatic Acid Phosphatase (Continued):

yield values for specific "prostatic" phosphatase in normal subjects ranging from 0.0 - 0.5 units/100 ml. serum. Clinically significant increases of total serum acid phosphatase occur in only about 40 per cent of cases of metastatic prostatic cancer, whereas abnormally high values for "prostatic" acid phosphatase are obtained in 90 to 95 per cent of such cases. Observations in cases under treatment indicate that changes in values for total serum acid phosphatase activity are due virtually exclusively to changes in "prostatic" phosphatase, and that increase in the latter may be demonstrable before the former has risen to abnormally high levels. It is claimed that increase in serum "prostatic" phosphatase occurs in up to 80 per cent of cases of prostatic cancer without demonstrable metastasis; however, some believe that abnormal values almost invariably point to metastasis, even though this may not be demonstrable. Slight increase in the specific enzyme has been reported in up to 30 per cent of cases of benigh prostatic hypertrophy, and a transient rise (1 to 1.5 units) may occur following massage or palpation of a normal prostate. Spuriously low values may be obtained in the presence of fever.

100 per cent inhibition of red blood cell acid phosphatase occurs but there is no inhibition of prostatic enzyme with 0.5 per cent formaldehyde. Other acid phosphatases are inhibited 12 to 60 per cent. Theoretically, the two common major acid phosphatases of the blood could be distinguished in this manner without need of a subtraction step. The procedure has not found wide use since hepatic, renal, and splenic sources are still capable of increasing the non-formaldehyde inhibited (prostatic) fractions.

70 to 80 per cent inhibition of erythrocyte acid phosphatase is inhibited by ethyl alcohol under conditions giving 90 to 100 per cent inhibition of prostatic acid phosphatase. However, it is of value to note that hepatic and splenic acid phosphatases are not inhibited by ethyl alcohol and so absence of inhibition may have significance when used in conjunction with other methods.

SERUM ALBUMIN
(Ultramicro)

PRINCIPLE: Serum albumin has the property of reacting with anions when on the alkaline side of its isoelectric point. The anionic dye, bromcresol green (3,3', 5,5' - Tetrabromo-m-cresolsulfonphthalein) binds tightly and specifically to albumin under these circumstances. There is a large difference in spectral absorption between the bound and unbound dye at 615 nm. Elevated bilirubin levels do not interfere either due to competitive binding or spectral absorbance; there is no appreciable effect due to hemolysis or moderate amounts of turbidity. Human and bovine albumins display little difference in their binding capacities.

SPECIMEN: 20 microliters of serum.

REAGENTS AND EQUIPMENT:

1. Bromcresol Green Indicator
   Available in set of 50 tubes of lyophilized dye by American Monitor Albumin Kit #1006. Item requires refrigeration.

2. Buffered Detergent Diluent
   Sufficient quantity supplied with American Monitor Albumin Kit #1006. Reagent must be refrigerated.

3. Standard Albumin Solution
   Albumin Standards should be included in the determination; any of the following are satisfactory:
   a). Bovine Albumin, Armour Pharmaceuticals; approximately 6.0 gm%. Convert nitrogen assay to Gm%. protein by: 1.0 mg. of protein nitrogen/ml. = 6.25 mg. protein/ml.

   b). Human Albumin (Fraction V or Mercaptalbumin); approximately 10-11 gm%. Convert nitrogen assay to Gm%. as for bovine albumin. This must be diluted into a usable range of concentration.

   c). Serum Albumin Standard, American Monitor; approximately 5.0 gm%. Supplied with Albumin Kit #1006 and/or American Monitor Albumin and Total Protein Standards #1010.

4. Ultramicro Dilutor
   Sample syringe set to take up 20 microliters of sample and dispensing syringe set to deliver 5.0 ml. of Buffered Detergent Diluent.

Serum Albumin (Continued):

5. <u>Gilford 300 N Spectrophotometer</u>
Instrument set at 615 nm. and fitted with standard evacuation cuvette.

COMMENTS ON PROCEDURE:

1. Very lipemic (milky) sera will contribute to the absorbance of the solution. Determine the value of the specimen plus diluent dispensed into a clean tube; read against water at 615 nm. Subtract this absorbance from the absorbance of the specimen plue dye.

2. Specimens with values above the highest albumin standard must be repeated after quantitative dilution.

3. Color development is complete almost immediately and is stable for 1 hour.

PROCEDURE:

1. Prepare a reagent blank by dispensing 5.0 ml. of Buffered Detergent from the dilutor into a tube of lyophilized dye. Mix well by inversion.

2. Using the microdilutor, take up 20 microliters of standard, control or specimen and dispense with 5.0 ml. buffered detergent into a series of lyophilized dye tubes. Mix well by inversion.

3. Determine the absorbance of all samples at 615 nm. against the reagent blank using the Gilford 300 N with the standard evacuation cuvette. See "Comments on Procedure" for extremely lipemic or elevated specimens.

4. Standards are plotted on linear graph paper. Determine the control(s) and test(s) using the calibration graph. If standards have proven to be linear, the concentration of controls and specimens may be calculated according to the following formula;

$$\text{Gm\%. of specimen} = \frac{\text{Conc. of Std.}}{\text{A. of Std.}} \times \text{A. of specimen}$$

NORMAL VALUES:

1. <u>First Week</u> (Full-term)
Albumin = 3.3 - 5.1 Gm%.

Serum Albumin (Continued):

2. 12 Months (Full-term)
   Albumin = 4.1 - 5.0 Gm%.

3. Four Years +:
   Albumin = 3.7 - 5.5 Gm%.

REFERENCES:

1. Rodkey, F: Clinical Chemistry, 11:478, 1965.

2. Rodkey, F.: Archives of Biochemistry and Biophysics, 108:510, 1964.

3. Dow & Pinto: Clinical Chemistry, 15:1006, 1969.

4. Ibbott & O'Brien: LABORATORY MANUAL OF PEDIATRIC MICRO-BIOCHEMICAL TECHNIQUES, Hoeber. (Normal Values)

CLINICAL INTERPRETATION:

Elevated serum albumin is seldom encountered. Elevated serum albumin usually signifies that the patient is dehydrated.

The most common cause for a low serum albumin in a hospitalized patient is excessive intravenous infusion of glucose in water.

Liver cirrhosis results in hypoalbuminemia because of a decreased synthesis by the pathologic liver. Malnutrition may also serve as a major cause for low serum albumin.

Low serum albumin may result from loss or lack of absorption related to gastrointestinal disease such as sprue, ulcerative colitis, villous adenoma, or protein-losing enteropathy. Prominent proteinuria associated with the various causes for nephrosis results in marked hypoalbuminemia.

Finally a large amount of albumin occurs in diffuse bullous dermatitis or an extensive burn with loss of albumin from the skin lesion.

SPECTROPHOTOMETRIC ALDOLASE ASSAY
(Bruns U. V. Method)

PRINCIPLE: The measurement of serum aldolase (Fructose-1, 6-diphosphate-0-glyceraldehyde-3-phosphate lyase) activity is performed rapidly and conveniently by spectrophotometric procedure. This is accomplished by coupling the aldolase reaction with that of a dehydrogenase acting upon one of the triosephosphates formed after splitting FDP. The latter reaction is accomplished by changes in NADH concentration which are measured spectrophotometrically at 340 nm.

1. $\text{FDP} \xrightleftharpoons{\text{ALD}} \text{GAP} + \text{DAP}$

2. $\text{GAP} \xrightleftharpoons{\text{TIM}} \text{DAP}$

3. $\text{DAP} + \text{NADH} + \text{H}^+ \xrightarrow{\text{GDH-GI}} \alpha\text{-Glycerophosphate} + \text{NAD}^+$

The disappearance of NADH is proportional to aldolase. Two moles of NADH are oxidized per mole of FDP hydrolyzed. See comments on calculation.

SPECIMEN: Specimen should be free of hemolysis - stable at 4°C. up to about 5 days. Need 0.2 ml. serum.

COMMENTS ON PROCEDURE:

The following abbreviations are taken from Bergmeyer, Methods of Enzymatic Analysis, Academic Press, New York, 1965:ALD, Aldolase; DAP, dihydroxyacetonephosphate, FDP, Fructose-1, 6-diphosphate; GAP, glyceraldehydephosphate; GAPDH, glyceraldehydephosphate dehydrogenase; GDH glycerol-1-phosphate dehydrogenase; a-GP, glycerol-1-phosphate; PGA, 3-phosphoglyceric acid; TIM, triosephosphate isomerase.

REAGENTS AND EQUIPMENT:
Biochemica Test Combination Kit Tc-D (Cat. No. 15974 TAAD).

1. Buffer - 0.056 M. Collidine Buffer; pH 7.4; 0.003 M. Monoiodacitrate; 0.003 M. FDP.

2. 0.020 M. NADH

Spectrophotometric Aldolase Assay (Continued):

3. 2.0 mg. GDH-TIM/ml.

4. Dilute reagent No. 1 in 100 ml. distilled water. Dissolve reagent No. 2 in 2.0 ml. distilled water - add to reagent No. 1. Add reagent No. 3 to above. Rinse bottle. This makes the Working Substrate (enough for 30 tests). Good for 4 weeks refrigerated.

5. DBG or Gilford 300 N
Equipped with thermostated cuvette at 37°C. with recorder.

PROCEDURE:

1. Add 0.2 ml. (200 lambda) of serum to 2.8 ml. of prepared Working Substrate. Place in 37°C. water bath for 6 minutes.

2. Take first reading (A_1) at exactly 6 minutes.

3. Take second reading (A_2) at 16 minutes (10 minute time difference) at 340 nm.

   Method linear with A/10 minutes up to 0.500. If greater, dilute serum 1:10 with saline. Multiply units by 10.

4. CALCULATIONS:

$$\frac{\Delta A}{\epsilon \times d} \times 10^6 \times \frac{TV}{SV} \times \frac{1}{Time} = \text{I.U./Liter or mU/ml.}$$

ϵ of NADH at 340 nm. = 6.22×10^3 Liter/Mole x cm.

$$\frac{\Delta A}{6.22 \times 10^3 \times 1} \times 10^6 \times \frac{TV}{SV} \times \frac{1}{Time} \times \frac{1}{2}$$ Since 2 Moles NADH are oxidized per Mole FDP hydrolyzed.

$$\Delta A \times \frac{1}{6.22 \times 10^3 \times 1} \times 10^6 \times \frac{3.0}{0.2} \times \frac{1}{10} \times \frac{1}{2}$$

$\Delta A \times F$ = mU/ml.

F = 121

NORMAL VALUES: 4 - 14 mU/ml.

Spectrophotometric Aldolase Assay (Continued):

REFERENCES:

1. Pinto, V. D., Kaplan, A., and Van Dual, P.: Journal of Clin. Chem., Vol. 15, No. 5, May, 1969.

2. Bruns, F.: Biochemische Zeitschrift, Bd. 325, S. 156 - 162, 1954.

3. Bergmeyer: METHODS OF ENZYMATIC ANALYSIS, Academic Press, 1965.

CLINICAL INTERPRETATION:

Elevated aldolase: (1) skeletal muscle necrosis, (2) myocardial necrosis, (3) liver necrosis. The adult level is from two to fourteen milliunits per ml. The newborn has four times the level of aldolase as the adult, and children have twice the level of the adult. The determination of serum aldolase has been found to be extremely valuable in differentiating primary diseases of skeletal muscle, such as pseudohypertrophic muscular dystrophy from disorders in which there is a neurogenic atrophy. Aldolase is present in many organs but is in highest concentration in the skeletal muscle. Skeletal muscle contains the enzyme as does the liver. The red cell contains approximately 150 times a greater amount than is present in the plasma and large amounts have also been found in platelets. When platelets disrupt during clotting, very little is released into the serum because it is bound by the fibrin clot. The main value for determining serum aldolase is in the assessment of skeletal muscle disease. Aldolase is elevated in patients with muscular dystrophy. It is also evidently elevated in acute muscular necrosis. Aldolase is not elevated in patients with myasthenia gravis. It is not elevated in patients with hypothyroidism. The serum aldolase activity in the newborn is twice that found in the umbilical cord blood and four times the adult level. Various explanations have been offered for the high activity in the neonatal period. One group feels that the elevated aldolase levels are related to the increased activity of the adrenal gland where the enzyme is present. Increased aldolase in the blood in rabbits occurs after ACTH is given. Another group feels that aldolase is increased because of the increased requirements of glycolytic enzymes in the newborn in response to their extrauterine life. The increase in aldolase in the newborn thus is related to adrenal cortical activity and is also partially due to the increased physiologic requirements of carbohydrate metabolism present at birth.

Spectrophotometric Aldolase Assay (Continued):

The enzyme is elevated in patients who have viral hepatitis. Some authors believe that the measurement of aldolase is more useful than that of the measurement of other enzymes in the assessment of patients with viral hepatitis. The activity of the enzyme is elevated in patients with acute viral hepatitis and parallels the activity of SGPT.

ALKALINE PHOSPHATASE
(Bessey-Lowry: Continuous Spectrophotometric)

PRINCIPLE: The enzyme alkaline phosphatase hydrolyzes the substrate p-nitrophenyl phosphate to yield phosphoric acid and p-nitrophenol. With an excess of substrate and defined conditions as to pH, temperature, and buffer molarity, the rate of reaction is constant and is proportional to the concentration of the enzyme. The rate of reaction can be determined by measuring the change in absorbance at 405 nm. which is the absorbance maximum for the reaction product p-nitrophenol.

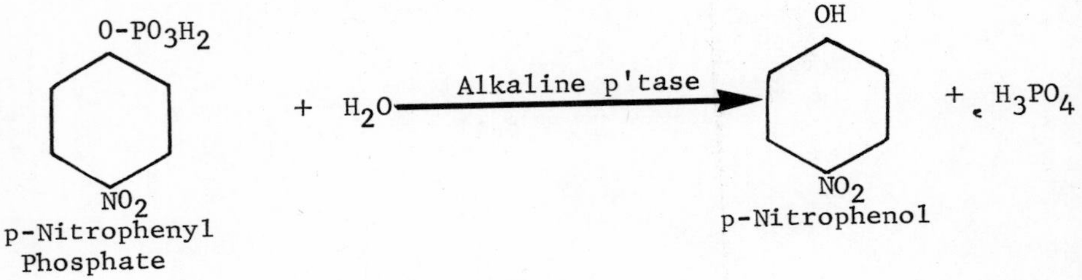

Elevated levels of serum alkaline phosphatase are usually due either to increased osteoblastic activity or to disease of the liver or bile ducts. The enzyme is normally elevated in pregnancy.

SPECIMEN: 25 microliters of serum. The anticoagulants EDTA and citrate may not be used, as they inhibit enzyme activity; heparin and oxalate - fluoride are not inhibitory for time periods up to 4 hours. Hemolysis does not interfere with the assay. Serum activity appears stable up to 16 months in frozen state. However, studies upon the stability at room or refrigerator temperatures have shown a 5 - 30% increase in enzyme activity within 24 hours, followed by decrease in activity within a few days.

REAGENTS AND EQUIPMENT:

1. 2-Amino-2-Methyl-1-Propanol Buffer, 0.625 M. pH 10.25
   This buffer is supplied as part of the SMA 12/60 Reagent System.

2. p-Nitrophenyl Phosphate
   The dry reagent is supplied as part of the SMA 12/60 Reagent System.

3. $MgCl_2$, 1.0 M.
   Solution is supplied as part of the SMA 12/60 Reagent System.

Alkaline Phosphatase, Bessey-Lowry Method (Continued):

4. Working Substrate
AMP buffer, 0.625 M., pH 10.25; p-Nitrophenyl phosphate, 5 mM.; $MgCl_2$ 2 mM. Pipette 1.0 ml. of $MgCl_2$ into 499 ml. of AMP buffer and mix. Transfer 1.0 gm. of p-Nitrophenyl phosphate into the above solution, and mix until the reagent is in solution. Prepare fresh daily. The 12/60 Working Substrate may be used.

5. Ultramicro Dilutor:
The sample syringe is set to take up 25 microliters of sample and the flush syringe is set to deliver 1.5 ml. of Working Substrate, (store dilutor filled with distilled water when not in use).

6. Spectrophotometer

A. Gilford 300 N, 405 nm.
This instrument is used when analyzing one sample at a time. The change in absorbance per minute is read from the Data Lister printout. The linearity of the reaction is visually monitored on the strip chart recorder. The instrument is used with the thermocuvette set at 30°C.

B. Gilford 222, 405 nm.
This instrument is used when analyzing four samples at a time. Temperature control is by means of a 30°C. circulating water bath. Rate of reaction is determined from the strip chart recorder set for 0.200 A. full scale and run at one inch per minute.

COMMENTS ON PROCEDURE:

1. Substrate exhaustion is rate-limiting; the reaction will be linear until this point.

2. Specimens with high enzyme activity may be monitored with the ratio switch of the Gilford 222 set at 0.5 or lower. The change in absorbance per minute must be multiplied by the appropriate factor.

3. One must be sure that the reaction mixture and cuvette chamber is at temperature before making a measurement.

Alkaline Phosphatase, Bessey-Lowry Method (Continued):

PROCEDURE:

1. With ultramicro dilutor, take up 25 microliters of serum, and flush with 1.5 ml. of Working Substrate. Mix well.

2. When using the Gilford 222, preincubate reaction mixture in 30°C. water bath to assure temperature equilibration. Preincubation is not necessary with the Gilford thermocuvette because the temperature equilibration and lag time are less than one minute.

3. Introduce reaction mixture into spectrophotometer, and record at least 2 minutes of linear reaction at 405 nm.

4. CALCULATION:

$$\frac{\Delta A}{\epsilon x d} \times 10^6 \times \frac{TV}{SV} \times \frac{1}{Time} = \text{IU/Liter or mU/ml.}$$

ϵ p-Nitrophenol at 405 nm. = 18.6×10^3 Liter/Mole x cm.

$$\frac{\Delta A}{18.6 \times 10^3 \times 1} \times 10^6 \times \frac{TV}{SV} \times \frac{1}{Time}$$

$$\Delta A \times \frac{1}{18.6 \times 10^3 \times 1} \times 10^6 \times \frac{1.525}{0.025} \times \frac{1}{1}$$

$\Delta A \times F$ = mU/ml.

F = 3279 or 3280

NORMAL VALUES: 30 - 85 mU/ml.

REFERENCES:

1. Bessey, O., Lowry, O., Brock, M.: "A Method for the Rapid Determination of Alkaline Phosphatase with Five Cubic Millimeters of Serum", J. Biol. Chem., 164:321, 1946.

2. Bowers and McComb: "A Continuous Spectrophotometric Method for Measuring the Activity of Serum Alkaline Phosphatase", Clin. Chem., 12:70-89, No. 2, 1966.

Alkaline Phosphatase, Bessey-Lowry Method (Continued):

3. Morgenstern, et. al: "An Automated p-Nitrophenyl Phosphate Serum Alkaline Phosphatase for the AutoAnalyzer", Clin. Chem., 11:876, 1965.

4. Henry, R. B.: CLINICAL CHEMISTRY: PRINCIPLES AND TECHNIQUES, Hoeber, pg. 491, 1965.

CLINICAL INTERPRETATION:

Elevated serum alkaline phosphatase is of two types, physiologic or pathologic.

Causes for physiologic elevation of serum alkaline phosphatase include:

1. Infant and children bone growth
2. Pregnancy

Conditions resulting in pathological elevations of serum alkaline phosphatase include:

1. Obstructive hepatobiliary tract disease or infiltrative lesions of liver
2. Osteoblastic lesions of bone
3. Primary or secondary hyperparathyroidism
4. Metastatic cancer or lymphoma lesion involving bone
5. Paget's disease
6. Healing bone fractures
7. Gastrointestinal lesions such as:
   - Peptic ulcer
   - Ulcerative colitis
8. Necrosis of lung
9. Neoplasms producing alkaline phosphatase isoenzyme of Regan
10. Intravenous use of plasma expander produced from human placenta
11. Acute infarction of the kidney
12. Acute infarction of the spleen

Causes for a low alkaline phosphatase include:

1. Hypophosphatasia
2. Magnesium deficiency
3. Cachexia
4. Pernicious anemia
5. Hypothyroidism

Alkaline Phosphatase, Bessey-Lowry Method (Continued):

6. Anticoagulants in sample:
   Fluorides
   Oxalates
   EDTA
7. High serum phosphate

<u>Cause For an Increase in Serum Alkaline Phosphatase</u>

Several etiologic factors result in elevations in serum alkaline phosphatase. The clinician usually considers the possibility of obstructive jaundice or the presence of a disease of bone as the primary causes for elevation in serum alkaline phosphatase. However, a number of other important diseases may contribute significantly to elevations in the serum alkaline phosphatase. Other causes for elevated alkaline phosphatase are neoplastic diseases in which a unique isoenzyme termed the Regan isoenzyme is produced by neoplastic cells, and gastrointestinal disease, such as ulcerative colitis in which the enzyme is liberated from the diseased colonic mucosa. Prominent increases in serum alkaline phosphatase may occur in patients receiving intravenous serum albumin. Commercial albumin may be prepared from blood derived from human placenta which contains alkaline phosphatase.

In considering elevations of alkaline phosphatase, one must always bear in mind the age of the patient since previously we have referred to normal elevations of alkaline phosphatase in growing children, and one must always remember that pregnant females in all trimesters of pregnancy normally have an elevated serum alkaline phosphatase.

Recently great interest has been present in rapid differentiation of both normal conditions and abnormal conditions which elevate serum alkaline phosphatase. Simple tests have been devised to quickly differentiate the isoenzymes of alkaline phosphatase in the Clinical Laboratory.

Liver alkaline phosphatase migrates with the serum beta globulins and beta lipoproteins. Kidney alkaline phosphatase migrates in the haptoglobin area. Bone alkaline migrates in the beta globulin zone. The intestinal alkaline phosphatase migrates in the beta globulin and beta lipoprotein zones. Patients with bone disease show alkaline phosphatase electrophoretic bands which migrate more slowly than the alkaline phosphatase derived from liver disease.

Alkaline Phosphatase, Bessey-Lowry Method (Continued):

Bone and kidney alkaline phosphatase are inhibited by urea. Placental and gastrointestinal alkaline phosphatase are not inhibited by urea. Liver alkaline phosphatase shows much inactivation by low urea concentrations less than 1.5 molar.

An important method to determine alkaline phosphatase isoenzymes is to incubate the enzyme at 56°C. for ten minutes. It has been found that the isoenzymes of alkaline phosphatase derived from liver and gastrointestinal sources are moderately heat stable, while alkaline phosphatase derived from the placenta is markedly heat stable, and in contrast, alkaline phosphatase derived from a bone source is heat labile. Another method for differentiating the isoenzymes of alkaline phosphatase is the utilization of phenylalanine to inhibit the action of alkaline phosphatase. Alkaline phosphatase from a placental source has been found to be both heat stable and inhibited by phenylalanine. The Regan isoenzyme which is produced by neoplastic cells such as bronchogenic carcinoma or other malignant lesions including lymphomas such as Hodgkin's disease. It is thought that these neoplastic cells produce this esoenzyme and secrete the enzyme into the blood stream which then causes the patient to present with an elevated serum alkaline phosphatase. The Regan isoenzyme is similar to the placental isoenzyme in that it is heat stable and inhibited by phenylalanine. Isoenzymes derived from a liver and bone source differ from intestinal, placental, and Regan isoenzyme in their reaction with phenylalanine. They are not inhibited by the action of phenylalanine.

When one is confronted with a bone disease in which an elevation of alkaline phosphatase is present, the pathophysiologic mechanism is injury to bone and in the reparative process the osteoblasts produce alkaline phosphatase which is secreted into the blood stream. Furthermore, an increase in serum alkaline phosphatase can signify the presence of other causes for osteomalacia. The outstanding causes for elevated alkaline phosphatase in bone lesions are hyperparathyroidism related to primary hyperplasia of the parathyroid glands or an adenoma, rickets, Paget's disease, metastatic carcinoma to bone from primary sources such as breast, lung, kidney, prostate, thyroid or primary malignancy of bone. In addition, trauma and fracture of the skeletal system with osteoblastic repair will give rise to alkaline phosphatase elevation. The determination of alkaline phosphatase in the work-up of a patient with vitamin D deficiency leading to rickets is one of the important aids in the assessment of this disease and an index to the severity of the disorder. Clinicians have noted that a progressive and rapid decline in elevated serum

Alkaline Phosphatase, Bessey-Lowry Method (Continued):

alkaline phosphatase is indicative of efficacy of therapy. In contrast, the serum alkaline phosphatase is consistently elevated when rickets is active. The enzyme begins to decline within several days after vitamin D therapy is instituted.

Extreme elevations of alkaline phosphatase are constantly seen in osteitis deformans (Paget's disease). The highest values for this enzyme have been seen in this condition compared with other bone diseases. Higher values are generally seen in younger individuals with Paget's disease, and patients who are elderly show slight elevations. The degree of elevation of the enzyme level can also be utilized in a prognostic way since polyostotic disease shows greater activity than monostotic. Slightly elevated levels are usually associated with quiescent stages of the disease.

Various other bone disorders cause elevation of the serum alkaline phosphatase. Metastatic carcinoma involving the bones and primary malignant neoplasms may elevate the serum alkaline phosphatase. In addition, the histiocytosis diseases and healing of bone fractures are associated with elevated alkaline phosphatase. In these conditions a response to appropriate therapy is usually associated with an initial rise in the enzymes shortly after therapy is begun which reflects the successful reparative process. Osteoporosis is usually associated with normal serum phosphatase.

<u>Hypophosphatasemia</u>

Several conditions will depress the level of serum alkaline phosphatase. The major cause for a low serum alkaline phosphatase is hypophosphatasia. In this condition, which is inherited, there is a marked reduction or absence in the alkaline phosphatase in myeloid leukocytes, the serum and tissues including bone. Clinically, the condition resembles rickets. The disease varies in severity. In patients who have a mild condition there may be survival into adult life. These individuals may have deformities of their bones with increased fragility resulting in frequent fractures. Hypercalcemia with calcinosis of the soft tissues may result. In contrast, the condition may be extremely severe at the time of birth and death may result shortly after birth or in early childhood. The diagnostic features in the blood are low leukocyte alkaline phosphatase and marked depression of the serum alkaline phosphatase. Pernicious anemia or hypothyroidism may also decrease serum alkaline phosphatase

AN ACCELERATED ION-EXCHANGE CHROMATOGRAPHIC PROCEDURE FOR THE DETERMINATION OF AMINO ACIDS IN PHYSIOLOGICAL FLUIDS

Klara Efron, M.D.

PRINCIPLE: A new Amino Acid Detection Program has begun at Stanford Clinical Laboratory, utilizing a new automatic Amino Acid Analyzer adjusted for clinical usage. Complete analysis of over 35 amino acids in blood plasma, urine, and cerebrospinal or amniotic fluids is accomplished in 6 hours compared with twice as long in other procedures. The system includes a single (33.5 x 0.9 cm.) column filled to a height of 27 cm. with UR-40 resin (Sondell Instruments, Palo Alto). The modified analytical procedure for physiological fluids represents one sequential run, using four different buffers for elution of the acidic, neutral, and basic amino acids. This reduces the analysis time, allows the use of one sample instead of two, and eliminates additional error. The high sensitivity (0.05 A full scale) permits the use of small samples (0.2 ml. of plasma ultrafiltrate for complete analysis). The preparation of the sample is also different from others. No picric or sulfosalicylic acid is added for protein precipitation. A special ultrafiltration system is used instead. The sample is placed into a small diameter dialysis tubing, and positive pressure (compressed nitrogen gas) is applied. The sample manifold is placed into a refrigerator at 4°C. and is connected to the nitrogen outlet of the instrument by means of a small diameter Teflon tubing. This process results in a clear solution containing the free amino acids. The rate of ultrafiltration is 1.0 ml. per one and one-half hours; 10 samples can be processed simultaneously.

SPECIMEN:

Collection of Blood Plasma

Collect 1.0 to 5.0 ml. of blood in a heparinized tube. Shake gently. Separate the plasma from the RBC by centrifugation at 2500 - 3000 rpm for 10 minutes at 4°C. Take off the supernatant and transfer into a chemically clean tube. Freeze immediately or start ultrafiltration.

Collection of Urine

Collect a 24 hour urine specimen (use a pinch of thymol as a preservative). The individual urine samples are refrigerated, and when the total 24 hour collection is completed, measure the total volume. Use a 2.0 to 4.0 ml. aliquot for storage or immediate ultrafiltration.

Determination of Amino Acids (Continued):

EQUIPMENT:

The Automatic Amino Acid Analyzer Model 335R-13 (Sondell Scientific Instruments, Palo Alto, California) used in our program consists of the following components: The Main Instrument (A); Tape-controlled Programmer (B); with Program (C); Log-linear Converter (D); Automatic Sample Injector (E) [Sample Injection Method and Apparatus, Patent No. 3,583,230]; Recorders (F & F_1) [Honeywell Electronic Model 194]; Integrator (G); [Series 200 Disc Integrator]; and Printer (H); [Disc Model 610 Automatic Printer]. (Figure 1.)

The components of the system include:

1. A short, water-jacketed column (33.5 x 0.9 cm.) filled with UR-40 Resin to a height of 27 cm. Special mounts eliminate end strain and minimize breakage of the column.

2. The UR-40 Resin (Sondell Scientific Instruments, Palo Alto, California) is a custom-designed sulfonated styrene co-polymer resin, 7.25% cross-linked, with spherical particle mean diameter of 14 $\pm$ 3 microns. It allows a shorter column, higher flow rates, without excessive pressure, resulting in shorter run times

3. The Head Space Eliminator [Chromatograph Column Head Space Reducer, Patent No. 3,487,038] automatically follows the resin eliminating void space, preventing backflow and channeling, protecting the column from excessive pressure and contraction.

4. The Circulating water bath provides the change in temperature required during the run. The time gradient (from low to high temperature) is 6 - 14 minutes.

5. The Reaction bath is kept at 100°C. The reaction coil is made of 22 gauge teflon tubing and has a 16 ml. volume. It maintains the column effluent and ninhydrin reagent for 10½ minutes at combined flow rate of 110 ml./hour.

6. The Colorimeter is equipped with 6.6 mm. light path cuvettes for 440 and 570 mμ. wavelengths and is regulated by the Log-linear converter.

7. The Flowmeter is electronically operated and records the buffer and combined (buffer + ninhydrin) flow time to 0.1 of a second.

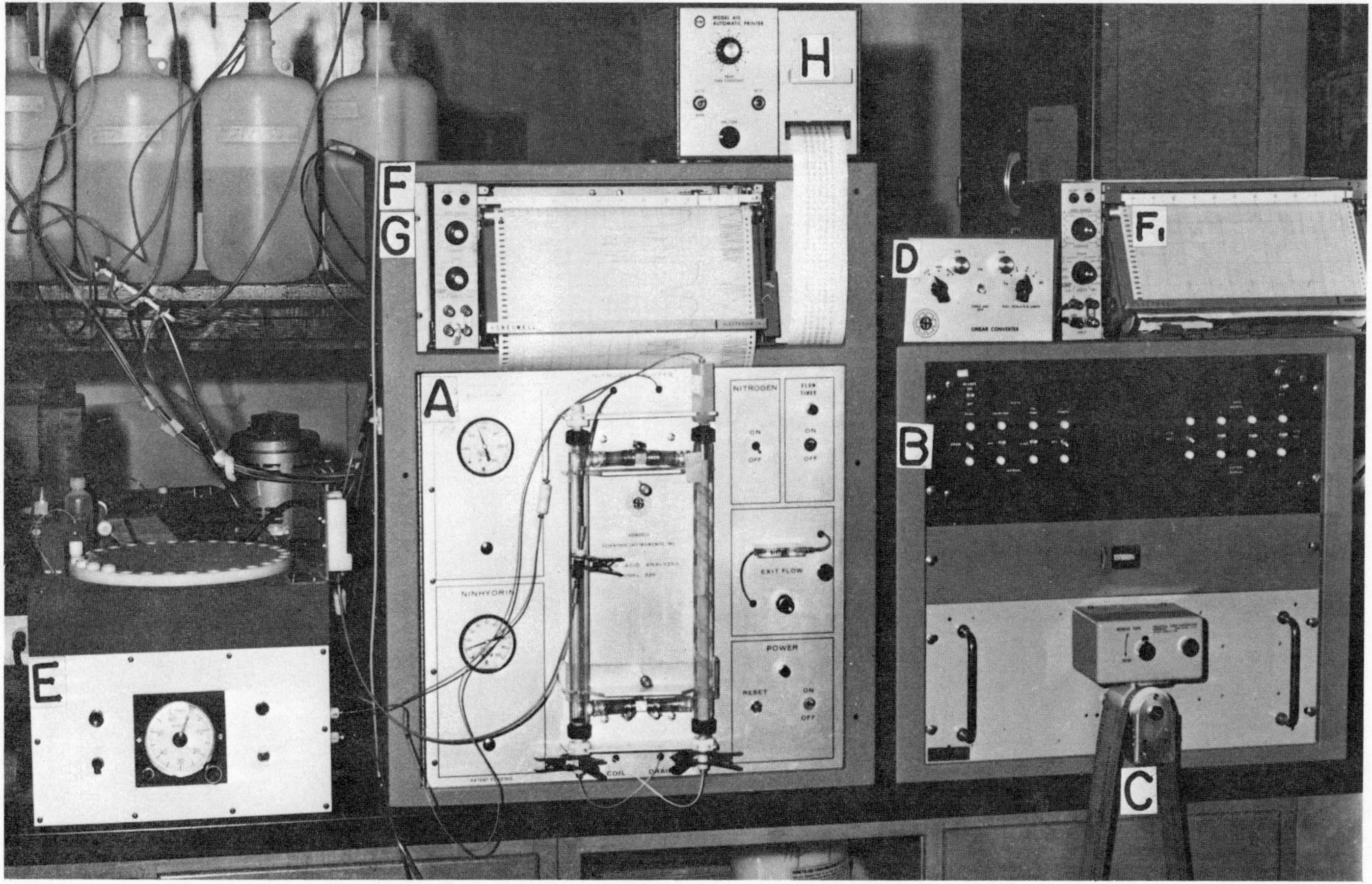

H
F
G
A
NITROGEN
ON
OFF
FLOW TIMER
ON
OFF
EXIT FLOW
POWER
RESET
ON
OFF
NINHYDRIN
COIL
E
D
LINEAR CONVERTER
F1
B
C

FIGURE 1.

Determination of Amino Acids (Continued):

8. The Set-Up for Deproteinization of the Sample
The technique which we developed for preparation of the sample involves a relatively simple piece of apparatus designed to produce an ultrafiltrate using positive nitrogen gas pressure. It consists of a nitrogen manifold, a test tube rack with a metal plate attached to the back of it. The plate holds copper tubing with ten individual outlets and valves, to the top of which a piece of rubber tubing is attached. Each piece of rubber tubing leads to a test tube (100 x 15 mm. for blood plasma and 150 x 17 mm. for urine), containing a connector that holds a dialysis tubing (BKH No. 25225, ¼ inch in diameter) bag, into which the sample is introduced. The general line of the manifold is connected by means of several feet of small diameter teflon tubing to the nitrogen outlet on the front panel of the Main Instrument (A). High purity compressed nitrogen gas (Liquid Carbonic Corporation) is used for this purpose.

REAGENTS:

The Classical Sodium Citrate Buffer System for amino acid determination in physiological fluids with some modifications, has been employed in our procedure. Four buffers are used in our system as eluants; another sodium citrate buffer is used as the sample diluent. The composition of each buffer is as follows:

	Sample Diluent	Buffer #1	Buffer #2	Buffer #3	Buffer #4
pH	2.2±0.03	3.17±0.01	4.24±0.01	4.15±0.01	5.36±0.02
Sodium Conc., N.	0.2	0.18	0.18	0.40	1.0
Sodium Citrate·$2H_2O$, Grams	19.6	70.60	70.60	156.90	45.7
Citric Acid, Grams	—	61.60	—	—	—
Concentrated HCl, ml.	16.5	35.00	29.00	65.00	8.8
Thiodiglycol ml.*	10.0	10.00	10.00	—	—
Pentachlorophenol, ml.**	0.1	0.40	0.40	0.40	0.1
Brij.-35, ml.	—	2.00	2.00	2.00	0.7
Sodium Chloride, Grams	—	—	—	39.70	353.0
Final Volume, Liters	1	4	4	4	4

* Pierce Chemical Company, 25% solution

** Stock Solution, 50 mg. in 10 ml. of 95% ethanol

Determination of Amino Acids (Continued):

6. Ninhydrin Reagent
This reagent is prepared by the method of Spackman, Stein, and Moore. A nitrogen atmosphere over the ninhydrin reagent is maintained by nitrogen supplied directly from a cylinder through a special regulator at a very low pressure of about 1 PSI.

7. NaOH, 0.2 N.
Used to regenerate the cartridges and column.

8. Standards

A. A synthetic mixture of amino acids which are or may be clinically significant. All amino acids were present in 0.1 μmole quantities, except phosphoserine, phosphoethanolamine, taurine, asparagine, sarcosine, cystathionine, β-alanine and anserine - 0.05 μmole; citrulline α-aminoadipic and α-amino-n-butyric acids - 0.25 μmole; creatinine - 0.6 μmole and urea - 1.5 μmole. (Sondell Amino Acid Calibration Standards)

B. Norleucine, 0.1 μmole added to each sample as an internal standard.

PROCEDURE:

1. Preparation of Samples:

A. Blood Plasma
Before placing the plasma into the dialysis tubing bag (refer to description for deproteinization of the sample), set the nitrogen gas regulator to 12 PSI. Barely open the valve, corresponding to the tube into which you want to introduce the plasma, and then turn off the nitrogen switch (on the front panel of the Main Instrument). Connect the rubber tubing to the connector of the corresponding tube, and open the dialysis tubing bag slowly.

Place the plasma into the bag using a long-nose Pasteur disposable pipette, and keep dilating the bag by slowly applying the nitrogen pressure; do not over-extend the bag; keep adding the sample. When addition of the sample is completed, connect the rubber tubing to the top of the connector; but the connector should just sit loosely on the top of the tube, to have an outlet for the excess of nitrogen.

Determination of Amino Acids (Continued):

Place the complete manifold in the 4°C. refrigerator and carry out the ultrafiltration.

A clear, colorless ultrafiltrate containing the free amino acids is received. The rate of ultrafiltration is 1.0 ml. per one and one-half hours as previously mentioned. Ten samples can be processed simultaneously. When ultrafiltration is completed, adjust the pH of the ultrafiltrate to 2.0 - 2.2 with 1 to 2 drops of 6.0 N. HCl. The sample is ready for analysis. 0.2 ml. to 0.5 ml. of plasma ultrafiltrate is used for a complete determination of the acidic, neutral, and basic amino acids. Results are expressed in micromoles of each amino acid/ml. of plasma.

B. Urine

The ultrafiltration of urine is carried out the same way as ultrafiltration of plasma. Take out a 1.0 to 2.0 ml. aliquot of the ultrafiltrate and measure the pH. To remove ammonia, add 4.0 N. NaOH to bring the pH to 11 or higher, place the urine ultrafiltrate under reduced pressure at about 60°C., and bring the volume down to almost dryness. Dilute the residue with 0.2 N. sodium citrate buffer, pH 2.2, to half of the initial volume of the ultrafiltrate. Adjust the pH to 2.2 with 6.0 N. HCl, and achieve the final volume by adding sodium citrate buffer, pH 2.2. The sample is ready for application. 0.3 to 0.5 ml. of urine ultrafiltrate is used for complete analysis. Results are expressed in μmole of amino acid/mg. creatinine, or μmoles of amino acid/24 hour urine specimen.

2. Application of the Sample

The sample is applied to the cartridge which is a part of the Automatic Sample Injector. This cartridge serves two functions:

a). It contains a resin that retains all material that is not wanted on the column.

b). From the moment the sample is applied to the cartridge, the amino acids are considered in solid state. The sample then may be safely stored.

Before application of the sample, regenerate the cartridge with 0.5 ml. of 0.2 N. NaOH. Equilibrate with the starting buffer, pH 3.2 using 0.5 ml. Follow this with 0.5 ml. sodium citrate buffer pH 2.2.

Determination of Amino Acids (Continued):

Apply sample with a micro-syringe and wash it in with 1 drop of sodium citrate buffer, pH 2.2.

Amino Acid analysis is carried out in agreement with the program.

3. Chromatographic Conditions

The initial column temperature is 30.5°C. At 76 minutes the temperature is raised to 60°C. and is kept at this level until the elution of arginine. The time gradient is 8 minutes.

The flow rates are: 67 ml./hour for the buffer and 43 ml./hour for the ninhydrin. The back pressure should not exceed 300 PSI.

Elution is started with sodium citrate buffer No. 1. A buffer change is made at 99 minutes, from buffer No. 2. to buffer No. 3 at 182 minutes, from buffer No. 3 to buffer No. 4 at 323 minutes. It takes 27 minutes for the buffer change to appear on the chart. At the conclusion of the run, at 359 minutes, there is a change from buffer No. 4 to buffer No. 1., and a temperature change from 60°C. to 30.5°C. Coil retention time is 10.5 minutes. Color development is observed at 570 and 440 mμ. wavelengths. The recorder chart speed is 15 cm./hour. Total time for complete analysis, including the elution of arginine, is 6 hours. Regeneration of the column is recommended every third run.

COMMENTS ON PROCEDURE:

A highly sensitive, specific, and fast analytical procedure for the determination of amino acids in small volumes of physiological fluids has been described.

Unlike other amino acid determination systems in operation now, our procedure represents an accelerated single-column sequential run for acidic, neutral, and basic amino acids present in physiological fluids. (Figures 2., 3., and 4. represent photographs of the patterns received with: a synthetic mixture of amino acids; a human plasma; and with human urine respectively.)

The baseline remains constant throughout the entire run. All components including the basic amino acids are well resolved, except for several overlaps. Sarcosine appears as a shoulder on asparagine and citrulline as a shoulder on the glycine peak. The aspartic acid-threonine separation is sufficient for quantitation, but can be improved by raising the initial temperature by 1°C. or increasing the pH of the starting buffer. Proline

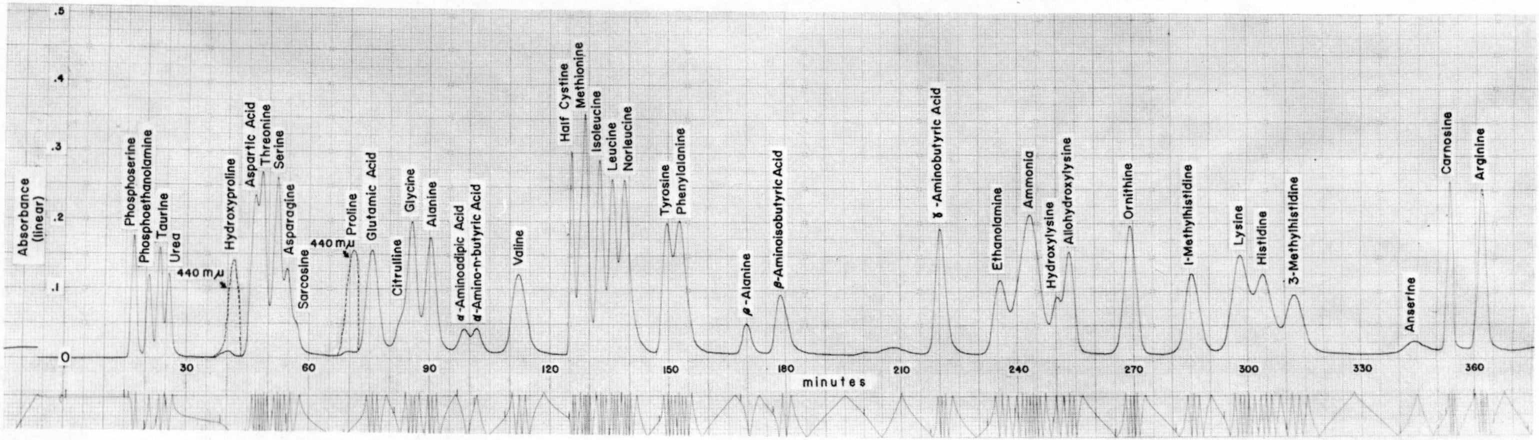

FIGURE 2.

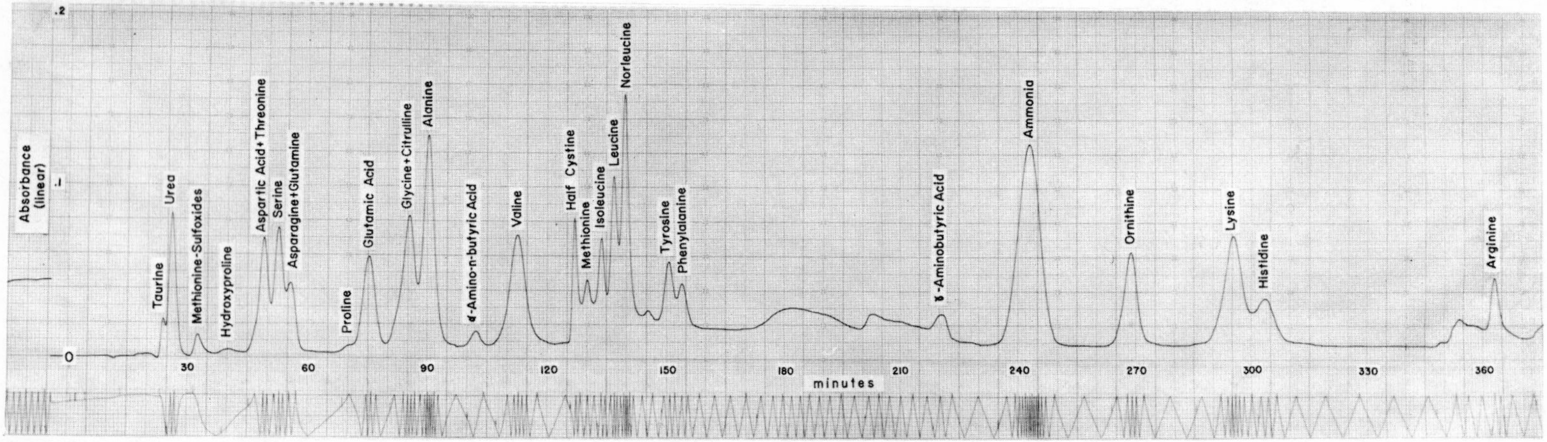

FIGURE 3.

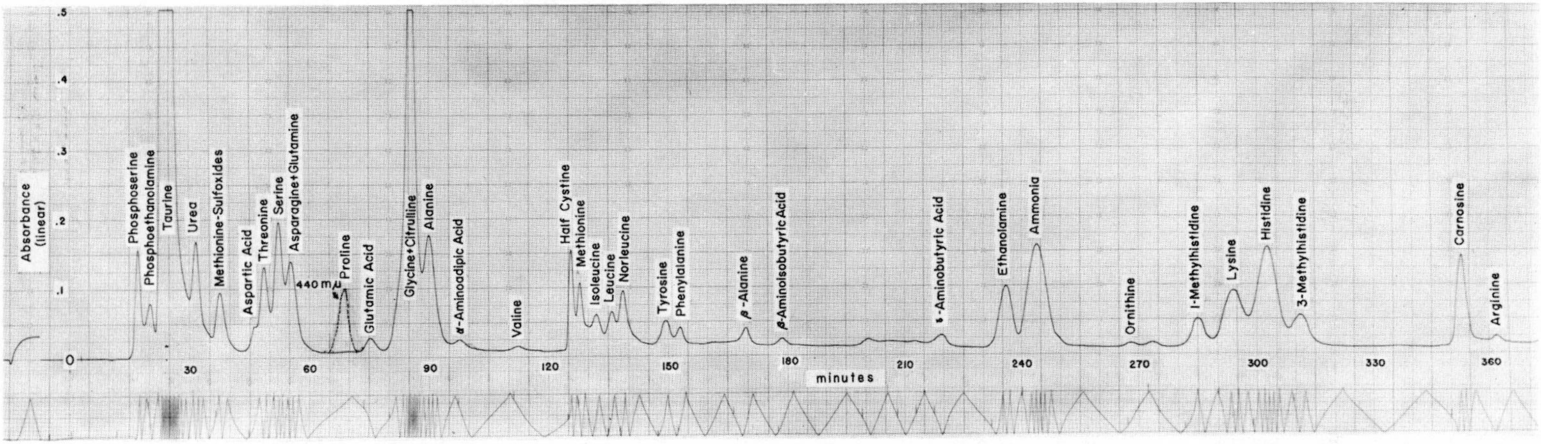
Absorbance (linear)
.5
.4
.3
.2
.1
0
30
60
90
120
150
180
210
240
270
300
330
360
minutes
Phosphoserine
Phosphoethanolamine
Taurine
Urea
Methionine-Sulfoxides
Aspartic Acid
Threonine
Serine
Asparagine+Glutamine
440 mμ
Proline
Glutamic Acid
Glycine+Citrulline
Alanine
α-Aminoadipic Acid
Valine
Half Cystine
Methionine
Isoleucine
Leucine
Norleucine
Tyrosine
Phenylalanine
β-Alanine
β-Aminoisobutyric Acid
γ-Aminobutyric Acid
Ethanolamine
Ammonia
Ornithine
1-Methylhistidine
Lysine
Histidine
3-Methylhistidine
Carnosine
Arginine

FIGURE 4.

Determination of Amino Acids (Continued):

and hydroxyproline have been recorded at 440 mμ. on a separate recorder (F_1) and the peaks transferred and shown by a dotted line. By increasing the sensitivity (0.2 A - 0.05 A full scale) which our system permits comfortably, most of the amino acids can be analyzed at a concentration level of 0.025 - 0.010 micromoles.

For quantitation, computing the area under each amino acid peak, an electronic integrator was used. Norleucine was added to each sample as an internal standard, and calculations for all amino acids have been based on norleucine equivalents. A norleucine equivalent is the ratio of peak area produced by unit molar quantity of norleucine to peak area produced by unit molar quantity of another ninhydrin-positive substance.

The color values obtained on repeated runs (over twenty) with synthetic amino acid mixtures and physiological fluids have been within $\pm$ 8% in the 0.02 - 0.10 micromole range.

The sensitivity of our system is such that 0.01 micromole (10^{-8} mole) of most amino acids can be determined. High sensitivity permits smaller samples (0.2 ml. of plasma ultrafiltrate for complete analysis), which is of great importance in pediatric practice.

The physical and chemical properties of our resin were carefully controlled, resulting in a uniform polymer matrix, lower back-pressure, and higher flow rates. These qualities allowed the use of a shorter column; they provided shorter analysis time and improved peak resolution. We would like to emphasize that care in packing of the column is important; trapped air will cause a channeling effect and increased back-pressure, which reduce buffer-resin contact and result in a decrease in the resolving capability of the resin column; therefore, it is quite worthwhile to pay special attention to this part, which is actually the heart of the whole system.

The buffers are regulated in a stepwise manner and are made up with deionized water. As the use of a detergent in the eluant has been found to permit faster flow rates without concomitant broadening of the peaks Brij.-35 was added. The composition of the buffers is basically the same as the one used by Spackman, Stein, and Moore, Benson and Patterson. Modification of buffer No. 4. (raising the sodium ion concentration to 1.0 N.) was necessary in order to accelerate the elution of arginine.

Determination of Amino Acids (Continued):

Higher pH values, higher sodium concentration of the buffers, and higher temperature increase the rate of migration of the amino acids; therefore, raising the temperature or pH of the starting buffer can be used in order to improve the aspartic acid-threonine separation.

Our results have been compared with data previously reported and also with results obtained earlier by these authors, using other methods such as: paper chromatography, high voltage electrophoresis paper chromatography, the Van-Slyke Ninhydrin-CO_2 Method for Total alpha-amino nitrogen, and chemical specific assays for the determination of certain amino acids.

The color values registered by our present automatic ion-exchange chromatographic system are in general agreement with our previous data and data reported by others. Thus, we feel strongly that for precise work these values must be determined for each system individually.

The addition of an internal standard (norleucine) to the unknown sample before analysis would indicate any deviation from the normal expected values (mechanical losses, chemical variation, equipment changes, etc.).

It is necessary to stress a few technical aspects in preparation of the sample. The presence of proteins in physiological fluids will interfere with amino acid analysis, causing poor separation and clogging of the column. Hamilton and Van-Slyke used 1% picric acid for deproteinization of plasma. Stein and Moore improved this method by removing the picric acid from the centrifugate before applying it to the column. Hamilton later used 3% sulfosalicylic acid, and the centrifugate was placed directly on the column. The above methods have been compared by Gerritsen with another one, without complete removal of protein, using high-speed centrifugation; they showed better recoveries and considered the results satisfactory.

Our technique represents an ultrafiltration procedure applying positive nitrogen gas pressure and no chemicals. A clear solution containing the free amino acids is obtained. The rate of ultrafiltration is 1.0 ml. per one and one-half hours, and 10 samples can be processed simultaneously.

Determination of Amino Acids (Continued):

REFERENCES:

1. Scriver, C. R.: "Hereditary Aminoaciduria", PROGRESS IN MEDICAL GENETICS, Vol. II. Ed. Arthur G. Steinberg and Alexander G. Bearn, Grune and Stratton, N. Y. and London, pg. 83, 1962.

2. Efron, M. L.: The New Eng. J. of Med., 272:1243, 1965.

3. Efron, M. L.: The New Eng. J. of Med., 272:1058, 1965.

4. Symposium on Treatment of Amino Acid Disorders, Am. J. Dis. Child, pg. 113, 1967.

5. Perry, T. L., et. al.: The Canadian Med. Assoc. J., 95:89, 1966.

6. McCully, K. S.: Am. J. of Path., 59:181, 1970.

7. Hamilton, P. B.: Anal. Chem., 35:2055, 1963.

8. Piez, K. A. and Morris, L.: Anal. Biochem., 1:187, 1960.

9. Benson, J. V. and Patterson, J. A.: Anal. Biochem., 13:265, 1965.

10. Efron, M. L.: PROCEEDINGS OF THE TECHNICON SYMPOSIUM ON AUTOMATION IN ANALYTICAL CHEMISTRY, New York, Sept. 1965; Medical Inc., 1966.

11. Linneweh, F. and Barthelmai, W.: Klin Wschr., 47:971, 1969.

12. Spackman, D. H., Stein, W. H., and Moore, S.: Anal. Chem. 30:1190, 1958.

13. Miller, E. J. and Piez, K. A.: Anal. Biochem., 16:320 - 326, 1966.

14. Hamilton, P. B. and Van Slyke, D. D.: JBC, 150:231, 1943.

15. Stein, W. H. and Moore, S.: JBC, 167:855, 1954.

16. Hamilton, P. B.: Ann. New York Acad. Sci., 102:55, 1962.

Determination of Amino Acids (Continued):

17. Moore, S. and Stein, W. H.: JBC, 192:663, 1951.

18. Benson, J. V., Jr. and Patterson, J. A.: NEW TECHNIQUES IN AMINO ACID PEPTIDE AND PROTEIN ANALYSIS, Ed. A. Niederwieser and G. Pataki, Ann Arbor Science Publishers, Inc., Ann Arbor, 1971.

19. Dickinson, J. C., Rosenblum, H., and Hamilton, P. B.: Pediatrics, 36:2, 1965.

20. Stein, W. H. and Moore, S.: JBC, 211:915, 1954.

21. Ackermann, P. G. and Kheim, T.: Clin. Chem., 10:32, 1964.

22. Tocci, P. M.: AMINO ACID METABOLISM AND GENETIC VARIATION, Chapter 34, Ed. William Nylan, McGraw-Hill, 1967.

23. Sackett, D. L.: J. Lab. & Clin. Med., 63:306, 1964.

24. Blackburn, S.: METHODS OF BIOCHEMICAL ANALYSIS, Vol. XIII, pg. 1-45, 1965.

25. Prockop, D. J. and Udenfriend, S. A.: Anal. Biochem., 1:228, 1960.

26. Troll, W. and Lindsey, J.: JBC, 215:665, 1955.

27. Gerritsen, T., Rehberg, M. L., and Waisman, H. L.: Anal. Biochem., 11:460, 1965.

28. Prescott, B. A. and Waelsch, H.: JBC, 167:855, 1947.

CLINICAL INTERPRETATION:

During the last ten - fifteen years, substantial progress has been made in the study of the primary biochemical errors in many hereditary and acquired disorders associated with defects in amino acid metabolism and transport.

These defects lead to mental retardation, neurological disorders, muscular difficulties, cirrhosis, kidney damage and, in some instances, early death in infancy and childhood. When enzyme-directed interruptions in the metabolic chain occur, abnormal amounts of amino acids are observed in physiological fluids. Aminoaciduria may be

Determination of Amino Acids (Continued):

a secondary manifestation of other disease. For instance, in severe liver disease, amino acid synthesis or degradation may be disturbed; in kidney disease, renal tubular absorptive mechanism may be damaged and large amounts of certain amino acids can be excreted in the presence of a neoplastic disease.

In many instances, diagnostic techniques and treatment are available now: low phenylalanine diet in Phenylketonuria, dietary treatment in Tyrosinosis, diet low in branched-chain amino acids (leucine, isoleucine, and valine) - for Maple Syrup Urine Disease, and many others. In other cases, where we do not have specific treatment yet, early diagnosis can be a useful tool in genetic counseling of the parents, in order to prevent the birth of another defective child; Hurler's Syndrome is an example of a disease in this category.

There are many other clinical conditions where the amino acid pattern is disturbed: In myocardial infarction, in hypertension, in arteriosclerosis, in psoriasis, and many more.

The importance of having in the clinical laboratory a precise, fast, and reliable procedure for amino acid analysis in physiological fluids cannot be overstated.

URINARY DELTA-AMINOLEVULINIC ACID (ALA)

PRINCIPLE: The quantity of lead in blood and the quantity of ALA (delta-aminolevulinic acid) in urine are the most reliable indices of lead poisoning. The effect of increased lead exposure is the declining ability of the body to adequately convert ALA, by the enzymatic action of ALA dehydrase, to porphobilinogen, as shown in the following mechanism:

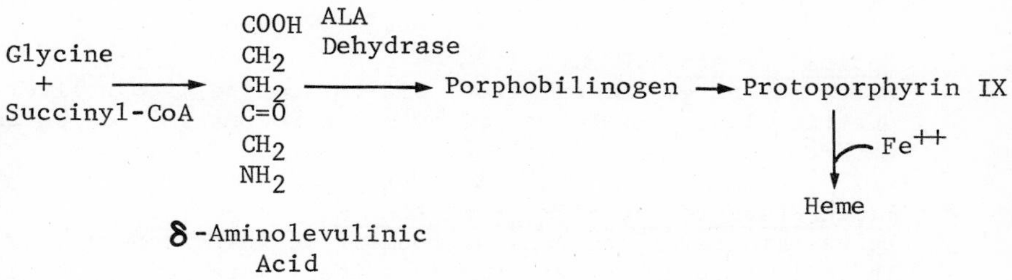

Since the function of ALA dehdrase is hampered, its substrate (ALA) increases and is excreted in the urine. No other heavy metal is known to affect this enzyme.

The method is a rapid, sensitive, and simplified determination using the Bio-Rad prefilled, disposable ion-exchange chromatography columns.

In order to isolate the ALA from the urine, the urine sample must first be passed through an anionic ion-exchange resin which permits both ALA and urea to pass through while porphobilinogen is retained on the resin. The effluent passing through the anionic resin must then be passed through a cationic ion-exchange resin which retains ALA while allowing urea to pass through. The ALA which has been retained on the cationic resin can then be eluted and collected for analysis.

After elution from the column, ALA is condensed with acetylacetone to form a pyrrole, 2-methyl-3-acetyl-4-(3-propionic acid) pyrrole. This pyrrole then reacts with Ehrlich's Reagent (p-dimethylamino-benzaldehyde) in acid solution to produce a cherry-red complex which can be measured spectrophotometrically.

Delta-Aminolevulinic Acid (Continued):

SPECIMEN: Only 0.5 ml. of a random urine sample is required for testing. It is advisable not to use the first voided morning specimen, late evening specimens after 8:00 p. m., or specimens obtained following excessive fluid intake. Add 0.1 ml. of concentrated glacial acetic acid to every 10 ml. of urine to keep the pH below 6.0. The urinary ALA content of samples stored in a dark refrigerator will remain constant up to 4 months, while frozen samples will remain constant up to a year.

REAGENTS AND EQUIPMENT:

1. Sodium Acetate 1.0 M.
   Weigh out 82 gm. of anhydrous sodium acetate into a liter volumetric flask. Dissolve and bring to volume with deionized water.

2. Stock Standard 100 micrograms ALA/ ml.
   Delta-Aminolevulinic Acid Hydrochloride (Calbiochem) M. W. 167.6 is used to prepare the solution. Weigh out 12.8 mg. of ALA · HCl into a 100 ml. volumetric flask. Dissolve and dilute to volume with deionized water. This standard has a concentration of 100 micrograms ALA/ml. Solution is stable under refrigeration.

3. Working Standard of ALA, 0.2 mg./100 ml.
   2.0 ml. of Stock Standard is diluted to 100 ml. with deionized water.

4. Perchloric Acid, 70 - 72%
   Available from J. T. Baker Chemical Company.

5. Acetylacetone, 2,4-pentanedione
   Obtained from Eastman Organic Chemicals, Rochester, N. Y.

6. p-Dimethylaminobenzaldehyde Acid Solution
   10 gm. of reagent grade p-dimethylaminobenzaldehyde (DMAB) is dissolved in 420 ml. of reagent grade glacial acetic acid. Solution is stored in an amber bottle in the refrigerator.

7. Fresh Ehrlich's Reagent
   To every 100 ml. of acid DMAB, add 19 ml. of 72% perchloric acid. Reagent is stable up to six hours. Prepare only the amount needed for the determination.

Delta-Aminolevulinic Acid (Continued):

8. Chromatography Columns
   Available in 20 ALA Column Units complete with reagents and Standard from Bio-Rad Laboratories.

9. Gilford Spectrophotometer 300 N.
   Equipped with an aspiration cuvette.

CALIBRATION CURVE: (To be set up with each determination)

1. Using culture tubes with screw caps, (20 x 150 mm.), aliquots of Working Standard are pipetted in the following manner:

Volume of Standard	Concentration of ALA	Corresponding Mg%. of ALA for 0.5 ml. of Urine
0.5 ml.	0.1 mg./100 ml.	0.2 mg./100 ml.
1.0 ml.	0.2 mg./100 ml.	0.4 mg./100 ml.
2.0 ml.	0.4 mg./100 ml.	0.8 mg./100 ml.
3.0 ml.	0.6 mg./100 ml.	1.2 mg./100 ml.
5.0 ml.	1.0 mg./100 ml.	2.0 mg./100 ml.

2. Dilute each of the Standards to 7.0 ml. with 1.0 M. NaAc.

3. BLANK - 7.0 ml. of 1.0 M. NaAc.

4. Add 0.2 ml. of acetylacetone to each tube and mix.

5. Place tubes in a boiling water bath for 10 minutes, remove, and cool to room temperature.

6. Add 7.0 ml. of fresh Ehrlich's Reagent to each tube, mix, and let stand for 15 minutes. The ALA colored product reaches a maximum in 6 - 15 minutes, and then remains stable for at least 15 minutes.

7. Using the Reagent Blank, set the Gilford Spectrophotometer to zero at 553 nm. and read the absorbance of the Standards.

8. Prepare a standard curve by plotting concentration of standards in Mg%. versus absorbance.

PROCEDURE:

1. Preparation of the Column
   Use one blue tinted column and one clear column for each sample to be run. Shake each column vigorously until all the

Delta-Aminolevulinic Acid (Continued):

resin is resuspended. Allow the resin to settle in each column. First remove the cap of the clear column and then snap off the tip and place in rack. Then remove the cap of the blue tinted column and then snap off the tip, placing the blue tinted column "piggy-back fashion" into the clear column. This two column set is referred to as the column unit. Each urine sample requires one column unit for the determination of ALA.

2. To each column unit, add 10 ml. of distilled water and allow to drain through both columns into the drain tray.

3. Pipette 0.5 ml. of random urine sample into the top of the column unit.
   CONTROL - 0.5 ml. of Supplemental Urine Control (Available from Hyland Laboratories.
   STANDARD RECOVERY - 0.1 ml. of Stock Standard corresponding to 2.0 mg%. ALA.

4. Wash three times with 30 ml. of deionized water, 10 ml. at a time. It is important that all column units are completely drained of the previous water wash before applying the next aliquot of water to the column unit. This prevents the column from overflowing.

5. Discard the blue tinted column, and place the clear column into a culture tube.

6. Elute with 7.0 ml. of 1.0 M. Sodium Acetate. Allow all of the sodium acetate to drain completely.

7. BLANK - 7.0 ml. of 1.0 M. Sodium Acetate.

8. Add 0.2 ml. of acetylacetone to each tube and mix.

9. Place tubes in a boiling water bath for 10 minutes, remove and cool to room temperature.

10. Add 7.0 ml. of fresh Ehrlich's Reagent to each tube, mix, and let stand for 15 minutes.

11. Read the absorbance of the colored product against the Reagent Blank on the Gilford Spectrophotometer 300 N at 553 nm.

Delta-Aminolevulinic Acid (Continued):

12. Read off Concentration of Delta-Aminolevulinic Acid in mg%. from the Standard Curve.

NORMAL VALUES: Children and Adults 0.00 - 0.54 mg%.

REFERENCES:

1. Mauzerell, D., and Granick, S.: J. Biol. Chem., 219:435, 1956.

2. Davis, J., and Andelman, S.: Arch. Environ. Health, Vol. 15, 1967.

3. Bio-Rad ALA Instruction Manual.

CLINICAL INTERPRETATION:

ALA is synthesized from glycine and succinyl-coenzyme A. Urinary excretion of ALA is greatly increased in lead poisoning. The major reason for the increased ALA in the urine in lead poisoning is the lack of conversion of ALA to porphobilinogen because of the inhibition of ALA dehydrase by lead.

ALA urinary excretion is also found in acute intermittent porphyria and has been found to be slightly increased in acute hepatitis, carcinoma of the liver and Porphyria Cutanea Tardea.

BLOOD AMMONIA DETERMINATION

PRINCIPLE: Potassium carbonate releases ammonia gas which dissolves in sulfuric acid on the acid dipped rod to form $(NH_4)_2SO_4$. Sulfate is exchanged for OH^- ion in the KOH of Nessler's reagent.

$$NH_4OH + 2(KI)_2\ HgI_2 + 3KOH \longrightarrow \underset{\text{(amber color)}}{HgNI} + 7KI + 4H_2O$$

REAGENTS:

1. Potassium Carbonate ($K_2CO_3 \cdot 1\frac{1}{2}H_2O$)
   Make a saturated solution. The amount of Potassium Carbonate required for a saturated solution depends a great deal on the temperature. It may require as much as 200 Gm./100 ml. water. The crystalline goes into solution easier than the anhydrous.

2. H_2SO_4 1 N. - ammonia free.
   14 ml. concentrated sulfuric acid/500 ml. distilled water.

3. Stock Nessler's Reagent
   Harleco (Koch-McMeekin)

4. Stock Ammonia Standard
   100 micrograms/ml. on the basis of "Nitrogen", 471.6 mg. $(NH_4)_2SO_4$/liter.

5. Working Standards
   Routinely run 1, 2, 3 micrograms/ml.
   Standard 1 - 1.0 ml. of stock + 100 ml. distilled water.
   Standard 2 - 2.0 ml. of stock + 100 ml. distilled water.
   Standard 3 - 3.0 ml. of stock + 100 ml. distilled water.

   Use 1.0 ml. of each of these for S_1, S_2, S_3, respectively.

PROCEDURE:

Preparation of Vials

The standards can be prepared before the blood is drawn since their time of diffusion over the first 20 minutes is not critical. Blood diffusion must be accurately timed as the Potassium Carbonate will start to denature proteins and liberate NH_3 if diffused too long.

Blood Ammonia Determination (Continued):

1. Blank - 1.0 ml. of distilled water with an Ostwald pipette.

2. Standard - 1.0 ml. of each of the 2 Working Standards with an Ostwald pipette. (200 microgram - 300 microgram standards.)

3. Pipette 1.0 ml. of saturated K_2CO_3 into the Blank, 2 Standards, and 3 vials for the test, 2 vials for the Control. A 5.0 ml. serological pipette can be used for this since the K_2CO_3 is in excess. Be careful not to wet the neck of the vials where it would contaminate a rubber stopper or a glass rod.

4. Dip the bead of the glass rod into the acid just enough to coat the bead and drain the excess acid against the side of the acid vessel. Do not leave the acid bottle open for long periods as ammonia will be absorbed from the atmosphere.

COLLECTION OF BLOOD:

1. Rinse the syringe with heparin (sodium salt - ammonia free).

2. If possible, draw the blood without using a tourniquet. If a tourniquet is necessary do not leave it on the arm while the sample is being withdrawn unless the veins are very poor. Treat Control person same as patient when obtaining blood.

3. Remove any air bubbles from the syringe immediately and stick the needle into a rubber stopper to prevent any leakage. Ice the blood immediately and keep iced until transfer.

PREPARATION OF THE UNKNOWN:

1. Mix the blood well in the syringe by rolling the syringe in the palms of your hand.

2. Attach a one inch length of small diameter rubber tubing over the syringe tip.

3. Flush a small amount of blood through the tubing and insert a 1.0 ml. Ostwald and load the pipette, which is held in one hand at a 45 degree angle, by gentle pressure on the syringe plunger with the other hand.

4. Blood is quickly introduced against the inside wall of the vial. Blow quickly to include the last drop, and stopper without delay with the prepared glass rods. Take care so the rod does not contact blood or standard solution either by touch or agitation.

Blood Ammonia Determination (Continued):

5. Mix the blood well again, flush out a few drops of blood and repeat sampling procedure. Three unknown vials are set up on a single sample of blood. The results should check within 5%. Do the same for Control specimen in the last two vials.

6. Place all 8 vials on the rotator and allow diffusion for exactly 20 minutes. Handle the vials carefully so that neither blood nor standard are shaken onto the glass rod.

7. Dilute Stock Nessler's reagent 1:10 with Ammonia-free water. Pipette 1.5 ml. dilute Nessler's into each of 8 cuvettes. (Nessler's may form a precipitate. Decant off supernatant fluid and use only if clear. An O.D. reading cannot be taken if there is any turbidity in the Nessler's.

8. Remove vials from the rotator after <u>exactly</u> 20 minutes. Carefully transfer glass rods to civettes, tilt and rotate cuvettes so rods are thoroughly washed with Nessler's.

9. Mix cuvettes, stopper and let stand 10 minutes for color development. Wipe off cuvettes with cloth while color development is taking place. Since the reading range is narrow, any lint or fingerprints may seriously affect the reading.

10. Read at 420 nm.

CALCULATIONS:

$$\text{Conc. Unk.} = \frac{\text{O.D. Unk.}}{\text{O.D. Std.}} \times \text{Conc. Std.} \times \frac{100}{1}$$

NORMAL VALUES: 75 - 125 micrograms%.

NOTE: NEVER USE ETHER OR ACETONE TO DRY PIPETTES OR CUVETTES.

REFERENCES:

1. Bessman: Modification of Seligson Method, <u>J. Clin. Investigation</u>, 34:622, 1955.

2. Seligson and Hirshara: <u>J. Lab. and Clin. Med.</u>, 49:962, 1957.

3. Nathan and Rodkey: <u>J. Lab. and Clin. Med.</u>, 49:779, 1957.

4. Moore and Stein: <u>J. Biol. Chem.</u>, 176:367, 1948.

Blood Ammonia Determination (Continued):

5. Bromberg, et al: J. Clin. Investigation, 39:322, 1960.

6. Roberts, Kathleen: J. Clin. Pathology, Vol. I.

7. Caesar: Clinical Science, 22:33-41, 1962.

CLINICAL INTERPRETATION:

Ammonia is present in a small quantity in the plasma. Part of the blood ammonia is derived from the gastrointestinal tract. Postprandial blood levels of ammonia are higher than that of the fasting state. An increase in blood ammonia occurs as a result of exercise derived from muscular contraction.

The most common cause for an increased blood ammonia is hepatic failure. It may result from the inability of the liver to utilize ammonia for urea production. The collateral portal circulation found in hepatic cirrhosis may reroute blood ammonia from the liver with subsequent cerebral encephalopathy. The blood ammonia may be enhanced by gastrointestinal hemorrhage associated with cirrhosis. This results from bacterial hydrolysis of nitrogenous substances. Furthermore, hypokalemia may cause an increase in blood ammonia because of increased ammonia production by the renal tubule.

A rare cause for hyperammonemia is hereditary hepatic ornithine transcarbamylase deficiency. Children with this enzyme deficiency develop high blood ammonia after a protein meal with subsequent cerebral encephalopathy.

SERUM AND URINE AMYLASE
(Iodometric Method by Caraway)

PRINCIPLE: Alpha-amylase is an enzyme secreted by the pancreas and salivary glands. It hydrolyzes starch to the disaccharide maltose. Starch, but not maltose, forms a blue colloidal complex with iodine in solution, and the intensity of this color is directly proportional to the concentration of the starch. The blue color produced by the starch substrate when combined with iodine, is measured after incubation with serum and compared to a blank. The decrease in color is proportional to the amylase activity.

One amylase unit is the amount of enzyme that will hydrolyze 10 mg. of starch in 30 minutes to a stage at which no color is given by iodine.

SPECIMEN: When serum determination is made, pipettable 0.1 ml. of serum should be obtained. A fasting sample is not essential. Urines can be analyzed in the same manner as serum. For adequate interpretation, either a 1 hour or a 24 hour collection of urine should be obtained; random specimens have little diagnostic value. Other body fluids such as bile, pancreatic secretions, duodenal drainage, and so on, may also be analyzed in the same manner, inasmuch as the substrate is well buffered.

REAGENTS:

1. Stable Buffered Starch Substrate pH 7.0
   Dissolve 13.3 gm. of anhydrous disodium phosphate and 4.3 gm. of benzoic acid in about 250 ml. of water. Bring to boil. Mix separately 0.200 gm. of Merck's solution starch (Lintner) in 5.0 ml. of cold water and add it to the boiling mixture, rinsing beaker with additional cold water. Continue boiling for 1 minute. Cool to room temperature and adjust pH to 7.0. Dilute to 500 ml. with water. Stable at room temperature and should remain water clear.

2. Stock Solution of Iodine, 0.1 N.
   Dissolve 3.567 gm. potassium iodate (KIO_3) and 45 gm. of potassium iodate in approximately 800 ml. of water. Add slowly and with mixing, 9.0 ml. of concentrated hydrochloric acid (12 M.) and dilute to 1000 ml. with water.

3. Working Iodine Solution, 0.01 N.
   Dissolve 59 gm. of potassium fluoride ($KF \cdot 2H_2O$) in approximately 350 ml. water in a 500 ml. volumetric flask. Add 50 ml. of

Serum and Urine Amylase (Continued):

the stock solution of iodine and dilute to the mark. Solution is stable for 1 - 2 months, when stored in a brown bottle in the refrigerator.

PROCEDURE:

1. Run normal and abnormal controls with each set of determinations. Do not use a blow-out pipette, as there is great amylase content in saliva.

2. Pipette 5.0 ml. (volumetric pipette) of starch substrate into each of 50 ml. graduated tubes marked "Test" and "Blank".

3. Place all the tubes in a water bath at 37°C. for 5 minutes to warm the contents.

4. Pipette exactly 0.10 ml. of serum into the bottom of the tube labelled "Test", and allow the reaction to proceed for exactly 7½ minutes. No serum is added to the "Blank" tube.

5. After 7½ minutes, remove the tubes from the water bath; add deionized water up to the 40 ml. mark. Immediately add 5.0 ml. (volumetric pipette) of working solution of iodine to each tube, then dilute to 50 ml. with deionized water. Mix well by inversion and shaking.

6. Measure the % Transmission of the "Test" and "Blank" without delay against water at 660 nm. in a spectrophotometer using 12 mm. cuvettes.

7. CALCULATIONS:

   Convert % T. readings to optical density.

$$\frac{\text{O.D. of Blank} - \text{O.D. of Test}}{\text{O.D. of Blank}} \times 800 = \text{Amylase units/100 ml.}$$

   "800" indicates that complete hydrolysis of the starch would correspond to a serum amylase activity of 800 units per 100 ml.

8. If the activity of the amylase in serum exceeds 400 units, the test is repeated using a 5-fold dilution of the serum with 0.9% NaCl. Final results are corrected by multiplying by 5.

Serum and Urine Amylase (Continued):

NORMAL VALUES: Serum: 40 - 160 units/100 ml.
Urine: 43 - 245 units/hour based on 6 to 24 hour collection.

REFERENCE: Caraway, Wendell T.: Amer. Jour. Clin. Path., 32:97, 1959.

CLINICAL INTERPRETATION:

The serum amylase is low in early infancy, is first demonstrable at two months of age, and attains the adult level at the age of one year.

Hyperamylasemia may be found in the following conditions:

1. Acute pancreatitis
2. Perforated peptic ulcer penetrating the pancreas
3. Intestinal obstruction
4. Pancreatic duct obstruction by carcinoma or calculi
5. Opiates, codeine or methylcholine spasm of sphincter of Oddi
6. Pancreatic stimulation by secretin or pancreozymin
7. Parotid gland disease, mumps, suppurative parotitis, and calculi in the parotid duct
8. Renal failure with oliguria or anuria, causing retention of amylase due to insufficient clearance by the kidney
9. Hepatic necrosis
10. Ectopic pregnancy
11. Macroamylasemia due to combination of amylase with an immunoglobulin
12. Ectopic production of amylase by extrapancreatic carcinoma
13. Hemolysis

Decreased amylase:

1. Citrate anticoagulant
2. Oxalate anticoagulant
3. Cachexia
4. Marked destruction of pancreas
5. Administration of glucose, insulin or cortisone

The commonest cause for a high serum amylase is acute pancreatitis. The destruction of pancreatic acinar tissue, with or without obstruction of flow of pancreatic secretion, results in the escape of the various pancreatic enzymes into the pancreas and into the peritoneal cavity. Absorption of pancreatic enzymes into the blood with subsequent prominent elevation of serum amylase occurs. In acute pancreatitis, the serum amylase invariably increases almost simultaneously

Serum and Urine Amylase (Continued):

with the onset of symptoms. It frequently rises above 500 Caraway units. The peak level of the enzyme is usually reached within 24 hours, after which there is usually a prominent drop in the enzyme level with a return to normal values within two to four days. An absence of an increase in the amylase level within the first 24 hours after the onset of symptoms is suggestive evidence against the diagnosis of acute pancreatitis. The serum amylase generally returns to normal within 24 - 48 hours.

ASCORBIC ACID DETERMINATION

PRINCIPLE: A protein-free filtrate of serum, plasma, or urine is prepared with metaphosphoric acid. The ascorbic acid content of the filtrate is determined from the decrease in color of added 2, 6-dichlorophenolindophenol. Ascorbic acid reduces the dye to its colorless form. Correction is made for any turbidity or background absorbance by measuring absorbance after adding excess ascorbic acid to decolorize the unreduced dye.

REAGENTS:

1. <u>Metaphosphoric Acid</u>, 3.0%
   Store in refrigerator. Stable at least 8 days.

2. <u>Stock Indophenol Dye Solution</u>
   100 mg. sodium salt of 2, 6-dichlorophenolindophenol per 100 ml. deionized water. Store in refrigerator. Can be used at least several months.

3. <u>Working Dye Solution</u>
   Dilute Stock solution 1:10 with deionized water. Prepare fresh from Stock Solution for each run. The reagent blank in the test should give an absorbance of about 0.67 with a spectrophotometer of high resolution such as a Beckman DU.

4. <u>Sodium Citrate Solutions</u>
   Use the following concentrations of sodium citrate dihydrate:

   For serum or plasma filtrates: 1.7%.
   For blank, standard, or urine filtrate: 4.37%.
   For blank, standard, or urine filtrate plus p-chloromercuribenzoic acid: 2.6%.

5. <u>p-Chloromercuribenzoic Acid</u>
   200 mg./100 ml. 0.05 N. NaOH.

6. <u>Standard</u>
   To 0.800 mg. ascorbic acid in a 100 ml. volumetric flask, add 40 ml. deionized water and then 3.0% metaphosphoric acid to volume. This standard is stable in the refrigerator for at least 1.0 month. Ascorbic acid is not stable in ordinary distilled water because of the copper and iron present.

Ascorbic Acid Determination (Continued):

PROCEDURE:

The following directions are for a cuvet needing no more than 3.5 ml. solution for measuring absorbance. If more is needed, the directions must be scaled upward.

1. Add 3.0 ml. 3.0% metaphosphoric acid to 2.0 ml. serum, plasma, or urine. Mix and filter or centrifuge.

2. Set up the following in cuvets:

REAGENT BLANK	- 1.2 ml. 3.0% metaphosphoric acid + 0.8 ml. water + 0.5 ml. 4.37% sodium citrate dihydrate.
STANDARD	- 2.0 ml. ascorbic acid standard (16 micrograms) + 0.5 ml. 4.37% sodium citrate dihydrate.
UNKNOWN	- 2.0 ml. filtrate + 0.5 ml. 1.7% sodium citrate dihydrate for serum filtrate (or 4.37% for urine filtrate).

3. Mix contents of each cuvet and then, working with one at a time, proceed as follows: Add 1.0 ml. Working Indophenol Solution, mix quickly and read absorbance at 520 nm. or with a filter in this region at 30 seconds against a water blank. This gives A_b, A_s, and A_x.

4. To each cuvet, add a small pinch of ascorbic acid crystals to decolorize excess dye, mix, and read absorbances again. These absorbances are A_{bb}, A_{sb}, and A_{xb}.

CALCULATION:

$$\text{mg. Ascorbic Acid/100 ml.} = \frac{(A_b - A_{bb}) - (A_x - A_{xb})}{(A_b - A_{bb}) - (A_s - A_{sb})} \times 0.016 \times \frac{100}{0.8}$$

$$= \frac{(A_b - A_{bb}) - (A_x - A_{xb})}{(A_b - A_{bb}) - (A_s - A_{sb})} \times 2$$

N. B. If $(A_x - A_{xb})$ is 0, or very nearly so, rerun the unknown on a dilution of the protein-free filtrate, dilution being made with 1.8% metaphosphoric acid. Appropriate modification in calculation must then be made.

Ascorbic Acid Determination (Continued):

NOTES:

1. Beer's Law.
The color obeys Beer's law on a Beckman DU Spectrophotometer at 520 nm. and with Klett filter Nos. 52 and 54.

2. Initial Dye Concentration.
The same absolute amount of dye is reduced by a given concentration of ascorbic acid whether the initial working indophenol reagent used contains 10 or 5.0 mg./100 ml., i.e., it is independent of indophenol concentration. Practical considerations thus dictate the choice of initial dye strength.

3. pH of Reaction Mixture.
Concentrations of sodium citrate are used to buffer the pH to about 4.0.

4. For Greater Accuracy with Urine.
Add 1.0 ml. p-chloromercuribenzoic acid solution to 3.0 ml. urine filtrate. Let stand 5 - 10 minutes and remove precipitate formed, consisting of excess reagent together with any insoluble complex formed by interaction with substances present, by centrifugation. Use 2.0 ml. supernate in test. Multiply final results by 4/3 to correct for this 3:4 dilution. p-Chloromercuribenzoic acid appears to decrease dye reduction by a standard by about 7.0%.

5. Serum vs. Plasma.
Most workers have used plasma rather than serum; the great instability of ascorbic acid in these samples almost makes it mandatory that plasma be used because of the greater speed with which plasma can be obtained from whole blood than serum from a clot.

6. Stability.
Practically all the vitamin C of whole blood is maintained in the reduced state, i.e., as ascorbic acid. It has been shown that serum or plasma and intact or hemolyzed erythrocytes inhibit oxidation of ascorbic acid. Ascorbic acid, however, is susceptible to oxidation by Cu^{++}, Fe^{+++}, and enzyme catalysis and these actions begin to overpower the opposing effect shortly after blood is drawn.

Serum or plasma is stable at room temperature no longer than 30 minutes and ascorbic acid completely disappears within

Ascorbic Acid Determination (Continued):

24 hours. Refrigeration apparently delays the oxidation but slight loss occurs in 3 or 4 hours. Ascorbic acid stays in the reduced state much longer in whole blood than in plasma presumably because of the greater concentration of sulfhydryl compounds in erythrocytes than in plasma.

ACCURACY AND PRECISION

The method appears to be fairly accurate for ascorbic acid in serum, plasma, and whole blood but not for urine. Reaction with sulfhydryl compounds, the chief source of non-specificity, is minimized by the presence of p-chloromercuribenzoic acid. The mean decrease in the apparent ascorbic acid content of plasma brought about by its use is 2.0%, of urine about 15%, and of erythrocytes about 50%. There is evidence that this treatment is quite successful in removing nearly all spurious reactants in the case of erythrocytes but only partially successful in the case of urine. Interference is also caused by phenolic compounds. It must also be remembered that this method will not distinguish closely related enediols but there is no evidence that this is of material significance with plasma or urine. The precision of the method is about $\pm$ 6.0%.

NORMAL VALUES:

The normal range for serum or plasma ascorbic acid is about 0.2 to 2.0 mg./100 ml. The Interdepartmental Committee on Nutrition for National Defense has adopted the following interpretation of plasma levels: Less than 0.1 mg./100 ml. indicates deficiency, 0.1 to 0.19 is low, and 0.2 to 0.4 is acceptable.

The normal range for whole blood is about the same as for plasma. Some of the published reports concerning the ratio of ascorbic acid in erythrocytes to that in plasma appear somewhat contradictory. This may be the result of the fact that at high levels (greater than ca. 1) plasma concentration is greater than erythrocyte concentration whereas at low concentration (less than ca. 0.6) the reverse is true.

REFERENCE: Henry, J. B.: CLINICAL CHEMISTRY: PRINCIPLES AND TECHNIQUES, Hoeber, pg. 715 - 719, 1965.

Ascorbic Acid Determination (Continued):

CLINICAL INTERPRETATION:

Seasonal changes occur in plasma ascorbic acid with higher levels during the summer months because of greater consumption. Levels of the vitamin decrease during pregnancy with lowest levels in the post-partum period. When an individual is receiving an ascorbic deficient diet, approximately four months transpire before the initial signs of scurvy occur. Deficient plasma and tissue ascorbic acid are associated with increased hemorrhagic tendency, increased susceptibility to infection, decreased wound healing and anemia.

Malabsorption may also predispose to decreased plasma and tissue ascorbic acid.

BARBITURATES AND SEDATIVES
(Gas-Liquid Chromatography)

PRINCIPLE: The specimen is acidified to form the free acids of the barbiturates, which are then extracted with CH_2Cl_2. The extract is then concentrated and injected into the GLC, where it is separated and recorded as peaks having characteristic retention times dependent upon the volatility and polarity of the compound. The amount of drug present is directly proportional to its peak height when this peak height is related to the peak height of the internal standard.

SPECIMEN: 1.0 ml. of serum, plasma or urine. If possible, avoid running urines. There is no restriction of the type of anticoagulant used. Hemolysis does not appear to affect the test. Specimen should be refrigerated and is stable in the frozen state indefinitely

REAGENTS:

1. Internal Standard Solution: Barbital
   a. STOCK: 100 mg./100 ml. in methanol; barbital (Merck or Mallinckrodt).
   b. WORKING: 0.2 mg./ml.
      Dilute Stock: 20 ml. q.s. to 100 ml. in deionized water.

2. Barbiturate Standard No. 1
   a. STOCK: 100 mg./100 ml. methanol. Dilute 100 mg. of each of the following individually in 100 ml. methanol:
      Butobarbitol (McNeill)
      Pentobarbitol (Abbott)
      Secobarbitol (Lilly)
      Meprobamate (Wallace)
      Phenobarbitol (Merck & Co. or Winthrop)

   b. WORKING STANDARD No. 1.

Barb.	Ml. of Stock	Final Concentration
Buta	2.0	2.0 mg%.
Pento	1.0	1.0 mg%.
Seco	1.0	1.0 mg%.
Mepro	3.0	3.0 mg%.
Pheno	4.0	4.0 mg%.

Volumetrically pipette the appropriate amount of each Stock Standard into dry 100 ml. volumetric flask. Evaporate to dryness under a stream of nitrogen in a warm sand

Barbiturates and Sedatives (Continued):

bath. Add 100 ml. of Blood Bank plasma or pooled serum. (Be sure that the plasma or serum has been analyzed and found negative for all barbs.) Let stand at room temperature for 24 hours, mixing occasionally. Aliquot and freeze. (The Standard may also be prepared by using deionized water to q.s. to 100 ml. instead of negative Blood Bank plasma.)

3. Barbiturate Standard No. 2
   a. STOCK: 100 mg./100 ml. methanol.
      Amobarbitol (Lilly)
      Glutethimide or Doriden (CIBA)

   b. WORKING STANDARD No. 2.

Barb.	Ml. of Stock	Final Concentration
Amobarb	1.0	1.0 mg%.
Doriden	1.0	1.0 mg%.

   Prepare in the same manner as Working Standard No. 1.

4. Methylene Dichloride (CH_2Cl_2) Reagent Grade.

5. Potassium Phosphate (KH_2PO_4)
   Saturated solution: 25 - 30 gm./100 ml. deionized water.

6. Chloroform ($CHCl_3$) Reagent Grade.

CHECK ALL REAGENTS AND STANDARDS AGAINST THE OLD AND MAKE SURE THEY ARE ACCEPTABLE BEFORE EXHAUSTING THE CURRENT SUPPLY!!

COMMENTS ON PROCEDURE:

1. Instrument used is a Hewlett-Packard Model 402 Gas Chromatograph with dual flame ionization detectors.

   A. Column: Silanized Glass Column 6' x 2 mm. ID

   B. Packing: 3.0% OV-17 on Gas Chrom Q and 5.0% OV-17

   C. Gas Readings:

Gas Readings:		Rotameter	Regulators on Tank
Helium	Flow Rate	2.5	40 psi
Hydrogen		2.5	8 psi
Oxygen		6.0	12 psi

Barbiturates and Sedatives (Continued):

D. Temperatures:
Column 206°C.
Injection Port 215°C.
Detector 250°C.

E. Chart speed: 40 inch/hour; ratio knob set on 4.

F. Paper: Varian #161A 11-5/16 inch width.

G. Alternate Column: 3' x 2 mm. ID
Packed with 3.0% PPE-20, Supelco Co., Bellafonte, Pa.
(A more polar liquid phase) to use when the identity of the material being chromatographed was doubtful. Used primarily to differentiate Pentobarbital and Phenacetin.

H. Sensitivity settings: Range 10 x 8 attenuation

2. 10 microliter Hamilton Syringe (Hamilton Co., Whittier, Calif.)

3. Quantitative determination is complicated by difficulties inherent in extraction and by absorption onto the GLC column. Use of an internal standard (chemically similar to compounds being analyzed) is the most satisfactory approach to this problem. Because this internal standard is added to the specimen prior to any of the reagents, the only volume measurements that need to be precise are:
   a. The 1.0 ml. of specimen.
   b. The 50 microliter of Internal Standard (equal to 1.0 mg%.)

   After these measurements are made, the determination is based completely on peak ratios.

4. Controls:
   At the present time, the only purchasable control is prepared by Hyland Laboratories. It is a urine specimen containing Pentobarbital reported by them as Phenobarbitol.

5. The F_x (Correction factor of sample relative to barbital) is the same whether you use a barbiturate standard q.s. with Blood Bank plasma or a barbiturate standard q.s. with deionized water.

6. At the present time we will not quantitate Doriden and Meprobamate, but rather just report their presence.

Barbiturates and Sedatives (Continued):

7. COLUMN PREPARATION FOR GLC

A. If packed, remove glass wool plugs from both ends; then take out packing.

B. Acid Wash: (Acid washing reduces the adsorptive properties of diatomaceous earth supports, glass wool and glassware, probably by removing of metal salts. -- The GLC of Steroids, J. K. Grant, p. 48 - 49)
   1). Rinse first with water, attaching one end of column to suction apparatus.
   2). Fill with dichromate cleaning solution.
   3). Let stand 10 minutes.
   4). Rinse well with water.
   5). Rinse with methanol.

C. Silanize: (Silanization converts the hydrophilic surface to a hydrophobic surface which will accept and retain a thin layer of stationary phase (OV-17 in our case). It also transform "active sites" (silanol groups) to less polar silyl ethers. -- The GLC of Steroids, J. K. Grant).
   1). Prepare solution:
       0.5 ml. dimethyldichlorosilane
       9.5 ml. toluene
   2). Using small beaker, pour solution into column.
   3). Let stand 20 minutes.
   4). Rinse with toluene.
   5). Rinse with methanol.
   6). Dry with stream of nitrogen by attaching tube to one end of column.

8. Column Conditioning -- No Flow Technique

A. Hook up column in oven by connecting one end to Injection Port and the other end to the dummy (do not attach one end to Detector).

B. Set Nitrogen flow to 40 ml./min. (normal setting).
Set temperature of oven to 205°C.
Let sit one hour.

C. Turn nitrogen flow off.
Raise oven temperature to 250°C.
Let sit two hours.

Barbiturates and Sedatives (Continued):

D. Turn temperature back to 205°C. (normal setting). After a few hours, raise nitrogen flow to 2.5. Let sit 15 hours.

E. Cool down oven (set oven control to fan only). Disconnect newly conditioned column. Put on new O-rings if necessary. Reconnect column; one end to Injection Port, the other to the Detector.

F. Run several plasma extracts through column as conditioner.

MAINTENANCE:

1. Change septum at end of day.
   a. Unscrew Injection Port.
   b. Remove old septum.
   c. Put in new septum.
   d. Replace Injection Port and screw in gently.

2. Weekly:
   a. Check ink supply in recorder pen.
   b. Check Chart Paper supply.
   c. Check gases.
   d. Check reagent and standard supply.

SPECIMEN PREPARATION:

1. Pipette volumetrically 1.0 ml. sample or standard or urine control, in duplicate, into 6.0 ml. stoppered centrifuge tubes.

2. Add 50 microliters of Internal Standard (use Drummond pipette) to all tubes.

3. To each add 0.5 ml. saturated KH_2PO_4 with a 2.0 ml. serological Pipet. Mix briefly.

4. Add 4.0 ml. volumetrically CH_2Cl_2. Stopper and invert 20 times. Centrifuge at moderate speed for 2 minutes.

5. Aspirate off the protein-containing aqueous layer.

6. Pipette 2.0 ml. volumetrically of the CH_2Cl_2 layer into a 5.0 ml. tapered tube and evaporate to dryness under nitrogen in a 40 - 50°C. sand bath.

Barbiturates and Sedatives (Continued):

7. Wash down the walls of tube (from top down) with approximately 0.5 ml. CH_2Cl_2 and re-evaporate.

8. Add 50 microliters of $CHCl_3$ (Chloroform) with Drummond pipette and mix gently.

9. Rinse Hamilton Syringe several times with chloroform. Rinse again with chloroform extract and inject approximately 1.0 microliter of this into GLC.

10. If the chromatogram yields a peak greater than full-scale deflection, change the electrometer sensitivity to the next higher setting and reinject the sample. The baseline will need adjustment when switching sensitivity.

11. Chromatograph the standard samples each day before and inbetween the test samples and determine the F.I.D. factors for the day.

12. CALCULATION:

    The F.I.D. response factors are calculated by chromatographing a standard solution which has been extracted from plasma. The weights ("W") injected are known. The peak heights ("H") are measured. The ratio H/W is then calculated for each peak. The correction factor ("F") is calculated by dividing the H/W of each peak by the Internal Standard (Barbital) H/W. These factors are relative to Barbital; i.e., the Barbital factor is arbitrarily set equal to 1.00.

    The weight injected of sample (x) can be calculated as follows:

$$W_x = \frac{Wa \cdot Hx}{Fx \cdot Ha}$$

    WHERE: Wa = Weight of Std. Barbital in mg%.
    Ha = Height of Std. Barbital
    Hx = Peak height of sample
    Fx = Correction factor of sample relative to Barbital

    W_x = Mg%. barbiturate.

    NOTE: See Item No. 5. under "Comments on Procedure", concerning reporting of Doriden and Meprobamate.

Barbiturates and Sedatives (Continued):

NORMAL VALUES:

THE MEANING OF BLOOD LEVELS OF BARBITURATES AND OTHER SEDATIVES

The values below are rough guides only. Very low blood concentrations of sedative drugs may be dangerous when the patient is debilitated from illness or injury or has taken other drugs.

Distinctions are often made between the short-acting barbiturates-those that act for less than three hours, and the long-acting-those that act for more than six hours. In general, coma is associated with lower levels of the short-acting barbiturates:

"Coma that may be associated with shock may be seen with blood levels ranging from 3.0 mg. per 100 ml. whole blood for short-acting barbiturates to 9.0 mg./100 ml. for long-acting barbiturates in fatal cases." (Todd-Sanford in I. Davidsohn and J. B. Henry, eds., CLINICAL DIAGNOSIS, 14th Ed., W. B. Saunders, Phila., 1969, pp. 588-589).

1. Blood Levels associated with average daily dosage schedules

   Six patients receiving daily doses of phenobarbital for psychiatric reasons had equilibrium concentrations of 1.3 to 3.3 mg./100 ml. plasma (T. C. Butler et al.: JPET 111:425, 1954). Twenty patients being treated for epilepsy received daily doses ranging from 50 to 400 mg. of phenobarbital. The serum levels attained after long term treatment (1 to 28 years) varied between patients from about 1.0 to 6.0 mg./100 ml. serum (Lous, P.: Acta Pharmacol. Toxicol 10:166, 1954).

2. Blood Levels associated with "sedative" effects

   Studies conducted on 40 healthy male prisoners rated effects on the scale shown below, and gave blood concentrations for degrees of depression.

Barbiturates and Sedatives (Continued):

	Impaired motor Performance	Sedated, calm, easily aroused	Comatose, depressed respiration	"Compatible with death" in ill or aged	"Usual lethal level"
		Mg./100 ml. Blood			
Pento	2	0.5 - 3.0	1.0 - 1.5	1.2 - 2.5	1.5 - 4.0
Seco	2	0.5 - 3.0	1.0 - 1.5	1.5 - 2.5	1.5 - 4.0
Amo	3	0.2 - 1.0	3.0 - 4.0	3.0 - 6.0	4.0 - 8.0
Buta	5	0.3 - 2.5	4.0 - 6.0	5.0 - 8.0	6.0 - 10
Pheno	10	0.5 - 4.0	5.0 - 8.0	7.0 - 12	10 - 20

(Parker, K. D. et al.: Clin. Toxicol., 3:131, 1970).

3. Blood Levels in a series of Untreated Cases at Autopsy

Pentobarbital (15 cases)	1.2 - 6.5 mg./100 ml. blood
Amobarbital (17 cases)	2.7 - 4.7 mg./100 ml. blood
Butabarbital (5 cases)	3.1 - 7.3 mg./100 ml. blood
Phenobarbital	15.1 - 16.4 mg./100 ml. blood

(Gillett, R. and Warburton, F. G.: J. Clin. Path., 23:435, 1970).

4. Blood Levels of Glutethimide (Doriden[R]) and of Meprobamate (Equanil[R]), Miltown[R]).

Sedative effects on human volunteers given therapeutic amounts of these drugs were rated as th CNS depression, and the blood levels achieved for this degree of depression were recorded.

Mild sedative effects were achieved with plasma levels of 0.4 to 0.8 mg./100 ml. for Glutethimide and 0.2 to 2.3 mg. per 100 ml. for Meprobamate (Parker, K. D.: Clin. Toxicol., 3:131, 1970).

In a series of comatose patients, the following ranges of plasma concentrations were detected:

Barbiturates and Sedatives (Continued):

Glutethimide (13 cases)	0.6 - 5.1 mg./100 ml. plasma (mean 1.9 mg./100 ml.)
Meprobamate (10 cases)	7.0 - 20.8 mg./100 ml. plasma (mean 12.9 mg./100 ml.)

(Bloomer, H. A. et al: Ann. Int. Med., 72:223, 1970).

REFERENCES:

1. Martin, H. F. and Driscoll, J. L.: "Gas Chromatographic Identification and Determination of Barbiturates". Anal. Chem., 28:345, 1966.

2. Anders, M. W.: "Rapid Micro Method for the Gas Chromatographic Determination of Blood Barbiturates". Anal. Chem., 38:1945, 1966.

3. McMartin, C. and Street, H. V.: "Gas-liquid Chromatography of Sub-microgram Amounts of Drugs". J. Chromatogr., 23:232, 1966.

4. Leach, H. and Toseland, P. A.: "The Determination of Barbiturates and Some Related Drugs by Gas Chromatography". Clin. Chem. Acta, 20:195, 1968.

5. Thompson, H. L. and Decker, W. J.: "Analysis of Blood", Am. J. Clin. Path., 49:103, 1968.

6. Parker, K. D., Wright, J. A., Halpern, A., and Hine, C.: "The Determination of Barbiturates". J. of Forensic Sci. Soc., 8:125, 1968.

7. Alber, L. L.: "All Purpose Gas-liquid Chromatographic Column for Pharmaceuticals". J.O.A.C., 52:1295, 1969.

8. Sunshine, I., Maes, R., and Faracci, R.: "Detection of Glutethimide and its Metabolites in Biologic Specimens". Clin. Chem., 14:595, 1968.

9. Sine, H. E., McKenna, M. J., Rejent, T. A., and Murray, M. H.: "Emergency Gas-liquid Chromatographic Determination of Barbiturates and Glutethimide in Serum". Clin. Chem., 16:587, 1970.

QUANTITATIVE ESTIMATION OF BARBITURATES
(by Ultra-violet Spectrophotometry)

PRINCIPLE: The ultra-violet spectrophotometric method is based on the change in the absorption spectra of barbiturates in strong alkali and in solution at pH 10.5. The changes in optical densities depend on the resonance form. When optical densities in pH 10.5 solution are subtracted from those in strong alkali, differences appear that are highly characteristic of all barbiturates, except the N-methyl and the thio-derivatives. By comparing the differences at various wavelengths with that at 260 nm., ratios are obtained that can differentiate among many of the commonly used barbiturates.

Also by alkaline hydrolysis, one can differentiate the duration of pharmacological action (whether it is a long, intermediate, or a short acting barbiturate).

SPECIMEN: 6.0 ml. whole blood, serum, plasma, urine, or gastric washing.

REAGENTS:

1. Ammonium Chloride, 16%

2. Sodium Hydroxide, 0.45 N. pH 13.0 - 13.5.
   To be titrated before use.

3. Phosphate Buffer, pH 7.0 (Clark and Lubs)
   To 50 ml. of 0.1 M. KH_2PO_4 add 30 ml. of 0.1 N. NaOH and dilute to 100 ml. with distilled water. Prepare on the day of use.

4. 0.1 N. NaOH

5. 0.1 N. HCl

6. 0.1 M. KH_2PO_4

7. Chloroform, Analytical Reagent
   Wash first with 0.1 N. Hydrochloric acid, then with 0.1 N. sodium hydroxide, and finally three or more times with water until the aqueous phase is neutral to litmus. Wash by shaking a large volume of Chloroform with the wash solution; allow the two phases to separate and discard the aqueous layer. May also use 1 x distilled $CHCl_3$.

Quantitative Estimation of Barbiturates (Continued):

8. Stock Barbiturate Standards, 40 mg./100 ml.
Weigh out 43.8 mg. of the Sodium salt of a barbiturate, (Phenobarbital, Amobarbital, or Secobarbital) into a 100 ml. volumetric flask, dissolve in water and make up to volume. The stock standards contain 0.4 mg./1.0 ml.

9. Working Barbiturate Standard, 4.0 mg%.
10 ml. of the stock standard are diluted to 100 ml. with water to give a concentration of 4.0 mg./100 ml.

PROCEDURE:

1. 6.0 ml. of blood, serum, or plasma are pipetted into a 125 ml. separatory funnel. Blank - 6.0 ml. of distilled water. Standard - 6.0 ml. of working standard (4.0 mg./100 ml.).

2. 90 ml. of Chloroform are added, and the mixture is shaken for 5 minutes.

3. The layers are allowed to separate and the chloroform extract is filtered into a 100 ml. volumetric flask, through a rapid flow filter paper. The use of a small funnel is advisable to minimize surface area. The filter paper is washed with 10 ml. of Chloroform, and the extracts are brought up to volume.

4. 80 ml. aliquot of the filtered Chloroform extract is put into another 125 ml. separatory funnel and 10 ml. of 0.45 N. NaOH is added.

5. The mixture is shaken for 5 minutes and the layers are allowed to separate.

6. The Chloroform layer is discarded, and the alkaline solution containing the barbiturate is placed into a small test tube, and centrifuged for 10 minutes to separate any residual Chloroform.

7. Place 3.0 ml. of the alkaline solution into a quartz cuvette. Determine the ultra-violet absorption characteristics in an ultra-violet spectrophotometer (Beckman DU Spectrophotometer) by reading the absorbances at the following wavelengths: 228, 232, 235, 240, 247, 249, 252, 255, 260, and 270.

8. Add 0.50 ml. of 16% Ammonium Chloride to the NaOH extract of the $CHCl_3$ which reduces its pH to 10.3.

Quantitative Estimation of Barbiturates (Continued):

9. Mix the solutions and then reread absorbances at the same wavelengths.

URINE OR GASTRIC WASHINGS:

A. 6.0 ml. sample is acidified with 0.1 N. HCL prior to the initial $CHCl_3$ extraction.

B. The $CHCl_3$ extract is washed with 10 ml. of pH 7.0 Phosphate Buffer.

C. 80 ml. of the $CHCl_3$ extract is then treated as in serum above.

HYDROLYSIS:

A. 5.0 ml. of the 0.45 N. NaOH extract is pipetted into a tube.

B. Place in boiling water bath for exactly 15 minutes, and then immediately put in ice.

C. 3.0 ml. is pipetted into a quartz cuvette, and is read at 260 nm.

D. 0.5 ml. of NH_4Cl is added, mixed, and then reread at 260 nm.

CALCULATIONS:

1. The Ammonium Chloride readings are multiplied by 7/6 to correct for the volume.

2. Plot the optical density reading of the standards and the test versus the different wavelengths. Compare the peaks.

3. Determine the difference ratios by the following formula:

$$\text{Difference Ratios} = \frac{\text{Optical density difference at a given wavelength}}{\text{Optical density difference at 260 nm.}}$$

4. Milligrams of barbiturate/100 ml. =

$$\text{Conc. barbiturate Std.} \times \frac{\text{Absorbance difference at 260 nm. of unknown}}{\text{Absorbance difference at 260 nm. of standard}}$$

5. Hydrolysis:

% Barbiturate remaining after hydrolysis = Absorbance difference at 260 nm.

Quantitative Estimation of Barbiturates (Continued):

$$\frac{\text{Absorbance difference at 260 nm. after hydrolysis} \times 100}{\text{Absorbance difference at 260 nm. before hydrolysis}}$$

% hydrolyzed = 100% - % Barbiturate remaining.

REFERENCES:

1. Sunshine, I.: STANDARD METHODS OF CLINICAL CHEMISTRY, 3:46-53, 1961.

2. Goldbaum, L. R.: ANALYTICAL CHEMISTRY, 24:1604-1607, 1952.

CLINICAL INTERPRETATION:

Refer to Barbiturates and Sedatives (GLC Procedure) on pages 61 and 62.

TOTAL BILIRUBIN
(Spectrophotometric Method)

PRINCIPLE: The yellow color of serum or plasma is due in part to bilirubin. This fact is the basis for the icterus index test. The icterus index is an imprecise and relatively nonspecific test; however, the simplicity of the test and its potential utility have led to numerous refinements and applications. The test described here is the direct spectrophotometric determination of total bilirubin. As a refinement of the icterus index, precision in this test has been achieved by the use of an ultramicrodilutor, a buffered diluent, and a spectrophotometer. Specificity has been improved by making corrections for hemoglobin and turbidity. This is accomplished by substracting the absorbance at 551 nm. from the absorbance at 461 nm. The latter wavelength is the peak for bilirubin; whereas, absorbance due to hemoglobin and turbidity is practically the same at these two wavelengths; by subtracting the absorbance at 551 nm. from the absorbance at 461 nm., any absorbance due to hemoglobin and turbidity is effectively cancelled out of the final answer. Thus, this method is a very simple, rapid, precise, ultramicro determination for bilirubin. The lack of specificity inherent in the method does not detract from its utility, particularly when it is restricted to its intended uses of monitoring neonatal jaundice and estimating bilirubin levels in adults.

SPECIMEN: 20 microliters of serum or plasma. Specimen should be processed as soon as possible or stored in a light-tight container at refrigerator temperature until assayed.

REAGENTS AND EQUIPMENT:

1. <u>Citrate Buffer</u>, 5% (W/V)
   Weigh out 50 gm. of Reagent Grade Sodium Citrate into a one liter volumetric flask. Dilute to volume with deionized water, pH = 8.8.

2. <u>Ultramicro dilutor</u>
   Set up to dilute 20 microliters of specimen with 1.0 ml. of sodium citrate.

3. <u>Gilford Spectrophotometer</u> 300N
   Instrument should be fitted with a microaspiration cuvette.

4. <u>Calibration Standard</u>
   An elevated Bilirubin Control with an Assay of around 20 mg%. Always check the assay, which varies with different lot numbers.

Total Bilirubin (Continued):

Several dilutions are made to give bilirubin concentrations of 2.0, 5.0, 7.0 15, and 20 mg%.

PROCEDURE:

1. Standardization. A calibration curve is made by diluting 20 microliters of each of the diluted standards with 1.0 ml. of sodium citrate. Set the machine to 0.000 absorbance at 551 nm. with a blank containing citrate buffer. Take the reading of the buffer at 461 nm. Take the absorbance readings of the standards at 461 nm. and 551 nm. Calculate the true absorbance of the bilirubin standards. Plot absorbance versus concentration on regular graph paper.

   A = (A of test at 461 nm.) — (A of Blank at 461 nm.) —

   (A of test at 551 nm.)

   The A of Blank at 551 nm. was preset to zero.

2. Using the dilutor, dilute 20 microliters of serum (Test) with 1.0 ml. of sodium citrate.
   Blank - Sodium Citrate
   Controls - Abnormal Control and Elevated Bilirubin Control.

3. Using the blank, set the machine to 0.000 absorbance at 551 nm. Take another reading at 461 nm.

4. Read the Test at 461 nm. and 551 nm.
   Calculate the bilirubin absorbance and refer to the curve.

NOTES:

1. Indirect sunlight lowers bilirubin values drastically. After standing for ½ hour under the influence of indirect sunlight, the values are lowered by an average of 7% of the original concentration.

2. The 5% sodium citrate maintains the pH of the solution, making absorbance readings more stable. To keep the effects of light on bilirubin to the minimum, read test very soon after each dilution.

NORMAL VALUES:	Adults	- less than 0.8 mg%.
	Children	- 0.1 to 0.7 mg%.
	Infants	- 5.0 mg%.

Total Bilirubin (Continued):

REFERENCES:

1. Henry, R. J., Golub, O. J., Berkman, S., and Segalove, M.: Am. J. Clin. Path., 23:841, 1953.

2. Chiamori, N., Henry, R. J., and Golub, O. J.: Clin. Chim. Acta, 6:1, 1961.

3. Gambino: MANUAL ON BILIRUBIN ASSAY, ASCP, 1968.

CLINICAL INTERPRETATION:

The three main causes for jaundice are hemolytic, hepatic and obstructive. The main reason for elevation of the non-conjugated bilirubin is hemolytic anemia. Conjugated bilirubin may rise in hemolysis if the liver is competent.

Other Etiologies for an elevated non-conjugated bilirubin are:

1. Presence of a large hematoma
2. Hemorrhagic pulmonary infarction
3. Utilization of a large number of blood bank units of blood in blood transfusions which are not fresh
4. Absence of liver glucuronyl transferase (Crigler-Najjar Syndrome)
5. Deficiency of liver glucuronyl transferase (Gilberts Syndrome)

There are many causes for liver disease responsible for a hepatic type of jaundice. Patients who have hepatic disease have elevation of conjugated and non-conjugated bilirubin. If the hepatic disease is due to infiltrative etiologies such as are seen in primary or metastatic cancer to the liver, granulomata, lymphoma, or cholestasis secondary to drugs, the conjugated bilirubin will be more elevated than the unconjugated bilirubin. Patients who have obstructive jaundice usually have an increased conjugated bilirubin. If liver disease develops after obstructive jaundice due to carcinoma of the head of the pancreas, choledocholithiasis, carcinoma of the bile ducts or sclerosing cholangitis, the unconjugated bilirubin will also become elevated. Patients with the Dubin-Johnson Syndrome also have an elevated conjugated bilirubin.

SERUM BILIRUBIN, DIRECT REACTING
(Ultramicro Method)

PRINCIPLE: Lathe and Ruthven have shown that a sharp distinction between the diazo coupling rate of conjugated and unconjugated bilirubin may only be achieved under the following conditions:

1. a low pH
2. a low concentration of diazotized sulfanilic acid

These two conditions are satisfied by the method presented, the reagents being optimal for the measurement of direct reacting bilirubin. Since standards of conjugated bilirubin are unavailable at the current time, the method must be standardized against a total bilirubin determination which uses the same reagents.

SPECIMEN: 40 microliters of unhemolyzed serum or plasma. Hemolysis will significantly lower the bilirubin result. For accurate results, specimen should be separated from cells as soon as possible, and stored in a light-tight container at refrigerator temperatures until assayed. Determination should be performed as soon as possible The instability of bilirubin when exposed to light and room temperature cannot be overemphasized.

REAGENTS AND EQUIPMENT:

1. Sulfanilic Acid, 0.2%
   Place 2.0 gm. of sulfanilic acid into a 1.0 liter volumetric flask, and bring to volume with 0.04 N. HCl. Stable indefinitely at room temperature.

2. HCl, 0.04 N.

3. Sodium Nitrite, Stock Solution, 5.0%
   Dissolve 12.5 gm. $NaNO_2$ in deionized water, and bring to volume in a 250 ml. volumetric flask. Store in dark bottle at refrigerator temperatures; reagent is labile at room temperature and when exposed to light.

4. Sodium Nitrate, Working Solution, 0.1%
   Place 2.0 ml. of the stock $NaNO_2$ into a 100 ml. volumetric flask, and bring to volume with deionized water. Prepare fresh solution weekly, and store in a dark bottle at refrigerator temperatures. Use fresh aliquots of this reagent with each set of determinations.

Serum Bilirubin, Direct Reading (Continued):

5. Brij.-Water, approximately 0.1%
   Add 1.0 ml. Brij.-35 to 1000 ml. deionized water.

6. Methanol, absolute
   This reagent is only required for standardization.

7. Ultramicro Dilutor
   500 microliter flush syringe set to deliver 350 microliters of sulfanilic acid reagent.

   100 microliter sample syringe set to take up 20 microliters of specimen.

8. Gilford Spectrophotometer 300 N
   Instrument should be fitted with a microaspiration cuvette.

COMMENTS ON PROCEDURE:

1. The presence of hemoglobin in the sample will significantly affect the accuracy of the diazo reaction. While the exact mechanism is unknown, hemolysis interferes with the coupling reaction resulting in a loss of azobilirubin, which is proportional to the amount of hemoglobin present. It has also been suggested that hemolysis interferes with the final reading of azobilirubin values, due to the increased rate of conversion of hemoglobin to methemoglobin as a result of the presence of nitrite in the test solution and its absence in the blank.

   Increased amounts of protein will proportionally decrease the amount of azobilirubin formed; this has been proposed as a partial explanation for the effect of hemolysis.

2. The instability of bilirubin when exposed to light and room temperature cannot be over emphasized.

3. When performing the direct bilirubin determination, timing of the readings is critical to reproducible results, particularly when the specimen has an elevated value.

4. Specimens with a direct bilirubin value greater than 20 mg%. must be repeated on dilution. Dilute with saline.

5. Standardization of the azobilirubin absorption values must be achieved by carrying the direct bilirubin procedure on to a

Serum Bilirubin, Direct Reading (Continued):

total bilirubin content. The addition of methanol will dilute the direct test volume in half; one must keep this in mind when assigning bilirubin values to the absorbance units.

6. Standardization

A. A series of specimens or control sera, or standards of known total bilirubin concentration (ranging in value up to 20 mg%.), are used.

Total Bilirubin Determination

B. Perform steps #1 through #4 of the direct bilirubin procedure on the above samples.

C. After a five minute time period, add 0.7 ml. of methanol to each tube, and vortex well. Allow five minutes of color development.

D. Determine absorbance value of specimen, and specimen blank against water at 540 nm. using the Gilford microaspiration cuvette.

E. Subtract absorbance value of specimen blank from that of the specimen.

F. Plot mg%. against absorbance on linear graph paper. From this curve, prepare a calibration chart for direct bilirubin. The absorbance values for a given mg%. should be multiplied by 2 to account for the difference in dilution between the direct and total methods.

PROCEDURE:

1. Prepare two microcentrifuge tubes or A.A. cups for each specimen to be analyzed. One tube is for sample color development, and one is reserved for the sample blank.

2. Using the ultramicrodilutor, draw up 20 microliters of freshly aliquoted 0.1% sodium nitrite, and flush this with 350 microliters of sulfanilic acid. Deliver into the color development tube, and vortex or mix thoroughly.

Serum Bilirubin, Direct Reading (Continued):

3. Take up 20 microliters of serum, and flush with 350 microliters of sulfanilic acid into the sample blank tube; dilute with an additional 350 microliters of sulfanilic acid. Vortex or mix thoroughly.

4. Take up a second 20 microliters of serum and flush into the tube with diazotized sulfanilic acid. Vortex and start timer.

5. At exactly one minute, determine absorbance of the azobilirubin against a Brij.-water blank at 540 nm. using the Gilford micro-aspiration cuvette. Determine the absorbance of the specimen blank against Brij.-water.

6. Subtract absorbance of specimen blank from absorbance of specimen. Determine the mg%. direct reacting bilirubin from the calibration chart. Repeat values greater than 20 mg%. on dilution with saline.

NORMAL VALUES: Less than 0.1 mg%.

REFERENCES:

1. Lathe, R. and Ruthven, C.: "Factors Affecting the Rate of Coupling of Bilirubin and Conjugated Bilirubin in the van den Bergh Reaction", J. Clin. Path., 11:155, 1958.

2. Jacobs, Henry, R., Segalove, M.: "Studies on the Determination of Bile Pigments", Clin. Chem., 10:433, 1964.

3. Amatuzio, D.: "The Rate of Coupling of Diazotized Sulfanilic Acid with Bilirubin", Archives of Biochemistry and Biophysics 86:77, 1960.

4. Meites, S. and Hogg, C.: "Studies on the Use of the van den Bergh Reagent for Determination of Serum Bilirubin", Clin. Chem., 5:470, 1959.

5. Gambino: MANUAL ON BILIRUBIN ASSAY, ASCP, 1968.

CLINICAL INTERPRETATION:

Refer to Interpretation section under Total Bilirubin - Spectrophotometric Method on page 70.

SERUM OR URINE BROMIDE DETERMINATION (Modified)

PRINCIPLE: Bromide is an important component of many patent medicines (particularly sedatives) and is toxic in high concentrations, as a result of its replacement of chloride. Bromide imparts a brownish color to gold chloride in solution. The density of the color is directly proportional to the concentration of the bromide, rendering the reaction suitable for the quantitative analysis of bromide. The minute quantity of bromide normally present in the blood (approximately 1.0 microgram/100 ml.) is not significant. The sensitivity of the method is adequate for measuring the level during bromide therapy in cases of suspected bromide poisoning

SPECIMEN: 0.100 ml. of a serum or centrifuged urine specimen.

REAGENTS AND EQUIPMENT:

1. Trichloroacetic acid, 10%
   Weigh out 100 gm. of trichloroacetic acid into a liter of volumetric flask. Dissolve and make up to volume with deionized water.

2. Acid Gold Chloride, 0.25%
   The 0.5% solution is available from HARLECO. Dilute 1:1 with deionized water.

3. NaCl, 0.06%
   Place 0.6 gm. NaCl in a 1000 ml. volumetric flask, and bring to volume with 10% TCA.

4. Stock Bromide Standard (1000 mg%.)
   Weigh out 1000 mg. of sodium bromide into a 100 ml. volumetric flask, and bring to volume with deionized water.

5. Working Bromide Standards (50 mg%. and 200 mg%.)
   50 mg%. = 5.0 ml. of Stock Std. diluted to 100 ml. with 0.06% NaCl in TCA.
   200 mg%. = 20 ml. of Stock Std. diluted to 100 ml. with 0.06% NaCl in TCA.

6. Disposable Micropipettes
   Wiretrol pipettes calibrated to contain and deliver 0.100 ml. of solution.

7. Eppendorf Pipette
   To deliver 0.500 ml. of solution.

Serum or Urine Bromide Determination (Continued):

8. Gilford Spectrophotometer, 300 N
   Equipped with a microaspiration cuvette.

CALIBRATION CURVE:

Dilute Stock Standards to give the following concentrations:

50 mg%. = 5.0 ml. of Stock Std. diluted to 100 ml. with 0.06% NaCl in TCA.
100 mg%. = 10 ml. of Stock Std. diluted to 100 ml. with 0.06% NaCl in TCA.
200 mg%. = 20 ml. of Stock Std. diluted to 100 ml. with 0.06% NaCl in TCA.
300 mg%. = 30 ml. of Stock Std. diluted to 100 ml. with 0.06% NaCl in TCA.
400 mg%. = 40 ml. of Stock Std. diluted to 100 ml. with 0.06% NaCl in TCA.

1. Pipette in duplicate into small disposable test tubes:
   BLANK = 1.0 ml. of 10% TCA + 0.100 ml. of deionized water using 100 lambda Wiretrol.
   STANDARDS = 1.0 ml. of 10% TCA = 0.100 ml. of standards using 100 lambda Wiretrol. Rinse two or three times with the solution. Mix all tubes well.

2. Using an Eppendorf pipette, deliver 0.500 ml. of the above solutions into 3.0 ml. AutoAnalyzer cups.

3. Add 0.500 ml. of 0.25 % Gold Chloride solution using an Eppendorf pipette.

4. Mix well, and read at 440 nm. on the Gilford 300N equipped with a microaspiration cuvette. Set the Blank to zero, and read absorbance of the standards. Read within 5 minutes.

5. Plot the concentration of Bromide versus absorbance on graph paper.

PROCEDURE:

1. Using small disposable test tubes, set up:
   BLANK = 0.10 ml. of water using 100 lambda Wiretrol.

   STANDARDS
   a. 50 mg%. = 0.10 ml. of Working Std.
   b. 200 mg%. = 0.10 ml. of Working Std.

Serum or Urine Bromide Determination (Continued):

TEST = 0.10 ml. of specimen.

2. Add 1.0 ml. of 10% to all tubes. Mix well and centrifuge.

3. Pipette 0.50 ml. of filtrate using an Eppendorf into 3.0 ml. AutoAnalyzer cup.

4. Add 0.50 ml. of 0.25% $AuCl_3$.

5. Cap, mix well, and read on the Gilford 300N equipped with the microaspiration cuvette. Set the machine at 440 nm., and set the Blank to zero.

6. Calculate the concentration from the standards.

$$\frac{\text{O.D. Unknown}}{\text{O.D. Standard}} \times \text{Conc. Std.} = \text{mg\%. Sodium Bromide}$$

NORMAL VALUES: For serum or urine:
Therapeutic level = 200 mg%.
Toxic level = 250 mg%.

REFERENCE: Natelson: TECHNIQUES OF CLINICAL CHEMISTRY, 3rd. Edition, Thomas, 1971.

CLINICAL INTERPRETATION:

Bromides are present in both organic and inorganic forms in various medicinals. They are utilized primarily for sedation and chemotherapy (Bromuridine). The drugs containing bromide are at times abused or ingested in overdosage accidentally. The minute quantity of bromide normally present in the blood is not significant (approximately 0.5 mg%.). Therapeutic levels are 100 - 200 mg%. and toxic levels are above 250 mg%. Hypochloremia without hyponatremia may occur in bromide poisoning with bromide replacing chloride in various body fluids.

BSP (BROMSULPHALEIN)

PRINCIPLE: The organic dye disodium phenoltetrabromphthalein sulfonate (sulfobromophthalein, bromsulphalein, BSP) is used to test the excretory function of the liver. BSP, injected intravenously at 5.0 mg. per kg. of body weight, is removed from the circulatory system by the parenchymal cells of the liver where it is conjugated with glutathione and then excreted by active transport into the bile. The test specimen is withdrawn 45 minutes after injection. The purple color of the BSP at alkaline pH, and its colorless state at acid pH, allow one to measure a serum level colorimetrically against a serum blank. The anion p-toluene-sulfonate is added to the alkaline buffer in order to negate the effect of protein on the spectral absorption curve of BSP. Specificity appears to be increased by controlling pH.

SPECIMEN: 180 microliters of fasting serum. Avoid hemolyzed specimens if possible.

REAGENTS AND EQUIPMENT:

1. Alkaline Phosphate Buffer, pH 10.6
   Place the following reagents into a 1 liter volumetric flask:
   24.4 gm. $Na_2HPO_4 \cdot 7H_2O$
   3.54 gm. $Na_3PO_4 \cdot 12H_2O$
   6.40 gm. sodium p-toluenesulfonate
   Dissolve in deionized water and bring to volume. Adjust to pH 10.6 - 10.7 with 1.0 N. NaOH or 1.0 N. HCl.

2. Acid Phosphate Buffer, 2.0 M.
   Into a 250 ml. volumetric flask, place 69 gm. $NaH_2PO_4 \cdot H_2O$. Dissolve and bring to volume in deionized water.

3. Stock BSP Standard, 10 mg%.
   Dilute 1.0 ml. BSP (5,000 mg%.) to 500 ml. with deionized water.

4. Brij.-Water, approximately 0.1%.
   Add 1.0 ml. Brij.-35 to 1000 ml. deionized water.

5. Serum Control, 20 mg%
   Add 20 ml. of 100% Stock BSP Standard to 80 ml. of pooled, BSP negative, non-icteric serum.

6. Calibration Curve
   Make the following dilutions volumetrically:

Bromsulphalein (Continued):

Ml. 10 mg%. dye	With Deionized water, dilute to following vol.	Conc. mg%.	Plot as % retention
--	no dilution	10.0	100%
5.0	10 ml.	5.0	50%
3.0	10 ml.	3.0	30%
1.0	10 ml.	1.0	10%
0.5	10 ml.	0.5	5%

Carry Standards through "Procedure" in same manner as a patient specimen.

7. Ultramicro Dilutor
   Flush syringe set at 0.9 ml., and sample syringe set to take up to 90 microliters.

8. Gilford 300 N
   Fitted with microaspiration cuvette.

9. Sequential Hamilton Dispenser, 1.0 ml.

COMMENTS ON PROCEDURE:

1. By minimizing the shift in pH (10.4 - 7.1), one reduces the interference due to turbidity hemolysis and bilirubin. With unbuffered pH changes, there is a shift in spectral absorbance curves of hemoglobin and bilirubin.

2. Time delays following additions of acid should be avoided, due to potential turbidity problems.

PROCEDURE:

1. Using the ultramicrodilutor, make duplicate dilutions of sample and control with alkaline buffer. One tube of each pair is used as a specimen blank, the other as the test sample. 90 microliters of sample is taken up in the dilutor and flushed with 900 microliters of buffer.

2. Using the sequential Hamilton dispenser, deliver 20 microliters of acid buffer into each of the tubes labeled "Blank".

3. Mix all tubes well, and determine absorbance at 580 nm. in Gilford microaspiration cuvette, using a Brij.-H_2O Blank in order to avoid inaccurate results due to carry-over in the

flow-through cuvette. When more than one test is to be read, all of the "Tests" (alkaline solutions) should be read first, and then, after thorough rinsing of the cuvette, all of the Blanks should be read.

4. Substract absorbance of specimen blank from the absorbance of the test sample. Take value of the patient from a calibration chart or calculate, using the standard, according to the following formula:

$$\% \text{ Retention} = \frac{\text{A. of sample}}{\text{A. of standard}} \times 10$$

NORMAL VALUES: Less than 5% dye retention

REFERENCES:

1. Seligson, D., et. al.: "Determination of Sulfobromophthalein in Serum", Clinical Chemistry, 3:638, 1957.

2. Davidsohn, I. and Henry, John: TODD-SANFORD: CLINICAL DIAGNOSIS BY LABORATORY METHODS, 14th Ed., pg. 687 - 689, 1969.

CLINICAL INTERPRETATION:

The determination of the rate of removal of bromsulphalein from the blood stream is an excellent one to ascertain hepatic function. The elimination of BSP dye from the circulation is accomplished by the liver and the dye is excreted into the bile. The dye is conjugated prior to excretion into the bile canaliculi. The removal rate is more rapid post-prandially because of increased blood flow. Thus, the test must be performed fasting.

The dye should not be injected outside of the vein. It is extremely irritating and may cause a falsely elevated value if injected outside of the vein. False elevations also occur when patients are extremely jaundiced because of the competition for excretion by bilirubin, following radiopaque contrast x-ray procedures with competition for excretion by dyes also eliminated by the liver, fever, and failure of the phlebotomist to obtain the 45 minute specimen from the opposite arm to that in which the dye was inserted.

Bromsulphalein (Continued):

False negative values are found in individuals who have a low serum albumin and lack an adequate transport mechanism, and injection of the dye in a non-fasting state.

Thus, BSP retention over 8.0% in 45 minutes signifies hepatic disease or extra or intrahepatic obstruction resulting in inhibition of dye elimination.

CALCIUM
(Clark-Collip modification of Kramer-Tisdall)

PRINCIPLE: Calcium is precipitated directly from serum as insoluble calcium oxalate. The precipitate is washed, redissolved with acid and heat. The resulting oxalic acid is titrated with standardized potassium permanganate. The calcium is calculated from the amount of permanganate used to titrate the liberated oxalic acid.

$$CaC_2O_4 + H_2SO_4 \longrightarrow CaSO_4 + H_2C_2O_4$$

$$5H_2C_2O_4 + 2KMnO_4 + 3H_2SO_4 \longrightarrow 2MnSO_4 + K_2SO_4 + 10CO_2\uparrow + 8H_2O$$

SPECIMEN: 2.0 ml. unhemolyzed serum. The serum should be separated from the cells as soon as possible. The serum may be preserved by freezing. Plasma from heparinized blood may be used in place of serum.

REAGENTS:

Use deionized water for all dilutions.

1. Sodium Oxalate, 0.1 N.
   a. Dry pure anhydrous sodium oxalate ($Na_2C_2O_4$) in an oven at 100 - 105°C. for 12 hours.
   b. Dissolve exactly 1.675 gm. $Na_2C_2O_4$ in approximately 500 ml. deionized water in a liter volumetric flask; add 1.25 ml. concentrated sulfuric acid, mix and dilute to volume.

2. Potassium Permanganate, approximately 1.0 N.
   a. In a 2.0 liter Florence flask, dissolve 32 gm. pure $KMnO_4$ in a liter of deionized water.
   b. Digest for several hours at or near the boiling point with a watch glass covered funnel in the neck of the flask to act as a condenser.
   c. Cool and let stand over-night. Filter through a sintered glass filter and store in a brown bottle in the dark for 4 - 5 weeks to allow it to stabilize. Do not use filter paper for filtering or allow the solution to come in contact with organic material. This solution slowly disintegrates.
   d. It is not necessary to standardize this solution.

3. Potassium Permanganate, 0.01 N.
   a. Make a 1:100 dilution of the 1.0 N. permanganate solution in deionized water, mix well, and let stand for days in a

Calcium, Clark-Collip (Continued):

brown bottle in the dark before standardizing.

b. Standardization

1). Make a 1:10 dilution volumetrically of the 0.1 N. sodium oxalate.

2). Volumetrically pipette 1.0 ml. of the 0.01 N. sodium oxalate into a conical centrifuge tube, add 2.0 ml. of approximately 1.0 N. sulfuric acid, heat to about 70 to 80°C. in a water bath for 1 minute and titrate with the approximate 0.01 N. permanganate solution. Using a 5.0 ml. burette graduated to 0.01 ml., add one drop of $KMnO_4$, and mix until pink color disappears. Continue titrating until the addition of one drop produces a faint pink which persists for 1 minute. The temperature should not go below 70°C. This titration should be done in triplicate.

3). Calculate the normality of the $KMnO_4$ by dividing the ml. of oxalate by the ml. of permanganate used in the titration and multiplying by the normality of sodium oxalate (0.01).

$$\text{N. of } KMnO_4 = \frac{\text{N. of } Na_2C_2O_4 \times 1.0 \text{ ml.}}{\text{ml. } KMnO_4 \text{ used}}$$

4). The normality of the $KMnO_4$ should be checked every day.

4. Ammonium oxalate, 4% (W/V)
Filter to remove any precipitate.

5. Sulfuric Acid, approximately 1.0 N.
Add 28 ml. concentrated H_2SO_4 to about 900 ml. water in a liter volumetric flask. Cool and dilute to volume.

6. Ammonium Hydroxide, 2%
Dilute 2.0 ml. concentrated ammonium hydroxide to 100 ml. with water.

7. Stock Calcium Standard, 0.1 mg. Calcium/ml.
Dissolve 0.125 gm. of highly purified calcium carbonate in 12.5 ml. of 1.0 N. hydrochloric acid in a 500 ml. volumetric flask. Dilute to volume with water.

COMMENTS ON PROCEDURE:

1. The deionized water should be checked periodically for the presence of calcium. This is most easily done by using a

Calcium, Clark-Collip (Continued):

40 ml. conical centrifuge tube and using 10 ml. water in lieu of the serum. Continue, as for serum calcium. If any calcium is demonstrable, discard all reagents made up with that water and find a purer source of water.

2. The sodium oxalate in a 0.1 N. solution is fairly stable, but the 0.01 N. solution is not and should be made up fresh each week.

3. Sources of error:
   a. All the excess oxalate must be removed by washing with ammonium hydroxide or too high results will be obtained.
   b. There must not be any precipitate in the 4.0% ammonium oxalate.
   c. Careless draining of tubes.
   d. Allowing temperature of the solution to drop below 70°C. during titration.
   e. The temperature of the solution during titration should be between 60° and 80°C.; if the solution temperature at the end point is greater than 80°, the end point is less definite.

PROCEDURE:

1. Test should be run in triplicate.

2. Place 2.0 ml. of serum in heavy walled, pyrex, conical centrifuge tube (12 ml. size). As controls, include at least one commerically assayed serum and a recovery consisting of 1.0 ml. unknown serum and 1.0 ml. of the standard $CaCO_3$ solution.

3. To each tube, add 2.0 ml. deionized water and mix by agitation.

4. Slowly add 1.0 ml. 4.0% ammonium oxalate while shaking. Again mix well by agitation.

5. Stopper and allow to stand at least 3 hours in a 35 - 50°C. water bath.

6. Centrifuge for 15 minutes at 2,000 RPM.

7. Decant the supernatant carefully, and allow tube to drain on filter paper for a minute or two.

Calcium, Clark-Collip (Continued):

8. Wipe lip of tube carefully, and wash down sides of tube with approximately 3.0 ml. of 2% ammonium hydroxide from a fine tipped wash bottle. Mix on vortex mixer. Make sure precipitate is completely broken up into fine powder, so that unbound oxalate is washed away.

9. Repeat Steps 6, 7, and 8 once more.

10. Again recentrifuge, drain, and wipe.

11. Add 2.0 ml. approximately normal sulfuric acid, and place tubes in a boiling water bath for 1 minute.

12. Titrate with previously standardized $KMnO_4$ using a 5.0 ml. microburette until a faint pink persists. The temperature should not fall below 70°C. during the titration.

13. CALCULATION:

    1.0 ml. of exactly 0.01 N. $KMnO_4$ is equivalent to 0.2 mg./ml. calcium.

    mg. Ca./100 ml. = ml. $KMnO_4$ used times normality of $KMnO_4$ x 1000.

    Derivation of above equation is:

$$\text{ml. } KMnO_4 \frac{\text{normality of } KMnO_4}{0.01} \times 0.2 \times 100/2$$

$$\text{mg. Ca./100 ml.} = \frac{\text{N. } KMnO_4 \times \text{ml. } KMnO_4 \text{ used}}{\text{ml. } KMnO_4 \text{ (used to titrate oxalate)}}$$

14. CALCULATION OF RECOVERY:

    a. Mixture conc. = $\frac{\text{mg\%. Ca.}}{2} + \frac{\text{mg\%. in serum}}{2}$

    $\text{mg\%. Ca.} = 2(\text{Mixture Conc.} - \frac{\text{mg\%. in serum}}{2})$

    b. % Recovery = The ratio between Calculated Recovery and the actual weighted concentration of Ca. x 100.

Calcium, Clark-Collip (Continued):

NORMAL VALUES: 9.0 - 10.5 mg%. in serum

REFERENCES:

1. Kramer and Tisdall: J. Biol. Chem., 47:475, 1921.

2. Clark and Collip: J. Biol. Chem., 63:461, 1925.

CLINICAL INTERPRETATION:

The causes for hypercalcemia that is, a value over 10.7 mg%. are hyperparathyroidism or metastatic cancer involving bone. Hyperparathyroidism may be caused by a parathyroid adenoma or hyperplasia of the parathyroid glands. The usual primary neoplasms which metastasize to bone and manifest hypercalcemia are carcinoma of lung, breast, thyroid, kidney, and testis. Hodgkin's Disease and other lymphomas may involve bone and present with hypercalcemia. Multiple myeloma patients may develop extreme hypercalcemia related to the extensive bone destruction. Certain neoplasms may produce parathyroid hormone and clinically manifest hypercalcemia. Lung and renal carcinomas are noted for this phenomenon. Two reasons exist for hypercalcemia in sarcoidosis. These are: 1). Involvement of bone by the lesion, and 2). Increased sensitivity to vitamin D with increased calcium absorption from the gastrointestinal tract. Excessive intake of vitamin D or calcium may cause hypercalcemia. Other rare causes are hyperthyroidism, and thiazide utilization.

The most common cause for hypocalcemia is accidental removal of the parathyroid glands during thyroidectomy for thyroid disease. Another common cause is hyperphosphatemia due to renal failure. Excessive infusion of intravenous fluids will decrease serum albumin and decrease protein bound serum calcium. Likewise other causes for low serum albumin will result in hypocalcemia, malabsorption with steatorrhea may result in loss of calcium with hypocalcemia. Acute pancreatitis is accompanied by a decline in serum calcium. Calcium soaps develop in the abdomen when adipose triglycerides are hydrolyzed to glycerol and fatty acids. Other unusual causes for hypocalcemia are rickets, Cushings Syndrome, and pseudohypoparathyroidism in which there is a renal tubular end-organ non-responsiveness to parathormone.

CALCIUM BY ATOMIC ABSORPTION

PRINCIPLE: Calcium atoms absorb resonant energy of a wavelength of 4227 Angstroms. Calcium content of blood or other body fluids may be directly determined on a specimen diluted with a lanthanum oxide solution and measured using an atomic absorption spectrophotometer.

SPECIMEN: Blood: 40 microliters of non-hemolyzed serum (or heparinized plasma) run in duplicate is required. The serum (or plasma) should be separated from the cells as soon as possible. The serum may be preserved by freezing. Blood anticoagulated with substances such as oxalate or EDTA is unacceptable since the calcium ions have been chelated. Urine: At least 1.0 ml. of well mixed urine is required. Urine specimens may be preserved by adding to them 3% of their volume, concentrated hydrochloric acid. If specimens are kept for a long period of time, let the sample warm to room temperature and then centrifuge before making the analysis. Because the calcium content of urine is so variable, dilution may be necessary.

EQUIPMENT AND REAGENTS:

1. Microdilutor
   Sample syringe set to pick up 40 microliters of specimen and flush syringe set to dispense 2.0 ml. of the lanthanum diluent.

2. Perkin-Elmer Atomic Absorption Spectrophotometer Model 403
   With a 3-slot burner head and calcium cathode lamp (15 minute warm-up period). Acetylene-air fuel mixture.
   a. Range: VIS
   b. Wavelength: 211
   c. Slit: 4 (1.0 mm., 13A)
   d. Flame: Reducing (yellow traces)

3. Stock Lanthanum Solution
   5% lanthanum in 25% (V/V) HCl. Weight out 58.65 gm. of La_2O_3 into a liter volumetric flask and wet it with deionized water. Add 250 ml. of concentrated hydrochloric acid very slowly until material is dissolved. This should be done with ice or under cold running water. Dilute to 1000 ml. with deionized water.

4. Working Lanthanum Solution
   Dilute stock lanthanum 1:100 with deionized water. Usually a large volume of 4 liters or so is made up at one time for convenience.

Calcium by A.A. (Continued):

5. Stock Calcium Standard
Fisher Scientific, 1.0 mg./ml. (1000 ppm.) calcium solution is used.

6. Working Standard
5.0, 10, and 15 mg%. standards are made by diluting the Stock Standard.

Final Concentration of Standard	Volume of Standard
5.0 mg%.	1.0 ml.
10.0 mg%.	2.0 ml.
15.0 mg%.	3.0 ml.

Dilute each to 1000 ml. with Working Lanthanum Solution.
NOTE: These dilutions correspond to patient samples diluted 1:50, making it unnecessary to dilute standards before use.

COMMENTS ON PROCEDURE:

1. Lanthanum oxide is necessary in small concentrations to protect the calcium determinations from interference from phosphorus and aluminum. Anionic chemical interferences can be expected if lanthanum is not used in samples and standards.

2. Calcium determinations require a reducing flame. An oxidizing flame permits formation of calcium oxides which are refractory to analysis.

3. This dilution (1:50) of serum prevents protein precipitation while allowing sufficient sensitivity for analysis.

PROCEDURE:

1. Dilution

   A. Dilute controls and specimens in duplicate or triplicate, using autoAnalyzer cups.

   B. Samples above 15 mg%. should be diluted as many times as necessary to bring the reading into the 0 - 15 mg%. range of linearity.

Calcium by A.A. (Continued):

2. <u>Instrument Set-Up</u>

A. Turn power to "ON". (Warm-up time is 15 minutes)

B. Check beam position over burner with white card. (Alignment of lamp should be necessary only when lamp has been moved or changed.)

C. Check amperage to lamp and adjust if needed. (Amperage requirements and limits are marked on lamps. Proper amperage setting will approach upper limit as lamp ages.)

D. Open acetylene cylinder and air valve on wall.

Acetylene: Do not run tank below 100 lbs. pressure. Seconary gauge on tank set to 15 psi.

Air: Wall gauge set at 60 psi.

E. Turn gases switch to "ON", acetylene and air switches up.

F. Fuel Flow Check: 8 psi. 45 on flow gauge.

G. Oxidant Check: 29 - 30 psi. 70 on flow gauge. (perform fuel check first, so that accumulated acetylene will be flushed out when followed by oxidant check).

H. Ignite: Have aspirator tip <u>in water</u>!

I. Check for "reducing flame" and adjust if needed, with acetylene. (thin whitish band should be present above center burner cones. If white band is absent, the flame is too hot; if white bands are too high, the flame is too cool).

J. Check Aspiration Rate: It should be 5.3 ml./minute. Turn knurled knob counterclockwise to reduce, clockwise to increase. Knob is very sensitive. Turn very slowly.

K. Set proper wavelength for test. Peak energy level as follows:

1). Set needle to pink area with GAIN.

2). Find point of maximum deflection using Fine Adjust knob on wavelength selector.

Calcium by A.A. (Continued):

3). Set needle to center of Red area with GAIN.

L. Check for proper SLIT, VIS or UV range.

3. Run

A. Aspirate water and set Automatic Zero.

B. Aspirate Blank. Check reading for absence of contamination. Reset autozero.

C. Run middle standard (10 mg%. Calcium). Set proper reading with "Concentration Potentiometer".

D. Check linearity: Depress "Absorbance" button. Read standards in series. Plotted graph of absorbance vs. concentration should be linear on regular graph paper.

E. Recheck Autozero with Blank (Concentration button depressed).

F. Recheck middle standard.

G. Run unknowns.

4. Instrument Shut-down

A. Aspirate water, then a few minutes of 0.1 N. HCl (for cleaning), then water again.

B. Turn GASES switch off. The flame will go out.

C. Turn main acetylene tank OFF and house air OFF.

D. Turn GASES switch back ON and hit the SENSOR OVERRIDE button. (This bleeds the lines to the Control unit).

E. When control unit gauges have dropped to zero, turn SENSOR OVERRIDE back OFF, and turn GASES switch to OFF.

NORMAL VALUES: 9.0 - 10.5 mg%.

Calcium by A.A. (Continued):

REFERENCES:

1. Perkin-Elmer Manual 403 Spectrophotometer.

2. Elwell, W. T., and Gridley, J. A. F.: ATOMIC ABSORPTION SPECTROPHOTOMETRY, 2nd. revised Edition, 1967.

3. Howe, Sister M. Martin: "Atomic Absorption Spectrophotometry, Theory, Instrumentation and Application", American J. of Medical Technology, Vol. 33, No. 2, March-April, 1967.

CLINICAL INTERPRETATION:

Refer to Interpretation section under Calcium Procedure (Clark-Collip modification of Kramer-Tisdall) on page 86.

SERUM CALCIUM (MICRO)
(Turner Fluorometer)

PRINCIPLE: This method is based on the formation of a fluorescent chelate complex between calcium and calcein in a strongly alkaline solution. The intensity of fluorescent light emitted by a sample under constant input light intensity is directly proportional to the concentration of the fluorescent compound.

SPECIMEN: At least 50 lambda of serum or plasma, if heparin is the anticoagulant used. Specimen should be unhemolyzed. However, some authors claim that hemoglobin up to 500 mg%. and bilirubin up to 20 mg%. do not interfere.

EQUIPMENT AND REAGENTS:

1. KOH, 0.8 N.
   In preparation of reagent use reagent grade KOH; make a correction in calculating amount of salt required according to assayed value. Use deionized water.

2. Stock Calcein, 100 mg%.
   Use the calcein prepared by G. Frederick Smith Chemical Co. Weigh out 10 mg. calcein on an analytical balance and dilute to 10 ml. volumetrically with 0.8 N. KOH. Store in refrigerator shielded from light. Good for three weeks.

3. Working Calcein, 7.0 mg%.
   Volumetrically dilute 0.7 ml. of stock calcein with 0.8 N. KOH to 100 ml. This solution is not stable to storage and should be prepared fresh each time.

4. Stock Calcium Standard, 50 mg%.
   In preparing the standard, use reagent grade anhydrous $CaCO_3$ which has been dried in the oven overnight and dessicated for at least 48 hours or until constant weight is reached.

   Dissolve 1.2485 gm. of $CaCO_3$ in approximately 5 - 10 ml. of 6.0 N. HCl and dilute to 1.0 liter with deionized water. Stable indefinitely at room temperature when tightly capped.

5. Working Calcium Standards
   Prepare the standards according to the following chart, using 25 ml. volumetric flasks:

Calcium, Turner Fluorometer (Continued):

Conc. Calcium	Ml. Stock Standard	Ml. Deionized Water
6.0 mg%	3.0	22.0
8.0 mg%	4.0	21.0
10 mg%	5.0	20.0
12 mg%	6.0	19.0

Mix flasks thoroughly and store at room temperature indefinitely.

6. Turner Fluorometer, Model III
   Prepare the fluorometer as follows:
   Filter Selection:
   Primary (right side) - 47B + 2A.
   Secondary (left side) - 10% + (2A - 12) or (2A - 15) (as required).
   Aperture: None. Range Selector: 10 X.

   For an activating wavelength other than 365 nm., it is best to allow a 45 minute warm-up of the instrument.

COMMENTS ON PROCEDURE:

1. Deionized water should be used throughout the procedure. Distilled water contains considerable amounts of calcium ions which, if present in large amounts (greater than 0.2 ppm.), does not give linear calibration curve. Usually blank when zeroed in with dummy gives 30 units on fluorometer dial (or less than 2.0 mg%. Calcium) due mostly to residual fluorescence of reagents.

2. 0.8 N. KOH is used to have an alkaline media where the fluorescence generated is by Ca^{++} only with minute amounts given by barium and strontium. Other ions do not interfere.

3. Fluorescence is affected by time, temperature, KOH and calcein concentrations. Deteriorated calcein solution gives high blanks and decreased fluorescence with added calcium (gives low spread).

PROCEDURE:

1. Push Fluorometer Power switch to "ON" position. Push the "START" switch up and hold for 3 - 4 seconds to actuate the circuit for ultraviolet lamp. Allow the instrument to warm up for at least 45 minutes.

Calcium, Turner Fluorometer (Continued):

2. Volumetrically pipette 5.0 ml. of working calcein into 12 X 75 mm. polystyrene tubes, preparing a tube for the blank, each control, each working standard, and a duplicate set of tubes for each unknown.

3. Using a 20 lambda micro-cap pipette, add 20 lambda of controls, standards, and duplicate unknowns to their respective tubes of working calcein.

4. Nothing is added to the "Blank" tube, which consists only of the working calcein.

5. Cap all tubes and mix by inverting ten times.

6. Zero the fluorometer dial with the dummy. Read the working calcein "Blank" and record the value.

7. Set the highest working standard at 90 units on the dial and obtain a value for the other working standards. (Difference between two standards should be approximately 20 dial divisions.)

8. Read the values for controls and unknowns and repeat standard readings at the end of the batch.

9. CALCULATION:
On linear graph paper, plot the dial units versus the standard concentrations and read off values for the controls and unknowns in mg%. On samples reading lower or higher than lower or higher standards, make dilutions using the same amount of calcein, but adding 10 or 40 lambda of sample or as required. Make corrections after graph readings either multiplying or dividing by dilution factor. There is a slight dilution error (approximately 0.2%) which for practical purposes can be neglected.

NORMAL VALUES: 9 - 10.5 mg%.

REFERENCES:

1. Wallach, Surgenor, Soderburg, Delano: <u>Anal. Chem.</u>, 31: 456-460, 1959.

2. Wallach and Steck: <u>Anal. Chem.</u>, 35:1035-1044, 1963.

3. Kepner and Hercules: <u>Anal. Chem.</u>, 35:1238-1240, 1963.

4. Operating and Service Manual, Model III Fluorometer, G. K. Turner Associates, 2425 Pulgas Avenue, Palo Alto, California.

Calcium, Turner Fluorometer (Continued):

CLINICAL INTERPRETATION:

Refer to Interpretation section under Calcium Procedure (Clark-Collip modification of Kramer-Tisdall) on page 86.

CARBON DIOXIDE CONTENT
(Micro Automated Procedure)

PRINCIPLE: The method determines carbonate and bicarbonate presen in the diluted sample. The sample is aspirated and then mixed wit a stream of acid diluent containing an anti-foam agent. The strea passes through mixing coils, where the released carbon dioxide ent the air phase. Emerging from the coils the stream feeds into a tr (liquid-gas separator), the liquid goes to waste, and the gas phas containing carbon dioxide is aspirated. This gaseous stream now segments a weak alkaline buffer reagent stream containing an appro riate pH color indicator(phenolphthalein). The carbon dioxide gas is absorbed by the alkaline solution, causing a decrease in pH whi is reflected by a decrease in color of the indicator. The indicat stream passes into a 15 mm. tubular flowcell, where the color is measured at 550 nm.

SPECIMEN: 50 microliters of serum or plasma. Blood specimen is centrifuged and serum specimen kept well capped. Analysis should be performed as soon as possible.

REAGENTS AND EQUIPMENT:

1. Technicon Equipment
   a. Recorder
   b. Colorimeter Model 1
   c. Proportioning Pump
   d. Sampler II fitted with 40/hour (2:1) cam

2. Microdiluter
   Sample syringe set to pick up 50 microliters of sample. Flush syringe set to dispense approximately 1.0 ml. of saline (CO_2 - free). Final dilution is 1:20. Saline reservoir should be fitted with a two-hole stopper: The air intake hole should be plugged with a cartridge of CO_2 adsorbant.

3. Stock Sodium Carbonate Standard, 50 mEq./L.
   Place 1.060 gm. dessicated, anhydrous sodium carbonate into a 200 ml. volumetric flask. Dissolve and bring to volume in CO_2 free deionized water. Store refrigerated.

4. Standard Curve
   Into 10 ml. volumetric flask, pipette the following volumes of stock 50 mEq./L. standard:

Carbon Dioxide Content (Continued):

Ml. Stock Standard	Ml. H_2O	Final Concentration
1.0	9.0	5 mEq./L.
2.0	8.0	10 mEq./L.
4.0	6.0	20 mEq./L.
6.0	4.0	30 mEq./L.
8.0	2.0	40 mEq./L.
Use undiluted		50 mEq./L.

Mix each flask well and store refrigerated. The Working Standards are prepared weekly to keep the absorption of CO_2 to a minimum.

5. Acid Diluent
Place approximately 500 ml. of deionized water into a one liter volumetric flask. Slowly add 2.8 ml. concentrated H_2SO_4, and mix by swirling. Add 1.0 ml. Antifoam B, and bring to volume with deionized water. The Antifoam B will settle out; before use each day, mix well.

6. Sodium Carbonate Reagent, 0.1 M.
Place 10.6 gm. anhydrous sodium carbonate into a one liter volumetric flask, and dissolve in approximately 500 ml. deionized water. Dilute to volume, and filter.

   To check molarity, see "Comments on Procedure".

7. Stock Phenolphthalein Indicator, 1%
Place 1.0 gm. phenolphthalein powder, A.C.S., in a 100 ml. volumetric flask, and bring to volume with methyl alcohol. Mix until dissolved, and store in an amber bottle.

8. Working Phenolphthalein Color Reagent
Into a four liter volumetric flask deliver: 12 ml. of 0.1 M. sodium carbonate reagent, 8.0 ml. stock phenolphthalein indicator, 2.0 ml. Brij-35. Bring to volume with distilled water. Protect this reagent from prolonged exposure to atmospheric CO_2; fill one liter amber storage containers to the top to exclude air.

9. Saline Diluent
Dissolve 9.0 gm. NaCl in one liter deionized water which is CO_2-free.

Carbon Dioxide Content (Continued):

COMMENTS ON PROCEDURE:

1. Standards and specimens are relatively unstable following dilution. No more than 3 - 4 cups should be uncapped on the wheel prior to sampling, for best results.

2. BASELINE ADJUSTMENT PROCEDURE
Baseline should be approximately 20% T. $\pm$ 5% T. If the baseline is greater than 25% T., add 0.1 M. Na_2CO_3 to the color reagent. (0.5 ml. will drop the baseline about 20% T.) If you add too much Na_2CO_3, you will get reduced sensitivity; start over with fresh color reagent. If the baseline reads less than 15% T., dilute the color reagent with H_2O. The 50 mEq./L. standard should read approximately 90% T.

3. If foaming in the trap occurs, add an additional ml. of Antifoam to the acid diluent.

4. To check the molarity of the sodium carbonate reagent: Place 20 ml. of 0.1 N. HCl into a beaker containing 3 drops of 1.0% methyl orange. Titrate with the sodium carbonate solution until the indicator changes from red to yellow.

   NOTE: The volume of sodium carbonate used in titration should be 10 ml., as the sodium carbonate solution is 0.2 N. (0.2 N. Na_2CO_3 = 0.1 M. Na_2CO_3).

PROCEDURE:

1. Using the microdiluter, prepare a single 1:20 dilution of all standards, controls, and specimens; cap each sample cup after making dilution, and mix. Keep dilute samples capped until shortly before sampling.

2. Set recorder at 100% T. with H_2O pumping through all lines.

3. Shake acid diluent reagent, and place all lines in appropriate reagent bottles. The baseline should read 20% T. $\pm$ 5% T. with all reagents running through. If baseline adjustment is necessary, see procedure under "Comments on Procedure".

4. Sample the following standards and specimens in this order: 5, 10, 20, 30, 40, 50, 10 mEq./L., Normal Control, elevated Control, and 10 mEq./L., followed by patient specimens.

Carbon Dioxide Content (Continued):

5. Plot the standard curve, and determine the values for controls and patient specimens.

NORMAL VALUES: Adults: 24 - 32 mEq./L.
Newborns: 19 - 27 mEq./L.

REFERENCES:

1. Technicon "N" Methodology (N-19)

2. Skeggs, Leonard: "An Automatic Method for the Determination of Carbon Dioxide in Blood Plasma", Am. J. Clin. Path., 33:181, 1960.

3. STANDARD METHODS OF CLINICAL CHEMISTRY, Vol. I, Academic Press, pg. 19 - 22, 1953.

CLINICAL INTERPRETATION:

Respiratory acidosis is associated with an increase in the carbonic acid, pCO_2 and HCO_3, of serum. It occurs because of rebreathing of an abnormally high percentage of CO_2 or in conditions where the elimination of CO_2 through the lungs is inhibited. The elimination of CO_2 is inhibited by pneumonia, emphysema, fibrosis, cardiac failure, excessive utilization of narcotics or mechanical failure of the lungs.

Metabolic acidosis with a deficit of alkali is associated with uncontrolled diabetes mellitus or renal insufficiency. In uncontrolled diabetes mellitus, a large accumulation of keto-acids and lactic acid is partly responsible for the acidosis. In addition, a large loss of base results in metabolic acidosis. Starvation or dieting also results in this type of metabolic abnormality.

Renal failure is characterized by retention of acid ions such as phosphate and sulfate and organic acids. In addition, the kidneys have lost their ability to conserve base and to produce ammonia. There is a decrease in HCO_3^-, H_2CO_3, and pCO_2 of serum.

In respiratory alkalosis, excess amount of CO_2 is eliminated from the blood by hyperventilation. This occurs in hysteria, fever, encephalitis, and drugs, which stimulate the respiratory center, such as salicylates. The serum H_2CO_3 or pCO_2 is decreased along with the HCO_3^-.

Carbon Dioxide Content (Continued):

Metabolic alkalosis results from vomiting or loss of excess HCl from the stomach. Potassium depletion from excess utilization of Cortisone or Diuretics, or Cushing's or Conn's hyperaldosteronism, will cause a metabolic alkalosis. With potassium loss, chloride is also lost, resulting in an increase in serum HCO_3^-, H_2CO_3, and pCO_2.

Technicon®
AutoAnalyzer®
Methodology

MICRO-CO_2

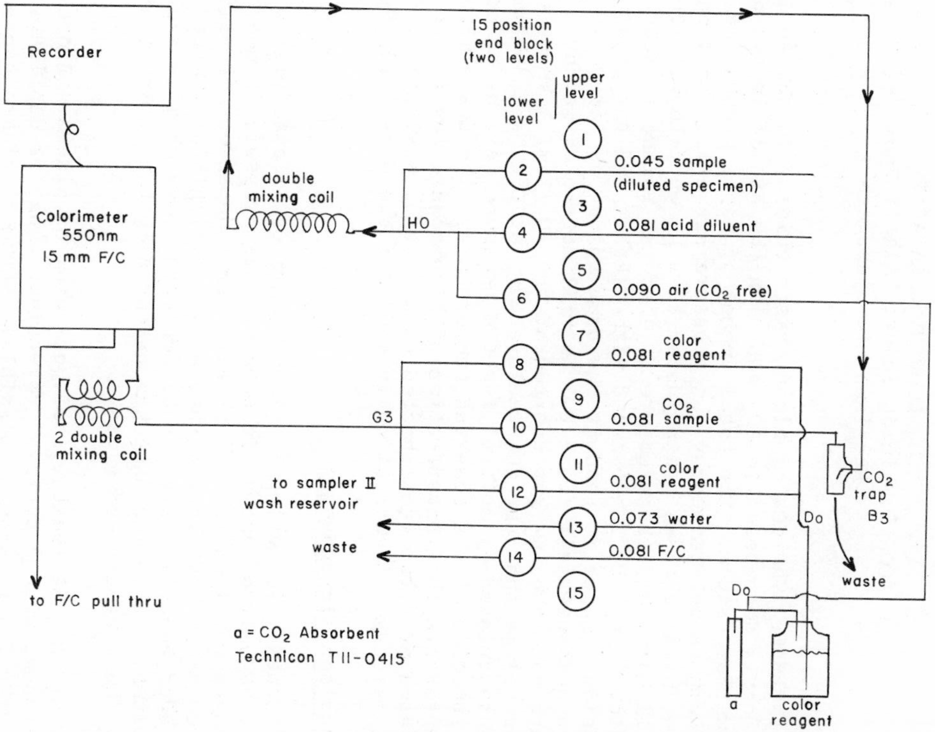

a = CO_2 Absorbent
Technicon T11-0415

Technicon Instruments Corporation
Tarrytown, New York

CO_2 COMBINING CAPACITY
(CO_2 Combining Power, "Alkali Reserve")
Natelson Microgasometer

PRINCIPLE: The CO_2 of blood exists chiefly in the form of HCO_3^-, which is decomposed to CO_2 gas with the aid of lactic acid and powerful vacuum. The total amount of gases present in a definite volume is determined by the pressure air exerts in a manometer. The released gases are adjusted to a known volume (liquid level to 0.12 Mark) and the pressure (P_1) of gas volume is read on the mano meter. The CO_2 is absorbed with alkali, and the pressure (P_2) of residual gases is read at the same volume as before (0.12 Mark). The pressure fall ($P_1 - P_2$) in millimeters of mercury, is the pressure that the CO_2 gas exerted at the (0.12 ml.) known volume Mark. The volume that the CO_2 would occupy at 0^oC. and 760 mm. of mercury is calculated by multiplying the pressure fall by a factor that is a function of the observed temperature. The CO_2 combining capacity represents the concentration of bicarbonate ion in the patient's plasma or serum equilibrated to a partial pressure of CO_2 of 40 mm. of mercury. A correction is made for the dissolved CO_2.

Natelson Microgasometer:

The theory for using Natelson apparatus is essentially the same as for Van Slyke manometric method. Gas pressure is measured under constant volume so that results are independent of atmospheric pressure. Temperature changes are negligible; a water bath is not needed because tests can be conducted very rapidly and the small volume of solution used has essentially the same temperature as surrounding air and glass.

SPECIMEN: At least 0.1 ml. of serum or plasma. Blood is collected as usual with a syringe, vacutainer tube, or capillary tube. Remove from cells as soon as possible. Serum or plasma can be used. Everything is done aerobically. For proper evaluation of CO_2 gas in blood, arterial plasma should be used and collection is anaerobic. It can be obtained from artery, capillary (finger stick), or "arterialized blood" from vein (keep arm warm for several minutes to arterialize blood before venous puncture). A significant amount of CO_2 is lost if sample is collected and centrifuged under oil. CO_2 content is also decreased (approximately 0.5 mmole/L.) if plasma is separated at room temperature. Solubility of CO_2 in serum is increased up to 8% in the presence of lipemia.

CO_2 Combining Capacity (Continued):

REAGENTS:

1. Na_2CO_3 Standard, 20 mEq./L.
   The AutoAnalyzer Standard may be used.

2. Lactic Acid, 1.0 N.
   90 ml. of 85% Lactic Acid is diluted to 1 liter.

3. Sodium Hydroxide, 3.0 N.
   12 gm. NaOH pellets are dissolved and made up to 100 ml.

4. Anti-foam, 10%
   Reagent No. 820 (Scientific Industries, Inc., 220-05 97th Ave., Queen's Village 29, New York). Shake before use.

5. Low Foam Detergent, 0.5%
   10 ml. of 5% Reagent No. 810 is diluted to 100 ml. (Scientific Industries). Used for cleaning machine.

COMMENTS ON PROCEDURE:

1. A leak in the apparatus is a source of great error.

   A. Keep stopcocks well greased and air tight; do not use grease with silicone if using alkali.

   B. If moisture collects in the manometer remove by adding small amount (approximately 0.5 ml.) of trimethylene glycol through stopcock. Small amounts left on walls do not affect the determination.

   C. Always leave the stopcock at the extraction chamber in the open position.

2. If any equilibration or alteration in partial pressure of CO_2 is necessary, changes are made before the blood cells are separated from plasma. Such true plasma represents the buffering action of plasma, serum, and red blood cells.

3. If blood is collected anaerobically and serum separated from cells before equilibration or change in partial pressure of CO_2 is made, this will represent only the buffering action of plasma or serum.

CO_2 Combining Capacity (Continued):

PROCEDURE:

Preparing Equipment and Solutions

1. Both stopcocks should be air tight and the apparatus clean (See Instruction Booklet #4).

2. Fill 4 vials with respective solutions (lactic acid, anti-foam, NaOH, and low foam) and mercury; use approximately 1.0 ml. of each.

3. Have a small cylinder of CO_2 (5%) and O_2 (95%) mixture, with pressure and regulator valve, ready for equilibrating samples.

4. Put at least 75 microliters of serum or plasma into a 10 x 75 mm. test tube with about 2 drops of mercury. Equilibrate samples for approximately 30 - 60 seconds; cover with parafilm.

Manipulating Gasometer for Tests

1. Always have a drop of mercury hanging from the tip of pipette to avoid introduction of air.

2. Draw up 0.03 ml. of equilibrated sample or standard followed with 0.01 ml. mercury.

3. Draw up 0.03 ml. Lactic acid and 0.01 ml. mercury.

4. Draw up 0.01 ml. Anti-foam and 0.01 ml. mercury.

5. Draw up 0.10 ml. water and mercury to 0.12 Mark. (Low foam reagent may be used instead of water.)

6. Close the reaction chamber stopcock and retreat with the piston until mercury is at the 3.0 ml. mark of the reaction chamber.

7. Loosen the clamping knob and shake for one minute.

8. Advance the piston until the top aqueous meniscus is at the 0.12 Mark.

9. Record the manometer reading (P_1) and the temperature.

10. Advance the piston until mercury is at the top of the manometer.

CO_2 Combining Capacity (Continued):

11. Hold the vial containing NaOH under the pipette and open the reaction chamber stopcock. Adjust the mercury if necessary until a drop is at the tip of the pipette.

12. Draw up 0.03 ml. NaOH and mercury to the 0.12 Mark.

13. Close the reaction chamber stopcock and retreat with the piston until mercury is at the 3.0 ml. Mark. Shake for 1 minute or let it sit for the same length of time.

14. Advance the piston until the aqueous meniscus is at the 0.12 Mark.

15. Record the manometer reading (P_2).

Rinsing Gasometer

1. Advance the piston until mercury is at the top of the manometer.

2. Open the extraction chamber stopcock and advance the piston to eject all solutions through the pipette tip.

3. Draw up 1.0 ml. Lactic acid to the stopcock level, then draw this back just below the 3.0 ml. mark and eject the Lactic acid wash.

4. Do the same with a water rinse.

Calculations

Gas pressure is measured:

1. At constant volume which is calibrated into the apparatus.

2. Under conditions of equilibrium between the gaseous and aqueous phases which is determined by the test method.

3. At the prevailing temperature. The temperature is the only variable. A set of factors based on temperature can therefore be determined for each gas. CO_2 Combining Power = (P_1 - P_2) x Factor. The CO_2 factors for temperature changes are calculated and can be found on a separate sheet in the Natelson Instruction Booklet #4, or in Peters and Van Slyke QUANTITATIVE CLINICAL CHEMISTRY, Vol. 2, 2nd. Edition, 1932.

CO_2 Combining Capacity (Continued):

NORMAL VALUES:

According to some authors CO_2 varies with age. Normal values vary with different authors. Average CO_2 (HCO_3^-) on venous blood with the test performed on serum.

1. Newborn and children: 18 - 27 mEq./L.
2. Adults: 21 - 28 mEq./L.

Reproducibility of method is about $\pm$ 1.5%

REFERENCES:

1. Instruction Booklet #4 Natelson Microgasometer Model 600. Scientific Industries, 220-07 97th Ave., Queen's Village 29, New York, New York.

2. Peters, J. P., and Van Slyke, D. D.: QUANTITATIVE CLINICAL CHEMISTRY, Vol. 2, Williams and Wilkins, Baltimore, 2nd. Ed., 1932.

3. Kolmer, J. A.: CLINICAL DIAGNOSIS BY LABORATORY METHODS, 2nd. Ed. pg. 97.

4. Davidsohn and Wells: TODD-SANFORD: CLINICAL DIAGNOSIS BY LABORATORY DIAGNOSIS, 13th Ed., pg. 471, 1962.

5. Behrendt, M.: DIAGNOSTIC TESTS IN INFANTS AND CHILDREN, 2nd. Ed., pg. 233, 1962.

CLINICAL INTERPRETATION:

Refer to Interpretation section under the Carbon Dioxide Content Procedure on page 99.

CATECHOLAMINES IN URINE

PRINCIPLE: The assay of catecholamines in normal human urine refers to the determination of two important amines, namely, norepinephrine and epinephrine. Norepinephrine and epinephrine are present in the urines as the free amines and as labile conjugates. The determination of these two compounds are significant for the diagnosis of pheochromocytoma in cases of hypertension.

The catecholamines are isolated from the urine and concentrated into a relatively pure extract. This done by absorption of the amines from the urine on aluminum oxide column at pH 8.5, and subsequent elution of these compounds with dilute acetic acid. The amines are oxidized by ferricyanides, in the presence of zinc ions, to adrenochrome and noradrenochrome. On the addition of strong alkali and ascorbic acid, they are further converted into the fluorescent compounds, adreno- and noradrenolutines, or commonly called the trihydroxyindoles.

SPECIMEN: A 24-hour urine should be collected in 15 ml. of 6 N. HCl, the resulting pH should be 3 or less. Under these conditions, the catecholamines are stable in urine for at least several days at room temperature and for years at $2^{o}C$. Volumes greater than 2500 ml. per day are difficult to handle and therefore, the liquid intake should be restricted to a minimum.

EQUIPMENT AND REAGENTS:

1. <u>Turner Fluorometer</u>. The instrument should contain the following:
   a. Primary filter - 405 (405 nm.)
   b. Secondary filter - 65A (495 nm.)
   c. Range selector - 30x
   d. Slit - None
   The machine should be warmed up for at least 30 minutes before taking the readings.

2. <u>HCl</u>, 6.0 N. Dilute 500 ml. concentrated hydrochloric acid to 1000 ml. with deionized water.

3. <u>HCl</u>, 2.0 N. Dilute 167 ml. concentrated hydrochloric acid to 1000 ml. with deionized water.

4. <u>Aluminum oxide</u> for chromatographic adsorption analysis. Available from British Drug House, England. It can be used right away without further treatment.

Catecholamines in Urine (Continued):

5. EDTA, 0.2 M. Dissolve (with heating), 37.2 gm. reagent grade disodium ethylenediaminetetraacetate ($Na_2C_{10}H_{14}O_8N_2 \cdot 2H_2O$) in water, cool and make to 500 ml.

6. NaOH, 5.0 N. Dissolve 100 gm. sodium hydroxide pellets in water and make to 500 ml.

7. NaOH, 0.5 N. Dilute 50 ml. of 5.0 N. NaOH to 500 ml.

8. Acetic Acid, 0.2 N. Dilute 5.8 ml. glacial acetic acid to 500 ml.

9. Acetic Acid, 1.0 M. Dilute 28.8 ml. glacial acetic acid to 500 ml.

10. Sodium Acetate, 1.0 M. Dissolve 68 gm. of reagent grade sodium acetate ($CH_3COONa \cdot 3H_2O$) in water and dilute to 500 ml.

11. Acetate Buffer, 1.0 M. pH 6.5. Adjust 400 ml. of 1.0 M. sodium acetate to pH 6.5 by the addition of approximately 6.0 ml. of 1.0 M. acetic acid with constant monitoring on a pH meter.

12. Acetate Buffer, 1.0 M. pH 3.5. Adjust 400 ml. of 1.0 M. acetic acid to pH 3.5 by adding approximately 33 ml. of 1.0 M. sodium acetate with constant pH monitoring.

13. Zinc Sulfate, 0.25%. Dissolve 0.25 gm. reagent grade $ZnSO_4 \cdot 7H_2O$ in deionized water to make 100 ml.

14. Potassium Ferricyanide, 0.25%. Dissolve 0.25 gm. reagent grade $K_3Fe(CN)_6$ in deionized water to make 100 ml. Store in a dark bottle in the refrigerator. Stable only for one month.

15. Ascorbic Acid, 1.0%. Dissolve 100 mg. ascorbic acid in 10 ml. deionized water just before use.

16. NaOH-ascorbic Acid Solution Mix 7 volumes of 5.0 N. NaOH and 3 volumes of 1% ascorbic acid immediately before use.

17. HCl, 0.01 N. Dilute 0.17 ml. 6.0 N. HCl to 100 ml. with deionized water.

18. Sodium Acetate, 0.2 M. Dilute 200 ml. of 1.0 M. sodium acetate to 1 liter with deionized water.

Catecholamines in Urine (Continued):

19. Sodium Phosphate, Dibasic, 1/3 M. Dissolve 47.32 gm. of reagent grade Na_2HPO_4 in deionized water and dilute to 1 liter.

20. Stock Standards:
    a. 1000 micrograms/ml. NE (M.W. 169.2) Dissolve 19.9 mg. 1-norepinephrine bitartrate monohydrate (M.W. 337.2) in 10 ml. 0.01 N. hydrochloric acid.
    b. 1000 micrograms/ml. E (M.W. 183.2). Dissolve 18.2 mg. 1-epinephrine bitartrate (M.W. 333.3) in 10 ml. of 0.01 N. hydrochloric acid.

21. Working Standards:
    Dilute 0.100 ml. of each stock standard to 100 ml. with 0.01 N. hydrochloric acid to give concentrations of 1.00 micrograms/ml. All reagents are stable at room temperature. Both stock and working standards are stable at 2°C. for at least 6 months.

PROCEDURE:

A. Preparation of Urine Eluate:

1. Filter 10% of the 24 hour sample (50-250 ml.) through two sheets of Whatman No. 1 filter paper. Transfer 50 ml. of aliquot for analysis to a beaker.
   CONTROL - Hyland Urine Chemistry Control. Reconstitute with 25 ml. of deionized water and run the whole amount.
   RECOVERY - Supplemental Urine Control with definite amount of working standard, usually from 3 to 10 micrograms of nor-epinephrine and or epinephrine.

2. Add 5.0 ml. Of 0.2 M. EDTA.

3. Using a pH meter, adjust pH of the urine to pH 8.4 - 8.5 with 5.0 N. NaOH. Add additional 1.0 ml. of 1/3 M. Na_2PO_4 to keep pH in that range while in the column.

4. Chromatographic tube to be used is 30X 2.0 cm. Add the aluminum suspended in water, to the tube until a column, 2.0 cm. is obtained. Allow the water to drain through, and it should flow by gravity at a rate of 1 - 2 ml. per minute. Wash the column with 5.0 ml. of 1/3 M. Na_2HPO_4 just before adding the urine

5. Chromatograph the urine without delay. Wash the column 3 times with 10 ml. of 0.2 M. NaAc and once with 10 ml.

Catecholamines in Urine (Continued):

of deionized water. Let it drain before eluting with acid.

6. Elute the catecholamines with 5.0 ml. of 0.2 N. acetic acid at a time. Allow to stand for 5 minutes before elution. Collect 25 ml. of eluate into 50 ml. graduated centrifuge tubes. The catecholamines are stable in this eluate for at least several weeks at 2°C. Centrifuge eluate before analysi

B. Formation of THI from NE and E at pH 6.5:

1. Pipette 0.5 ml. of eluate to each of 3 test tubes:
   a. Blank
   b. Sample
   c. Sample + Standard

   Set up a tube for Reagent Blank (1.0 ml. of water) and duplicate tubes for 0.5 ml. of norepinephrine and epinephrine Working Standard (External Standards).

2. To tubes a and b, and the external standards, add 0.5 ml. of water. To tube c, add 0.5 ml. of norepinephrine internal standard.

3. To all tubes, add 3.0 ml. of acetate buffer, pH 6.5.

4. Pipette 0.1 ml. of $ZnSO_4 \cdot 7H_2O$ to all tubes and standards excluding the blanks.

5. Pipette 0.1 ml. of $K_3Fe(CN)_6$ to all tubes and standards excluding the blanks. Let stand for 2 minutes.

6. Add 1.0 ml. of NaOH - Ascorbic acid mixture to all the tubes. To the blank, add 0.1 ml. $K_3Fe(CN)_6$. Mix and read in the fluorometer right away, using Coleman microcuvettes.

7. Set the machine, so that the external standard should read around 70 fluorescence units. Read fluorescence units of the urine samples.

CALCULATIONS:

Sample fluorescence - Blank fluorescence = Corrected Sample Fluorescense.

Flourescence of Sample + Standard - Sample Fluorescence = Standard Fluorescence.

Catecholamines in Urine (Continued):

The catechol urine excretion in micrograms per day is calculated as follows:

$$\frac{\text{Reading of 0.5 ml. of Sample at pH 6.5}}{\text{Reading of 0.5 ug. of NE Internal Standard}} \times 25 \times$$

$$\frac{\text{Urine volume}}{\text{Vomume of liquot analyzer}} = \text{micrograms catecholamines per 24 hours.}$$

C. Oxidation at pH 3.5.

1. Prepare a second series of tubes exactly as described above, except use 0.5 ug. E instead of NE in the internal standard tube.

2. Add 3.0 ml. of acetate buffer, pH 3.5 to each of the tubes and oxidize the same way as in the pH 6.5 series. At this pH, E is oxidized quantitatively, but NE remains unchanged.
   Primary filter - 405 (405 nm.)
   Secondary filter - 2A-15 (520 nm.)

PROCEDURE (for total Catecholamines):

1. Transfer 25 ml. of urine into a 125 ml. Erlenmeyer flask and add 6.0 N. H_2SO_4 to pH 2.0. Cover the mouth of the flask with a small funnel, and heat in the boiling water for 20 minutes.

2. Cool to room temperature, add EDTA, pH to 8.5, and run through the column the same way as for the free catecholamines.

NOTES:

Quenching lowers the readings. To minimize this problem, analyze with very dilute solutions. Also, the eluates should always be centrifuged to get rid of any fine particles, which inhibit the formation of the trihydroxyindoles.

REFERENCES:

1. Crout, J. Richard: STANDARD METHODS OF CLINICAL CHEMISTRY, 3:62 - 79, 1961.

2. Jacobs, S. L., Sobel, C., and Henry, R. J.: Journal Clin. Endocrin. & Metab., XXI, 305 - 314.

Catecholamines in Urine (Continued):

CLINICAL INTERPRETATION:

Two distinct substances, epinephrine and norepinephrine, are secreted by the adrenal medulla. Because they are generically related to catechol and are amines, they have been termed catecholamines.

Studies of the metabolic fate of administered catecholamines have provided evidence of several pathways for the degradation of these compounds. Only about 4% of intravenously infused catecholamine is recovered as such in the urine. The transformation to urinary metabolites involves enzymatic O-methylation and enzymatic oxidative deamination. The scheme is usually O-methylation followed by oxidation, although methylation may follow rather than precede the action of the amine oxidase. In either case the end product is the same, 3 methoxy-4-hydroxymandelic acid (Vanillylmandelic acid, VMA). When O-methylation occurs first, the important urinary metabolites, metanephrines, are formed. A portion of the adrenal medullary hormones, as well as their metabolites, is excreted in the urine. Thus one may assess the state of adrenal medullary function by assaying the urine for the hormones or their metabolites.

The levels of catecholamines in blood are extremely low and large amounts of blood are thus required for analysis. Urinary catecholamines provide a more distinct and reproducible separation of normal persons from patients with pheochromocytoma than does the assay of plasma. VMA is of value primarily as a confirmatory test with the urinary catecholamines, although in rare instances the VMA may be elevated with an apparently normal catecholamine excretion.

The determination of urinary catecholamines correlates with the clinical state; however, there are situations which may lead to elevation of catecholamines not related to pheochromocytoma. These are: (1) Vigorous exercise prior to urine collection can result in an increase in the output of catecholamines. (2) increased output of catecholamines has been reported in cases of progressive muscular dystrophy and myasthenia gravis. Medications which may lead to fluorescent urinary products may lead to spuriously elevated results. Medications which have this effect are antihypertensive drugs of the alpha methyl dopa configuration, tetracycline antibiotics, the B Vitamin complex, adrenaline and adrenaline-like drugs used in asthmatic seizures. Other medications and materials reported to lead to spurious results are: carbon tetrachloride, erythromycin, hydralazine, quinine, quinidine, methenamine and formaldehyde. Patients should be off such medication for at least one week before specimens are collected for catecholamine determinations. It is essential

Catecholamines in Urine (Continued):

that the urine be kept acid during the entire collection. When the collection is completed, the pH should be tested with narrow range pH paper and should be less than 3.

NORMAL RANGE:

Random Sample: Up to 18 ug./100 ml. urine.
24 Hour Sample: Up to 103 ug./24 hours.

The analysis for VMA is useful both as a confirmation of elevated catecholamine excretion and in cases of unexplained hypertension with normal catecholamine levels. VMA analysis may be of particular value in cases of neuroblastoma which produce high VMA levels. All medications and foods which may give rise to phenoxy acids in urine, i.e., coffee, fruits (bananas) and substances containing vanilla, be excluded from the diet before and during collection periods. Anileridine, aspirin and methocarbamal are medications reported to interfere with this test.

NORMAL RANGE: 0.7 to 6.8 mg./24 hours.

Homovanillic acid (HVA) is an important urinary metabolite in cases of neuroblastoma.

NORMAL RANGE: Up to 15 mg./24 hours.

CEPHALIN - CHOLESTEROL FLOCCULATION TEST

PRINCIPLE: A cephalin-cholesterol emulsion is flocculated slightly or not at all by serum with normal proteins. Sera with altered proteins, as found in parenchymatous liver disease and certain other diseases, cause flocculation. Flocculation is caused by the precipitation of globulin and cholesterol. Cephalin acts as an emulsifying agent. Albumin inhibits flocculation. A positive test results from an increase in globulin and/or decrease in albumin or a change in its anti-flocculation properties.

SPECIMEN: 0.4 ml. unhemolyzed serum. Test should be set up on fresh serum as the stability of the test is unpredictable. Frozen serum should not be used.

REAGENTS:

1. 0.9% NaCl (W/V)
   In a liter volumetric flask dissolve 9.0 gm. NaCl in approximately 300 ml. deionized water. Dilute to volume.

2. Cephalin - Cholesterol Stock Ether Solution (Difco Bacto-Cephalin Cholesterol Antigen)
   Reconstitute one Difco vial volumetrically with 5.0 ml. anesthetic ether. If turbidity persists, add one drop of distilled water. Stable for months under refrigeration if protected against evaporation.

3. Cephalin - Cholesterol Emulsion
   Warm 35 ml. distilled water to 65 - 70°C. in a 50 ml. Erlenmeyer flask calibrated at 30 ml. with a marking pencil; add 1.0 ml. Stock Ether solution slowly with stirring. Raise temperature slowly to boiling and allow to simmer until final volume is 30 ml. Cool to room temperature before using. The emulsion should be milky, translucent, and have no trace of ether. It is stable when refrigerated for two weeks.

COMMENTS ON PROCEDURE:

False positives may result from:
1. Exposure to light during the incubation of the test.
2. Dirty glassware (particularly heavy metal or strong acid contamination).
3. Serum contamination with bacteria.

Cephalin-Cholesterol Flocculation Test (Continued):

PROCEDURE:

1. Set up each test in duplicate. Add 0.2 ml. serum in a 12 ml. tube (unscratched, conical-tipped centrifuge tube) containing 4.0 ml. saline. Mix gently.

2. Set up an emulsion control with each run, consisting of 4.0 ml. saline. If a 4+ positive control serum is available, the test should be included.

3. Add 1.0 ml. cephalin - cholesterol emulsion to each of the above tubes. Cover each well with parafilm and mix gently and thoroughly.

4. Place tubes in dark at 25°C. ± 3°C.

5. Read the reaction at 24 hours and again at 48 hours as follows:

Negative	No Flocculation
1+	Minimal Flocculation
2+	Definite Flocculation
3+	Considerable Flocculation but definite residual turbidity in the supernate
4+	Complete Flocculation with clear supernate

If more than minimal Flocculation occurs in the "Emulsion Control" or the 4+ positive control does not display a 4+, the test should be repeated with fresh emulsion and fresh serum specimens.

REFERENCE: Henry, R. B.: CLINICAL CHEMISTRY: PRINCIPLES AND TECHNIQUES, Hoeber, pg. 559 - 560, 1964.

CLINICAL INTERPRETATION:

A three or four-plus flocculation in forty-eight hours is abnormal and positive. Flocculation may occur as a result of decrease in serum albumin which stabilizes the emulsion. The low serum albumin may result from hepatic disease, malnutrition, nephrosis, or malabsorption. Alpha-globulin may stabilize the emulsion and a decrease in this globulin will cause a flocculation as will an increase in gamma-globulin.

Positive reactions occur in sera heated at 56°C. for thirty minutes or if the reaction mixture is exposed to light during the test period.

The clinical value of the test is in hepatocellular damage with decreased serum albumin and increased gamma-globulin.

SERUM CERULOPLASMIN DETERMINATION

PRINCIPLE: Ceruloplasmin concentration is determined from the rate of oxidation of p-phenylenediamine at 37°C. and at pH 6.0. The rate of appearance of the purple oxidation product (Wurster's red), which has an absorption peak at 520 - 530 nm. is measured spectrophotometrically or photometrically.

SPECIMEN: 0.1 ml. serum or plasma.

REAGENTS AND EQUIPMENT:

1. Beckman Model Du Spectrophotometer
   Equipped with Thermospacers as the compartment temperature must be kept constant at 37°C.

2. Acetate Buffer, 0.1 M. pH 6.0
   Add 10 ml. 0.1 M. acetic acid (0.57 ml. glacial acid plus water to 100 ml.) to 200 ml. 0.1 M. sodium acetate (1.36 gm. CH_3COO-Na · $3H_2O$ per 100 ml.). The pH must be 5.95 - 6.00.

3. Sodium Azide, 0.1% in 0.1 M. Acetate Buffer
   The pH must be 5.95 - 6.00. Store in refrigerator.

4. p-Phenylenediamine · 2HCl, 0.25% in 0.1 M. Acetate Buffer
   Recrystallize commerical salt as follows: Dissolve in water, add Darco charcoal, warm in water bath at 60°C. with occasional mixing, and filter. Add acetone to filtrate until turbidity appears, refrigerate for several hours, filter off the p-phenylenediamine · 2HCl (PPD), and dry the crystals in the dark in a vacuum desiccator over anhydrous $CaCl_2$. Store in brown bottle. To prepare the 0.25% reagent, dissolve 12.5 mg. in 3.0 ml. acetate buffer and, using narrow range pH paper, adjust the pH to 6.0 by adding 1.0 N. NaOH dropwise from a 0.2 ml. serologic pipette (approximately 0.1 ml. required). Add acetate buffer to a final volume of 5.0 ml. This reagent can be used up to about 2 hours after preparation if kept in the dark.

PROCEDURE:

1. Set up the following in cuvets with 1.0 cm. light path:
   BLANK: 1.0 ml. azide reagent + 1.0 ml. buffer + 1.0 ml. PPD reagent.
   TEST: 2.0 ml. buffer + 1.0 ml. PPD reagent.

2. Place cuvets in compartment and allow 5 minutes for temperature equilibration.

Serum Ceruloplasmin Determination (Continued):

3. Add 0.1 ml. serum (heparinized or oxalated plasma is satisfactory) to each tube from a TC pipette, effecting mixing in the process.

4. Read absorbance of the Test against the Blank at 530 nm. at exactly 10 minutes and again at 40 minutes after addition of serum.

CALCULATION:

Ceruloplasmin Units = (A at 40 min. - A at 10 min.) x 1000

COMMENTS ON PROCEDURE:

1. Artificial Standard
Of a number of dyes studied, pontacyl violet 6R (Du Pont) possesses an absorption curve closest to that of PPD oxidized by ceruloplasmin. The curves are not identical, however, and the concentration of dye given as a 400 unit standard was established for a Klett No. 54 filter and may not be valid for other photometers. It is not valid for a spectrophotometer at 530 nm.

2. Effect of Light
Catalytic oxidation of PPD is increased by exposure to light. The test, therefore, must be carried out in the dark.

3. Variation in pH
In the technic presented, the optimal pH for serum ceruloplasmin activity is at 6.0 and is fairly sharp. At pH 6.0 and above, the rates, after the lag phase, are linear to 60 minutes. At pH 5.8 and lower, however, a decrease in rate occurs during the 30 to 40 minute period. The cause for this is unknown.

4. Variation in Temperature
In the method presented, Arrhenius plots of log of activity against reciprocals of absolute temperature give straight lines between 22 and 45°C. with a slope of —3700. This gives an activation energy, u, of 17,000 cal./mole and a Q_{10} (temperature coefficient) between 30 and 40°C. of 2.45.

5. Lag Phase
Evidence indicates that PPD oxidized by ceruloplasmin is again reduced by the ascorbic acid present in the serum until all the ascorbic acid is used up. A lag period greater than 10 minutes

has never been observed by our laboratory in the method presented, i.e., a linear rate is established by 10 minutes.

6. Hemolysis
   It has been reported that minimal hemolysis does not interfere. Hemoglobin has been added in our laboratory to sera to a concentration as high as 200 mg./100 ml. and no interference was observed.

7. Stability of Samples
   Stability at room temperature is somewhat variable, some sera showing no degradation in 2 days, others decreasing up to 15% in 24 hours. Samples are stable at least 2 weeks at $4^{o}C$. or in the frozen state. Ultraviolet light of 253.7 nm. wavelength inactivates the oxidase activity of ceruloplasmin by causing a splitting off of the copper.

ACCURACY AND PRECISION

The accuracy of any method for determination of the oxidase activity of ceruloplasmin is restricted by the absence of any reference ceruloplasmin standard. Oxidase activity can be plotted against ceruloplasmin concentration determined by immunochemical analysis but this is not feasible for most laboratories. The situation is further complicated by the fact that there are at least two, and possibly four or five ceruloplasmins in serum. In any event, the enzymatic activity observed in a method is governed not only by the enzyme concentration but also by the concentrations of various ions. Ferrous ion, and to a lesser extent ferric and other cations at low concentration, enhance the oxidase activity of purified ceruloplasmin. There is also evidence that albumin inhibits.

The precision of the test (95% limits) is about $\pm$ 5%.

NORMAL VALUES: The 95% adult limits are 280 to 570 Units

REFERENCE: Henry, J. B.: CLINICAL CHEMISTRY: PRINCIPLES AND TECHNIQUES, Hoeber, pg. 500 - 503, 1965.

CLINICAL INTERPRETATION:

Copper is present in two forms in the serum. Five per cent is free or loosely bound to albumin. Ninety-five per cent is firmly bound to alpha-two globulin. Ceruloplasmin is a blue protein and

Serum Ceruloplasmin Determination (Continued):

acts as a ferroxidase. The chief aspect of determining ceruloplasmin is in the evaluation of Wilson's disease. The serum ceruloplasmin is markedly decreased in Wilson's disease. Loss of the alpha-two globulin in nephrosis will also result in a low ceruloplasmin.

There are many causes for an elevated serum ceruloplasmin. These are pregnancy, utilization of oral contraceptives, Hodgkin's disease, hyperthyroidism, hepatic disease, tissue necrosis as is seen in myocardial infarction, and acute inflammatory states. Ceruloplasmin is an acute phase reactant and is elevated in active disease and returns to normal with successful treatment of the disease. The serum ceruloplasmin is a useful test to determine the activity of Hodgkin's disease.

The elevation of ceruloplasmin in pregnancy has been attributed to estrogens. When administered alone, estrogens can produce a markedly increased ceruloplasmin. The rise in plasma ceruloplasmin in women taking oral contraceptives is most probably due to the estrogen component.

The green color of plasma secondary to elevated ceruloplasmin is more prominent when a large volume of plasma is present as in a blood-bank unit. It is not as noticeable and infrequently seen in a clinical laboratory test tube or pilot tube attached to a blood-bank unit.

The presence of elevated levels of plasma ceruloplasmin in rheumatoid arthritis is unexplained. The green plasma color is easier to detect if there is a reduction in yellow plasma pigments, which are primarily carotenoids, bilirubin and heme. These yellow pigments are decreased in rheumatoid arthritis, and observation of the green color is easier.

CHLORIDES
(Manual Titration: Schales & Schales)

PRINCIPLE: Chloride ion in centrifuged urine, centrifuged cerebrospinal fluid or a protein - free filtrate is titrated with a standard solution of mercury ion, forming undissociated, but soluble, $HgCl_2$. The end-point is signaled by a violet-blue color resulting when excess Hg^{++} forms a complex with the indicator diphenylcarbazone.

REAGENTS:

1. Folin-Wu precipitating Reagents:
   2/3 N. Sulfuric Acid
   10% Sodium Tungstate

2. Diphenylcarbazone Indicator Solution
   Dissolve 100 mg. s-diphenylcarbazone (Eastman Kodak) in 100 ml. 95% ethanol. Store in dark bottle in refrigerator. Prepare fresh monthly.

3. Standard Mercuric Nitrate Solution (10 mEq. Hg^{++}/ Liter)
   With a few hundred ml. deionized water in a 1 liter volumetric flask, add 20 ml. 2.0 N. Nitric Acid. Add 1.6 gm. reagent grade mercuric nitrate ($Hg(NO_3)_2$), dissolve, and dilute to volume. Stable indefinitely.

4. Standard Sodium Chloride Solution (10 mEq. Cl/Liter)
   Dry reagent grade Sodium Chloride in an oven at 110°C. overnight. Cool and weigh out exactly 584.4 mg. Dissolve in deionized water, and transfer with rinsings to a 1 liter volumetric flask. Dilute to volume and mix. This solution is stable indefinitely.

5. Standardization of Mercuric Nitrate Solution
   Transfer 2.0 ml. of Standard NaCl solution (#4 above) to a small beaker, add 4 drops of indicator (#2 above), and titrate with the mercuric nitrate solution using a microburette. The strength of the $Hg(NO_3)_2$ solution is adjusted so that 2.0 ml. are required in the titration, either by adding more mercuric nitrate or by dilution with 0.04 N. Nitric Acid.

NOTES ON PROCEDURE:

1. When a protein-free solution is titrated, the end-point should be quite distinct and the resulting color fairly stable.

Chloride

Due to the capacity of protein to bind heavy metals such as mercury, titrations carried out directly on serum will tend to be higher (up to 15 mEq./L.), and the end-point of the titrations will be less distinct and the color change less stable.

2. The pH of the solution just prior to reaching the end-point should be between 1.5 and 3. If the pH is too low, there is a loss of sensitivity, and the titer will be too high. If the pH is too high, the titer will be too low. Thus, it is necessary to acidify alkaline urines before titrating to pH 3.0.

3. Diphenylcarbazone is also an acid-base indicator, giving a salmon-pink color at roughly pH 6.0 or higher. This color should disappear as the titration proceeds and the solution becomes progressively more acid. If, however, the salmon-pink color is very intense on addition of the indicator (due to alkalinity or presence of amphoteric proteins), the color may be removed by adding a drop of 1.0 N. nitric acid before beginning the titration.

4. The fresh diphenylcarbazone solution will have a weak reddish-orange color. As it deteriorates, it will turn cherry-red or yellow and should no longer be used.

5. Adjust urines by diluting with distilled water, so that roughly 2.0 ml. of dilution are needed to react with 2.0 ml. of mercuric nitrate solution. Multiply result by dilution factor.

PROCEDURE:

1. Prepare a Folin-Wu protein - free filtrate by adding in the following order:
   1.0 ml. serum
   7.0 ml. deionized water
   1.0 ml. 2/3 N. H_2SO_4
   1.0 ml. 10% sodium tungstate

   Allow the mixture to stand for approximately 10 minutes and then filter.

2. Place 2.0 ml. Folin-Wu filtrate of control serum or serum specimen in a small beaker.

3. Add 4 drops of diphenylcarbazone indicator.

Chlorides - Schales & Schales (Continued):

4. Using a microburette calibrated in 0.01 intervals, titrate the filtrate with mercuric nitrate solution. The first drop of excess mercuric nitrate turns the solution from colorless to an intense blue-violet.

5. If only small amounts of serum or CSF are available, use 0.2 ml. specimen, add 1.8 ml. distilled water, acidify with a drop of 1.0 N. nitric acid, and proceed with titration.

6. CALCULATION:

If the mercuric nitrate has previously been standardized so that 2.0 ml. exactly equals 2.0 ml. of the standard NaCl solution, the chloride in the unknown is equal to the ml. of mercuric nitrate used in the titration times 50.

It is generally more convenient to calculate as follows:

Titrate 2.0 ml. of the standard NaCl solution.

$$\text{Factor} = \frac{100}{\text{ml. } Hg(NO_3)_2} \quad \frac{(10 \text{ mEq. } Cl^-/\text{Liter} \times 10 \text{ (dilution factor))}}{\text{needed to titrate standard}}$$

Factor x ml. $Hg(NO_3)_2$ used to titrate unknown = mEq./L. Cl^- in unknown.

mEq//L. x 0.058 x 100 = mg%. NaCl

NORMAL VALUES:

Serum or plasma: 98 - 106 mEq./L.
Cerebrospinal Fluid: 122 - 132 mEq./L.
Urinary Chloride: 170 - 250 mEq./24 hours

REFERENCES:

1. Schales and Schales: J. Biol. Chem. 140:827,

2. Henry, J. B.: CLINICAL CHEMISTRY: PRINCIPLES AND TECHNIQUES Hoeber, pg. 406, 1964.

Chlorides - Schales & Schales (Continued):

CLINICAL INTERPRETATION:

Hyperchloremia is found in patients with metabolic acidosis as in renal failure. It may also occur in dehydration or when inappropriate excess administration of saline occurs.

Hypochloremia occurs with severe vomiting or excess removal of gastric contents by gastric suction. It is also seen in respiratory acidosis and metabolic alkalosis. If inappropriate excess intravenous glucose is administered, or if inappropriate ADH occurs, hypochloremia will be present.

A decrease in CSF chloride occurs when the CSF protein rises due to various causes. Tuberculous meningitis is one of the usual diseases which causes low CSF chloride. The explanation for the decrease in CSF chloride is the Donnan Equilibrium, low serum chloride due to inappropriate ADH, or consumption of chloride by organisms.

SWEAT CHLORIDE DETERMINATION

PRINCIPLE: A filter paper disc, pre-weighed within a clean plastic vial is used to collect the sweat produced by philocarpine iontophoresis. After plating the skin with 0.2% philocarpine nitrate which stimulates sweat production, the area is covered with the filter paper disc. The disc is completely enclosed by water-proof tape and left on the arm for 30 minutes. The filter paper disc, saturated with sweat is returned to the vial and weighed immediately before evaporation can occur. The sweat NaCl is then eluted with deionized water and a chloride determination run on the chloridometer. An increased concentration of Na^+ and Cl^+ in sweat above normal ranges is indicative of cystic fibrosis.

REAGENTS AND EQUIPMENT:

1. Pre-weighed Discs: Using clean dry forceps, insert a Whatman #42 filter paper disc (5.5 cm. diameter) into a clean wide-mouth plastic vial with a tight fitting cap. Weigh the container with paper accurately (to four decimal places), and from this point on handle the container only with forceps or tissue paper and not with the hands. Place the container into a plastic jacket and lebel it with the dry weight.

2. Double-strength Acid Reagent
   Nitric Acid, Conc. 12.8 ml.
   Acetic Acid, glacial 200 ml.
   Distilled water qs. 1000 ml.

3. Sodium Chloride Standard, 1.0 mEq./ml. NaCl
   Prepare a 1:100 dilution of the Chloride Standard used for serum chlorides.

NOTES ON PROCEDURE:

Under no circumstances should the filter paper or the clean vial be touched with hands or moistened in any way until the final weighing. This is to avoid any NaCl contamination, or weight change not due to actual sweat collection.

PROCEDURE:

1. When the vial is returned to the laboratory with the filter paper saturated with sweat, again weigh the vial and paper disc. The difference in the two weighings represents the amount of sweat collected.

Sweat Chloride (Continued):

2. Volumetrically add 4.0 ml. deionized water to the container, making sure that the filter paper is submerged in the fluid. Allow it to elute for at least 30 minutes. Do not attempt to hasten elution by shaking, as the filter paper will disintegrate, making it extremely difficult to pipette.

3. To each of three titration cups add 2.0 ml. Double Strength Acid Reagent.
   Add for the Blank: 2.0 ml. water (run in duplicate)
   Std: 2.0 ml. of the 1.0 mEq./ml. NaCl Standard (run in duplicate)
   Sweat: 2.0 ml. of the sweat eluate.

4. To each cup add 4.0 drops of gelatin indicator.

5. Titrate on the Chloridometer using the following procedure.
   a. Turn Direct Read-out to "BLANK" position.
   b. Set titration range switch to "MEDIUM" position.
   c. Titrate Blank, Standard and Sweat dilution, recording time in seconds.

6. CALCULATION:

$$\frac{\text{Unknown-Blank}}{\text{Standard-Blank}} \times \frac{4.0 + \text{weight of sweat}}{\text{weight of sweat}} = \text{mEq./Liter of sweat}$$

NORMAL VALUES:

Children under 6 years	8 - 40 mEq./L.
Adults (including Heterozygotes	30 - 70 mEq./L.
Clinical cases of cystic fibrosis	70-140 mEq./L.

REFERENCES:

1. O'Brien: LABORATORY MANUAL OF PEDIATRIC MICRO AND ULTRA-MICRO TECHNIQUES, 3rd. Edition, page 84.

2. Gibson, L. E. and Cooke, R. E.: Pediatrics, 23:545, 1959.

CLINICAL INTERPRETATION:

The dermination of sweat chloride is important in the diagnosis of cystic fibrosis of the pancreas also known as mucoviscidosis. The disease is present in infants and children. It is characterized by bulky diarrhea, malnutrition, obstructive jaundice, diabetes mellitus, and chronic pulmonary infections. Mucous secretions accumulate in pancreas, intestine, biliary tree, and lung leading to obstruction and infection. A failure of mucolytic function exists systemically. In addition, an abnormality in sweating occurs with excessive loss of Na^+ and Cl^-. During childhood, the sweat chloride should be below 40 mEq./L. Patients suffering from mucoviscidosis exhibit sweat Cl^- levels far above 40 mEq./L. Because of this tendency to lose an excessive amount of sodium chloride in the sweat, children with this disorder may develop shock during hot weather from the loss of sodium chloride. Two other conditions which cause excessive sodium chloride in the sweat are hypothyroidism and adrenal cortical insufficiency.

ULTRAMICRO CHLORIDE DETERMINATION
(Chloridometer)

PRINCIPLE: The therory of operation of the chloridometer is based on established principles of coulometric generation of reagent and of amperometric indication of the endpoint. A constant direct current is passed between a pair of silver generator electrodes in the coulometric circuit, causing an oxidation of the silver wire to silver ions (Ag^+) and their release into solution at a constant rate. The Ag^+ will combine with Cl^- to form the insoluble AgCl. As soon as all of the Cl^- in the acid solution have combined with the Ag^+, the presence of free silver ions causes a rise in current which flows through the indicator electrodes and the Meter-Relay (amperometric circuit). This stops the timer which was activated with the first release of silver ion. The current to the generator electrodes is also shut-off at this time. Since the release of silver ion is constant, the amount of chloride precipitated is proportional to the elapsed time.

SPECIMEN: 20 microliters of serum or plasma; the specimen is run in duplicate.

REAGENTS AND EQUIPMENT:

1. Buchler-Cotlove Chloridometer (with Buchler Direct Reader)
   The instrument is set at the medium titration rate. The microampere meter (red needle) is set at 10 microamps. The direct read-out auxillary has been adapted by exchanging the 5,000-ohm 4-W rheostate with a 50,000-ohm 2-W rheostate; this allows use of the read-out at low as well as medium and high titration settings. The Standard adjust vernier was replaced with a ten turn digital knob pot. A reading of 20 units is approximately equivalent to 1.0 mEq./L. The blank time delay was left unchanged from original purchase.

2. Ultramicrodilutor
   Sample syringe set to take up 20 microliters, and flush syringe set to deliver 4.0 ml. of nitric-acetic acid reagent.

3. Nitric-Acetic Acid Reagent, 0.1 N. HNO_3, 10% Glacial Acetic Acid
   To 900 ml. of deionized water, add 6.4 ml. of concentrated HNO_3 and 100 ml. of glacial acetic acid. Do not store reagents in polyethylene containers. Use glass containers and glass fittings; never use rubber fittings. Sulfide and other sulfhydryl groups precipitate silver ions.

Ultramicro Chloride Determination (Continued):

4. Gelatin Reagent
   Dry, prepared mixture (available through Buchler Instruments); consists of gelatin, thymol blue, and thymol in the weight ratio of 60:1:1. 6.2 gm. is contained in each vial. Place the contents of one vial into 1 liter of hot water, and heat gently with continuous swirling until solution is clear. Dispense into small aliquots, and store in refrigerator. Stable for at least 6 months. To liquefy gelatin before use, insert container in hot water briefly. Gelatin is stable for 24 hours at room temperature

5. Sodium Chloride Standard, 100 mEq./L. Cl^-
   Dissolve 5.8450 gm. of dessicated, reagent grade sodium chloride in deionized water, and bring to volume in a 1 liter flask.

COMMENTS ON PROCEDURE:

1. Chloridometer On-Off switch must be Off even when A.C. line is disconnected, as this switch also controls the mercury battery which is part of the indicator amperometric system.

2. All electrodes must be clean and shiny-silver in color at all times. Use silver polish on a thin strip of cloth to clean. Rinse well with water.

3. The silver wire anode electrode should be positioned so that it is as long as the other electrodes. When advancing silver wire, make sure that good electrical contact is made at the binding post at the wire spool. If it is not, titration values will not be obtained.

4. If silver wire becomes very thin, trim it off and advance new portion of wire. Polish this new section to remove any oily insulation coating on the wire which would prevent generation of silver ions. Rerun blanks and standards before using.

5. The gelatin reagent contains thymol blue as a pH indicator in the acid reagent (indicating that the gelatin has been added), and thymol is included as a preservative. The gelatin itself is used to give a smooth indicator current by preventing the reduction of silver chloride.

Ultramicro Chloride Determination (Continued):

6. The titration is started as soon as possible after the serum is added to the vial and the needle of the ampere meter falls below the shut-off point. This gives a more uniform time lapse before the titrations are started.

7. Best results are obtained when minimum time elapses between removing the vial for specimen and replacing it again.

8. Titrate a couple of "Standard" aliquots first, in order to condition the electrodes. Then proceed with the Reagent Blank titration.

9. Trouble-Shooting Procedure:

   A. To check the error of the instrument and the dilution technique: Make 20 dilutions of the 100 mEq./L. Standard. Titrate as samples with the Mode Selector on Blank. Record the time. Calculate the coefficient of variation. At the time the method was set up, the range was 0.32 - 0.43.

   B. To check the component error of the Chloridometer, dispense with a dilutor, 4.0 ml. of a 1:200 dilution of 100 mEq./L. Standard into 20 vials. Titrate as samples with Mode Selector on Blank. Record the time. Calculate the coefficient of variation. Range was 0.20 to 0.28.

PROCEDURE:

1. Set the Titration Rate switch on Medium, the Blank Time Delay on "0", and Mode Selector on Blank. Clean electrodes with silver polish then wash well with distilled water. Set the Microampere Meter (red needle) at 10 Microamps.

2. Using a microdilutor, dilute 20 microliters of standard or specimen with 4.0 ml. nitric acid reagent. Add 2.0 drops of gelatin.

3. Condition the electrodes with 2 "Standard" vials. When ready to titrate, turn the Titration Switch from position No. 2 to position No. 1. As soon as the black needle falls below the red, turn the switch back to position No. 2. After each titration, rinse electrodes with distilled water and blot dry with a Kimwipe.

Ultramicro Chloride Determination (Continued):

4. Titrate 3 "Blank" vials (containing 4.0 ml. reagent and 2.0 drops of gelatin), record and average the readings within 0.2 seconds. Using the averaged blank time, set the Time Delay Dial. This interval will be automatically subtracted from the final mEq./L. The Time Delay resets itself for the next titration.

5. Turn Mode Selector on Standard or Unknown. Set Digital Read-Out to 0.000 for each vial, standard and test. Titrate 2 "Standard" vials. After the second titration, set the "Standard Adjust" Dial to read 100 mEq./L. The third Standard titration should read 100 ± 0.5 on the Digital Out-Put. Each 20 units on the Standard Adjust Dial is equal to a change of approximately 1.0 mEq./L. Always check the Standard Setting with an additional standard, either the 130 or 70 mEq./L.

6. Proceed with sample titration in the same manner. The specimen run in duplicate should be within 3.0 mEq./L.

7. As an alternative means, the readings in seconds may be used to calculate the unknowns. The Direct Read-Out Auxillary must be Off or in Blank position.

8. CALCULATIONS:

$$\frac{\text{Test in Seconds} - \text{Blank in Seconds}}{\text{Standard in Seconds} - \text{Blank in Seconds}} \times \begin{array}{c}\text{Conc. Std.}\\ \text{in mEq./L.}\end{array} = \begin{array}{c}\text{Unknown}\\ \text{in mEq./L.}\end{array}$$

NORMAL VALUES:

1. Adults: 98 - 106 mEq./L. in serum. (Males tend to be somewhat higher than female
2. Newborn: 93 - 117 mEq./L. in serum.

REFERENCES:

1. Siggaard-Andersen: Technical Bulletin of the Registry of Medical Technologists, 37:240, September, 1967.

2. Cotlove: J. Lab. and Clin. Medicine, 51:461, 1958.

Ultramicro Chloride Determination (Continued):

CLINICAL INTERPRETATION:

Hyperchloremia is found in patients with metabolic acidosis as in renal failure. It may also occur in dehydration or with inappropriate excess administration of saline.

Hypochloremia occurs with severe vomiting or excess removal of gastric contents by gastric suction. It is also seen in respiratory acidosis and metabolic alkalosis. If inappropriate excess intravenous glucose is administered, or if inappropriate ADH occurs, hypochloremia will be present.

TOTAL CHOLESTEROL DETERMINATION
(N-Methodology, Technicon AutoAnalyzer)

PRINCIPLE: The serum proteins are precipitated and the total Cholesterol (esterified and free) extracted by 2-Propanol. Propanol is excellent as an extraction reagent, as it precipitates the serum proteins in a finely divided state, without the use of heat, and extracts total Cholesterol in a minimal amount of time. The addition of the iron color reagent to an aliquot of the clear extract produces a stable color with a density proportional to the amount of Cholesterol present. Formation of color depends upon the amount of heat present when acetic and sulfuric acids are combined with the Propanol extract of Cholesterol. For this reason, the mixture is diverted through a heating bath (95°C.) just before passing into the colorimeter. Iron, added to the color reagent in the form of Ferric Chloride, is critical for achieving maximum color production. Color production also depends upon use of the optimal ratio between the concentrations of acetic and sulfuric acids. The source of error found by researchers to give the greatest variation in color reaction is the manner of mixing the color reagent and the extract. Because this should be consistent and gentle, the AutoAnalyzer method has been found to be ideal. All forms of Cholesterol yield the same amount of color production.

SPECIMEN: A fasting, non-hemolyzed serum specimen is preferred. Bilirubin does not appear to have any adverse effects.

REAGENTS AND EQUIPMENT:

1. Ferric Chloride ($FeCl_3 \cdot 6H_2O$), 0.05% in 95% Acetic acid
   Put 50 ml. distilled water into a liter volumetric flask. Add about 500 ml. glacial acetic acid (J. T. Baker; conforming to Dichromate test) with gentle swirling. Add 0.5 gm. Ferric Chloride ($FeCl_3 \cdot 6H_2O$), and mix until dissolved. Bring to volume with glacial acetic acid.

2. Cholesterol Color Reagent
   Place 1000 ml. 0.05% Ferric Chloride reagent in a 2 liter Erlenmeyer flask. Add 430 ml. concentrated sulfuric acid (Specific Gravity 1.84) in 150 ml. aliquots, swirling gently and allowing air bubbles to escape between additions. Cap with aluminum foil and allow to cool. Stable for at least 1 month.

3. 2-Propanol, 99%
   J. T. Baker or Mallinckrodt Spectrophotometric Grade. Redistillation is not necessary.

Total Cholesterol Determination (Continued):

4. Stock Cholesterol Standard (125 mg%.)
Place 625 mg. Cholesterol (Eastman Organic Chemicals) in a 100 ml. beaker, and add about 50 ml. 2-Propanol. Mix well and heat gently on hot plate, putting the Cholesterol into solution. Cool and transfer carefully, with rinsing, into a 500 ml. volumetric flask. Bring to volume with 2-Propanol.

5. Working Cholesterol Standards
Prepare dilutions 2-Propanol, using volumetric glassware according to the following chart:

Stock Standard (1.25 mg./ml.) in ml.	Dilution volume isopropanol	Actual Conc. mg%.	Graph value = mg%. 40 x actual conc.
1	100 ml.	1.25	50
2	100 ml.	2.50	100
4	100 ml.	5.00	200
6	100 ml.	7.50	300
8	100 ml.	10.00	400
10	100 ml.	12.50	500

The concentration of the standards compensates for the 1:40 sample dilution when the 2-Propanol extract of serum is made. Hence, they may be plotted as 50, 100, 200, 300, 400, and 500 mg%. Cholesterol.

6. Technicon AutoAnalyzer
Technicon Sampler II set at 40/hour
Technicon module with special acidflex and solviflex tubing is used throughout the manifold.
Technicon colorimeter with 520 nm. filter, and #1 aperture.
Technicon heating bath at 95°C.

NOTES ON PROCEDURE:

1. The operator is advised to wear glasses or safety goggles in the event that a manifold connection should come loose during the run and spatter reagent.

2. A generous flow of water through the waste sink will minimize the corrosive effects of the acid on the plumbing.

3. NEVER use the high speed setting on the pump, even with water-wash.

Total Cholesterol Determination (Continued):

4. Always pump water through the system before starting run. At this time check system for leaks, pinched tubing and blockages. Then drain water completely before starting acid reagents through. The intermixing of acid and water causes excessive heat and pressure build-up, which could cause a connection to come loose.

5. Do not stop the pump with pressure on the manifold tubing, as the tubing will distort on standing and result in improper proportioning of reagents.

6. In the event of a nipple connection leaking during the run, immediately stop pump by cutting power. With a hemostat clamp off the solviflex tubing between the 0.056 sample line and the "HO" fitting. Then transfer reagent lines to an empty beaker and release pump rollers, allowing reagent to flow back into beaker. When reagent stops flowing back, disconnect line from heating bath inlet nipple and dip it into a beaker. Inspect manifold and reinsert or replace leaking nipple, being careful not to get acid on your fingers. Then remove hemostat and flush air through manifold. Check for further leaking with a water-wash followed by an air flush. Then reconnect to heating bath, continue to flush air to remove reagent in heating bath. Finally restart reagents. Suppress the desire to keep a close watch on the manifold unless you are wearing safety goggles!

7. Always have bicarbonate of soda on hand!

8. Always use the <u>same type</u> of 2-Propanol in standards, serum extracts, and as a reagent.

PROCEDURE:

1. <u>Preparation of extract</u>: Into a screw-top tube containing 3.9 ml. 2-Propanol, add 0.1 ml. control or serum specimen while mixing on the vortex. Cap with a <u>Teflon-lined</u> cap, and mix immediately for 20 seconds to produce a finely divided precipitate. Let sit for 15 minutes, mix again, and centrifuge. Keep capped until ready for use.

2. Turn on colorimeter and recorder; check for proper filters and aperature.

3. Attach Cholesterol manifold, using acidflex tubing for all connections between modules. Be sure that the drain is connected

Total Cholesterol Determination (Continued):

Wash well with water, and then disconnect water-wash and allow all fluid to be expelled from the system.

4. Start pumping of reagents. When baseline has been established (set for about 80% T.), begin dispensing the first standards on sampler plate. Dispense specimen into cups on plate only shortly before it is to be picked up. In this way, evaporation of solvent may be kept to a minimum.

5. When run is completed, lift reagent lines carefully, wiping acid reagent from the outside of the tubing with paper towels. Allow system to clear entirely of acid reagent before submerging pick-up lines into water-wash. Store manifold with water in the tubing.

6. Plot a curve from the standard peaks on the recorder chart, and connect the points with straight lines. Determine Mg%. of Cholesterol in control sera and specimens by reading from the curve.

NORMAL VALUES: 160 - 280 mg%.

REFERENCES:

The automation of the colorimetric procedure for cholesterol has been based on the methods of:

1. Zlatkis, Zak and Boyle: J. Lab. & Clin. Med., 41:486, 1953.

2. Zak, Dickenman, White, Burnett & Cherney: Amer. J. Clin. Path., 24:1307, 1954.

3. Rosenthal, Pfluke & Buscaglia: J. Lab. & Clin. Med., 50:318, 1957.

4. Leffler: Amer. J. Clin. Path., 31:310, 1959.

CLINICAL INTERPRETATION:

Hypercholesterolemia occurs in a diverse group of conditions. It is present consistently in uncontrolled diabetes. Diminution of the serum cholesterol occurs with insulin treatment of the diabetes. Changes in serum cholesterol are less marked in diabetic children.

Total Cholesterol Determination (Continued):

The nephrotic syndrome due to various causes is characterized by hypercholesterolemia. The hypoalbuminemia results in an attempt for the liver to resynthesize albumin with concommitant increased synthesis of hepatic cholesterol esters. In addition, plasma lipoprotein lipase deficiency is present in nephrosis which leads to hypercholesterolemia.

Obstructive jaundice leads to increased cholesterol ester synthesis and concommitant reflux of cholesterol into the blood stream. Hypercholesterolemia is thus a common biochemical feature of obstructive jaundice.

Untreated myxedema is frequently associated with hypercholesterolemia. When thyroxin is administered, the serum cholesterol returns to normal.

Patients on long term cortisone as an immunosuppresive agent, exhibit an increase in blood lipids including hypercholesterolemia.

Finally hereditary lipoprotein abnormalities with premature arteriosclerosis is associated with hypercholesterolemia in Types II, III, IV, V.

Hypocholesterolemia is a common abnormality in hyperthyroidism exactly opposite to hypothyroidism. The exact cause for the low blood cholesterol in anemia, especially pernicious anemia, is not known. When severe hepatic disease is present, cholesterol ester synthesis is markedly impaired and the blood cholesterol is low. Severe systemic infections result in a hypermetabolic state with hypocholesterolemia.

Plasma cholesterol levels are lower in children and females. Malnutrition and malabsorption may result in marked hypocholesterolemia.

Technicon ®
AutoAnalyzer ®
Methodology

Total Cholesterol

Flow diagram

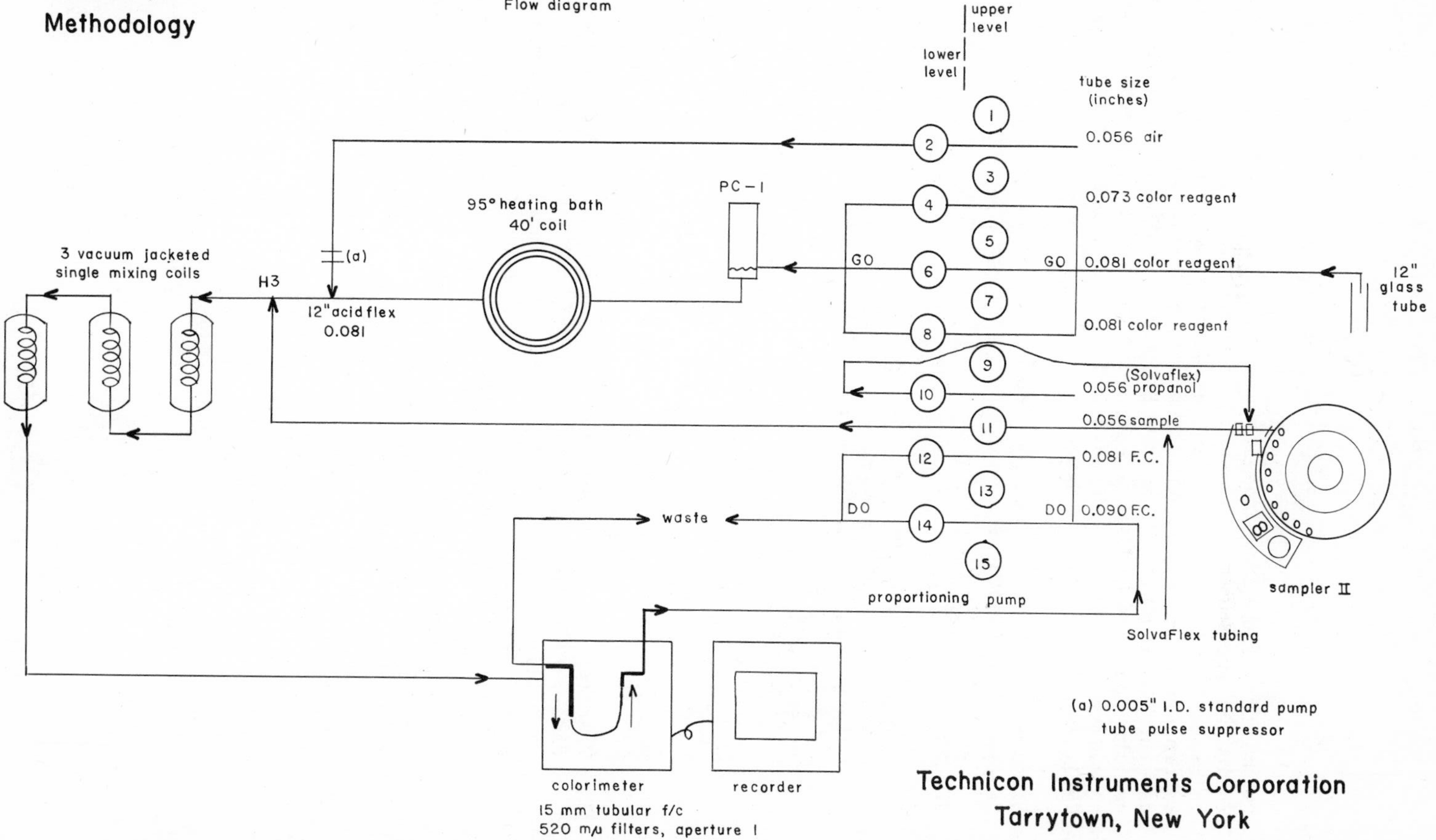

(a) 0.005" I.D. standard pump tube pulse suppressor

Technicon Instruments Corporation
Tarrytown, New York

CHOLESTEROL ESTERS

PRINCIPLE: The serum proteins are precipitated and cholesterol including its esters are extracted by a warm solution of Ferric chloride and Acetic acid. In the determination of Cholesterol Esters, an Ethanol - Ether extract is prepared and the free Cholesterol is precipitated as the Digitonide. The mixture is dried and the Cholesterol Esters extracted with Petroleum ether. Finally an aliquot is taken to dryness and color development is induced with $FeCl_3$, Acetic Acid, and H_2SO_4.

SPECIMEN: 0.4 ml. of <u>unhemolyzed</u> serum or plasma specimen. See "Comments on Procedure".

REAGENTS:

1. <u>Ethanol - Ether Solvent Mixture</u>
   3 volumes of 95% ethanol + 1 volume ether. Ethanol denatured with methanol is satisfactory if 5.0% water is added.

2. <u>Digitonin Solution</u>
   A 0.5% solution is added to 95% ethanol.

3. <u>Petroleum Ether</u>
   B. P. 30 - 60°C.

PROCEDURE:

1. Transfer 0.4 ml. serum or plasma to a dry, 10 ml. volumetric flask. Add 5.0 ml. ethanol - ether solvent mixture and bring to boiling point for a few seconds by immersion in 60°C. water bath. Let flask come to room temperature, add solvent mixture to volume, mix, and filter through a dry Whatman 1 filter paper.

2. Transfer 2.5 ml. filtrate to large test tube (25 x 150 mm.), add 0.5 ml. digitonin solution, mix, and let stand at least 10 minutes.

3. Evaporate to dryness in a water bath at 60 - 80°C. This evaporation can be hastened by a stream of air or N_2.

4. Extract the esters from the residue by adding approximately 5.0 ml. petroleum ether, bring to boil in 60°C. water bath, allow to cool, stopper, centrifuge, and decant supernate into a large test tube. Repeat extraction and combine extracts.

Cholesterol Esters (Continued):

5. Evaporate extract to dryness under air or N_2 at 60 - 80°C.

6. Add 5.0 ml. $FeCl_3$ - acetic acid reagent, mix, and set aside for 10 - 15 minutes. Heat tubes in 60°C. water bath for 2 minutes with occasional shaking. Allow to cool to room temperature and centrifuge. The supernate and precipitate should be straw color.

7. Set up the following in large test tubes (25 x 150 mm.):
BLANK: 5.0 ml. $FeCl_3$ - acetic acid reagent.
STANDARD: Place 1.0 ml. Working Standard in tube, bring to dryness at 60°C., then add 5.0 ml. $FeCl_3$ - acetic acid reagent. Evaporation of $CHCl_3$ can be hastened by a stream of air or N_2.
UNKNOWN: 5.0 ml. supernate.

8. Add 3.0 ml. Concentrated H_2SO_4 to each tube from a buret and mix immediately.

9. After 10 minutes, read Absorbance at 560 nm. or with a filter with nominal wavelength in this region vs. a water blank.

CALCULATION:

Mg. Cholesterol esters (as free cholesterol)/100 ml.

$$= \frac{A_x - A_b}{A_s - A_b} \times 0.2 \times \frac{100}{0.1}$$

$$= \frac{A_x - A_b}{A_s - A_b} \times 200$$

COMMENTS ON PROCEDURE:

1. <u>Beer's Law and Color Stability.</u>
The absorption peak occurs at 560 nm. The color obeys Beer's Law in a Beckman DU and in a Klett photometer with a No. 54 filter. Color reaches a maximum in 5 minutes and is stable for at least one hour.

2. <u>Water in the Concentrated H_2SO_4</u>
The reaction is performed in absence of water and great care must be taken that there is no more than a trace of water contamination.

3. <u>Purity of Acetic Acid</u>
It has been claimed that the color intensity varies inversely with the purity of the acetic acid.

Cholesterol Esters (Continued):

4. Temperature during Color Development
Addition of concentrated H_2SO_4 to the $FeCl_3$ acetic acid reagent results in evolution of considerable heat, which hastens the attainment of peak color intensity and increases the degree of color formed.

5. Precipitation of Free Cholesterol as the Digitonide
Digitonin reagents employed for precipitation of free cholesterol have varied from aqueous and 50% ethanolic to 95% ethanolic. The presence of water is required. If the reagent is prepared from absolute ethanol or ethanol denatured with methanol, 5% water must be added. Precipitation is complete within 10 minutes. Inclusion of any digitonin in the final color would result in an orange-red rather than purple because of color produced with digitonin alone, but this causes no interference at 560 nm.

6. Interference by Hemolysis
The presence of 1.0 gm. of hemoglobin per 100 ml. serum or plasma causes approximately a 10% positive error. This method, therefore, is not directly applicable to whole blood.

7. Serum vs. Plasma
Plasma obtained with an anticoagulant such as heparin, which does not cause a water shift, has a cholesterol concentration identical to serum. The level in oxalated plasma, however, is significantly lower (up to 15%).

8. Stability of Samples
On standing, free cholesterol decreases and cholesterol esters increase as a result of enzymatic action by cholesterol esterase. The optimal pH for this activity is at 8, and heat inactivation at 55 - 60°C. for 30 minutes, destroys the enzyme.

NORMAL VALUES:

Free Cholesterol:	22 - 30% of Total Cholesterol
Cholesterol Esters:	70 - 78% of Total Cholesterol

REFERENCE: Henry, J. B.: CLINICAL CHEMISTRY: PRINCIPLES AND TECHNIQUES, Hoeber, pg. 856 - 859, 1965.

Cholesterol Esters (Continued):

CLINICAL INTERPRETATION:

Approximately seventy per cent of Total serum Cholesterol exists in the ester form. Esterification of Cholesterol occurs in the liver with the hepatic cell being responsible for this process. Thus, a decrease in serum Cholesterol esters is found in hepatic disease. A decrease below fifty per cent is found especially in marked necrosis of the liver.

An increase in serum Cholesterol Esters occurs in obstructive hepatobiliary disease. A reflux of Cholesterol Esters into the blood from the liver due to biliary obstruction occurs. In addition, biliary obstruction stimulates hepatic cell membrane synthesis of Cholesterol Esters.

SERUM OR PLASMA CHOLINESTERASE
(Acholest Test Paper)

PRINCIPLE: Cholinesterase, an enzyme found in the tissues of all animals, formed in the liver, hydrolyzes acetylcholine to choline and acetic acid.

$$\text{Acetylcholine} \longrightarrow \text{Choline} + \text{Acetic Acid}$$

Acetylcholine stimulates the nerve impulses in the parasympathetic nervous system. When cholinesterase is low, acetylcholine accumulates, resulting in continuous stimulation of the parasympathetic system with undesirable symptoms.

The ACHOLEST TEST PAPER is a rapid, simplified screening test for the determination of plasma or serum cholinesterase. The kit consists of a test strip impregnanted with a special substrate and a control strip (comparative color strip). The time required to reach the endpoint with the control strip is a measure of cholinesterase activity, described as "increased", "normal", "suspicious", or "decreased".

Cholinesterase activity is another diagnostic test for liver disease. It is very useful in detecting poisoning by organic phosphate insecticides or some drugs as in anesthesia.

SPECIMEN: 0.1 ml. of non-hemolyzed plasma or serum is required. Samples are stable for several days at $4^{o}C$. and for weeks in the freezer.

REAGENTS AND EQUIPMENT:

1. ACHOLEST, cholinesterase Test Paper
   Bottle I.

2. Comparative Color Strip
   Bottle II.

3. Clean Slides

4. 0.05 ml. Wiretrol Disposable Pipettes

5. Stop-Watch

6. Tweezers

Serum or Plasma Cholinesterase (Continued):

The whole kit (for 30 determinations) is available and supplied by E. Fougera and Co., New York)

PROCEDURE:

1. Using Wiretrol disposable pipettes, pipette 0.05 ml. of plasma or serum on each of two clean slides. Label one slide "Test" and the other one "Control".

2. Cut the "Test" Strip and the "Control" Strip in half.

3. Using tweezers, place the Test Strip and Control Strip on the respective slides.

4. Cover with clean slides to ensure complete saturation of the paper.

5. Set the stop-watch. The Test Paper turns green, gradually developing into a yellowish color, which is the color of the Control Paper.

6. Cholinesterase activity is measured by the time required for the Test Paper to reach the color of the Control Paper.

7. From the moment of contact of the serum or plasma and Test Paper, to the point where the comparative tone of color has been reached, the following time values have been established:

Minutes	Activity of Plasma or Serum Cholinesterase
Below 5	"Increased"
5 - 20	"Normal"
20 - 30	"Suspicious"
30 and Longer	"Decreased"

NORMAL VALUES: 5 - 20 Minutes

REFERENCES:

1. STANDARD METHODS OF CLINICAL CHEMISTRY, 3:93 - 98, Academic Press, 1961.

2. STANDARD METHODS OF CLINICAL CHEMISTRY, 4:47 - 56, Academic Press, 1963.

Serum or Plasma Cholinesterase (Continued):

CLINICAL INTERPRETATION:

A rapid test to qualitatively determine the plasma level of cholinesterase. The test paper known as Acholest Test Paper is utilized as a simple screening method for plasma cholinesterase. Determination of this enzyme is important in suspected decrease due to poisoning by organic phosphate insecticides. Enzyme levels also are low in hepatic disease due to various causes. If syccinyl choline is utilized before surgery in patients with hepatic disease when there are low enzyme levels, patients may suffer undesirable "overdose" type effects. Thus, the determination of plasma cholinesterase may be extremely useful in the surgical patient. The lab kit consists of a test strip impregnanted with a special substrate and a control strip. The time required to reach the endpoint with the control strip is a measure of cholinesterase activity which is reported as normal or decreased. The plasma cholinesterase acts on the substrate to liberate choline and acetic acid. The acetic acid changes the color of the test strip containing an acid base indicator

CONGO RED TEST

PRINCIPLE: The dye, Congo Red, is a derivative of benzidine and naphthionic acid and has an unexplained affinity for amyloid. When a measured amount of the dye is injected intravenously, a greater proportion of the dose is removed from the blood in a given time by patients with amyloidosis than by normal persons.

REAGENT:

Congo Red, 1.0% Aqueous Solution, Sterile.

Dissolve 0.3 gm. biologically tested Congo red in 30 ml. warm, distilled, pyrogen-free water. Shake until completely dissolved. Filter through a fine filter paper and transfer the filtrate to a rubber-capped vaccine bottle. Pierce the cap with a fine hypodermic needle and autoclave for 15 minutes at 10 lbs. pressure. Remove the hypodermic needle from the cap immediately on taking the bottle from the autoclave. Allow to cool. This solution must be freshly prepared (on the afternoon of the day prior to the test); any unused solution must be discarded.

PROCEDURE:

1. Sterile Congo red is injected intravenously as a 1.0% aqueous solution in an amount of 1.0 ml. for each 10 lbs. body weight; the injection should take 1 or 2 minutes. To avoid lipemia the patient should be fasting. Note the time of completion of the injection.

2. Exactly 4 minutes later withdraw about 10 ml. of blood from the opposite arm, avoiding hemolysis. Place in a centrifuge tube and allow to clot.

3. Exactly 1 hour after completion of the injection withdraw a second sample of venous blood as in Step 2. Collect a sample of urine at the same time.

4. Allow 1½ to 2 hours for clotting and clot retraction to proceed to completion. Then centrifuge and transfer the serum to clean tubes.

5. Set up two centrifuge tubes, marked 4 minute and 60 minute samples. Into each pipette exactly 2.0 ml. of serum from the respective blood samples.

Congo Red Test (Continued):

6. To each tube add 8.0 ml. absolute ethanol; stopper and shake vigorously for 30 seconds.

7. Centrifuge for 10 minutes at 2800 to 3000 rpm.

8. Transfer clear supernatants to correspondingly marked spectrophotometer tubes.

9. Read the optical density of each tube at 510 nm. setting the zero with absolute ethanol (use blue-green filter with photoelectric colorimeter).

CALCULATION:

$$\%\ \text{Dye absorbed} = 100 - \left\{\frac{\text{O.D. of 60 minute sample}}{\text{O.D. of 4 minute sample}} \times 100\right\}$$

NORMAL VALUES: Less than 40% Retention.

REFERENCE: Lynch, M., Raphael, S. and et. al.: MEDICAL LABORATORY TECHNOLOGY AND CLINICAL PATHOLOGY, 2nd. Ed., Saunders, 1969.

CLINICAL INTERPRETATION:

If the Congo red dye absorbed or retained in the body is more than 90%, the diagnosis of secondary amyloidosis is strongly indicated. If 60 to 90% is retained the diagnosis of secondary amyloidosis is suggestive. A retention of less than 60% is negative for secondary amyloidosis. Do not perform the test in primary amyloidosis or when the nephrotic syndrome is present since less plasma proteins are present which bind the Congo red dye. Thus, more dye is lost into the tissues. In addition, more Congo red dye is lost into the urine because of its combination to albumin.

SERUM COPPER
(Atomic Absorption Spectrophotometry)

PRINCIPLE: In the determination of copper in serum, the serum is treated with 20% trichloroacetic acid, and a minimal amount of 8% butyl alcohol. The solution is digested in the 90°C. water bath for 15 minutes, to separate the metals from the protein, cooled to room temperature and centrifuged. The supernatant is then analyzed for copper, by atomic absorption spectrophotometry.

Copper is present in the blood in two forms. The direct reacting fraction is loosely bound to albumin and represents 5% of total copper. The indirect reacting fraction, which represents 95% of total copper, is a protein complex, called ceruloplasmin.

SPECIMEN: UNHEMOLYZED, pipettable 2.0 ml. of serum is needed for the determination, as hemolysis tends to elevate the values. There is general agreement that there is a diurnal variation in the serum copper; so to be consistent, it should be drawn in the morning. Serum copper is stable for two weeks in the refrigerator, and for several months in the freezer.

REAGENTS AND EQUIPMENT:

1. Trichloroacetic acid, Baker's Analyzed Reagent Grade, 20% (W/V). Weigh out 200 gm. of Reagent Grade Trichloroacetic acid, into a one liter volumetric flask. Add deionized water slowly, and bring the crystals into solution. Bring up to volume with deionized water.

2. Butyl Alcohol, Baker's Reagent Grade, 8% (V/V). Measure 80 ml. of butyl alcohol, and pour into a one liter volumetric flask. Dilute and bring up to volume with deionized water.

3. Copper Stock Standard I, 0.5 mg./ml. Weigh out 0.3937 gm. of Reagent Grade $CuSO_4 \cdot 5H_2O$ into a 200 ml. volumetric flask. Dissolve in deionized water and bring to volume. This represents 0.500 mg. Cu^{++} / ml. or 50,000 micrograms/100 ml.

4. Copper Stock Standard II, 1000 micrograms/100 ml. Pipette 2.0 ml. of Stock Standard I into a 100 ml. volumetric flask and dilute to volume with deionized water.

Serum Copper (Continued):

5. Copper Working Standards
The Working Standards are made up by diluting definite amounts of Stock Standard II to 50 ml. with deionized water.

Vol. of Stock Std. II	Total Volume	Concentration
2.5 ml.	50 ml.	50 ug%.
5.0 ml.	50 ml.	100 ug%.
7.5 ml.	50 ml.	150 ug%.
10.0 ml.	50 ml.	200 ug%.
12.5 ml.	50 ml.	250 ug%.

6. Perkin-Elmer Model 403 Atomic Absorption Spectrophotometer
Equipped with a three-slot burner head.

PROCEDURE:

1. Using volumetric pipette (special washing), pipette 2.0 ml. of serum into a 12 ml. glass-stoppered centrifuge tube.
BLANK - 2.0 ml. of deionized water
STANDARDS - 2.0 ml. each of the 50, 100, 150, 200, and 250 ug%. Working Standards are pipetted into respective tubes
CONTROLS - a. Normal Control sera.
b. Reference Standard (100 ug./100 ml.)

2. Add 1.7 ml. of 20% TCA and 0.3 ml. of 8% butanol to all tubes. Stopper, and mix well on the Vortex mixer.

3. Digest all tubes in the 90°C. water bath for 15 minutes.

4. Cool, centrifuge, and analyze the supernatant for copper, using the atomic absorption spectrophotometer.

5. Copper cathode lamp should be warmed up for at least 15 minutes.

6. INSTRUMENT SETTINGS
Range -- U.V.
Wavelength -- 324.7
Slit -- 4
Fuel -- acetylene (oxidizing flame)
Sensitivity -- 0.15 ug./ml. $Cu.^{++}$ for 1% absorption

7. Set the machine for concentration mode. Set to zero with the Blank. Run Standards, and check for linearity of the curve. Run Controls and Unknowns. Report in micrograms/100 ml.

Serum Copper (Continued):

NORMAL VALUES: Male 68 - 134 micrograms/100 ml.
Female 84 - 143 micrograms/100 ml.

REFERENCES:

1. Olson, A. D., and Hamlin, W. B.: Serum Copper and Zinc by Atomic Absorption Spectrophotometry. <u>Atomic Absorption Newsletter</u>, 7:4, 1968.

2. Perkin-Elmer: ANALYTICAL METHODS FOR ATOMIC ABSORPTION SPECTROPHOTOMETRY, September, 1968.

3. Piper, K. G., and Higgins, G.: <u>Proceedings of the Association of Clinical Biochemists</u>, 4:7, August, 1967.

CLINICAL INTERPRETATION:

Copper is present in two forms in the serum. Five per cent is free or loosely bound to albumin. Ninety-five per cent is firmly bound to alpha-two globulin. The variations of serum copper and ceruloplasmin are essentially the same and will found in the Interpretation of the Ceruloplasmin Determination on page 118.

CREATINE PHOSPHOKINASE

PRINCIPLE: Determination of the enzyme creatine phosphokinase (CPK, adenosine triphosphate, creatine phosphotransferase) in serum is of value in the investigation of skeletal muscle disease, possibly in the detection of carriers of muscular dystrophy, and in the diagnosis of suspected myocardial infarction. The CPK value is also elevated with muscle trauma, such as multiple intramuscular injections, and with severe exercise. CPK is found in highest activity in skeletal muscle, heart muscle and brain tissue, with minimal quantities found in lung, kidney, liver and red cells. For this reason, the CPK is helpful in interpreting elevated SGOT values in the presence of hepatic problems. The CPK activity also rises much more rapidly than SGOT following myocardial infarction.

This enzyme catalyzes the reversible formation of adenosine triphosphate and creatine from adenosine diphosphate and creatine phosphate. CPK activity in serum is determined by the procedure in which adenosine triphosphate, liberated by the action of the enzyme, is linked to the reduction of nicotinamide-adenine dinucleotide phosphate and the formation of reduced nicotinamide-adenine dinucleotide phosphate followed spectrophotometrically at 340 nm.

1. $\text{ADP} + \text{CP} \longrightarrow \text{ATP} + \text{Creatine}$

2. $\text{ATP} + \text{Glucose} \xrightarrow{\text{Hexokinase}} \text{Glucose-6-Phosphate} + \text{ADP}$

3. $\text{Glucose-6-Phosphate} + \text{NADP} \longrightarrow \text{6-Phosphogluconate} + \text{NADPH}$

Increased reaction rate is achieved by including optimal amounts of Mg^{++} and glutathione as activators. AMP is included to inhibit the activity of any myokinase which may be present.

SPECIMEN: 50 microliters of unhemolyzed serum. The enzyme is extremely heat labile, and should be stored at refrigerator temperatures until one is prepared to make the determination.

REAGENTS AND EQUIPMENT:

The Reagents for the test are the Biochemica Test Combination supplied by Boehringer Mannheim Corporation: UV System CPK Single Test (Cat. #15790 - for 20 determinations). Volumes have been modified for use in this laboratory.

1. Pipette 2.5 ml. of Solution 2 into each Bottle No. 1. Prepare enough working substrate for one day run.

Creatine Phosphokinase (Continued):

Concentrations in the test volume:
a). 0.1 M. Triethanolamine buffer, pH 7.0, 20 mM. glucose
b). 10 mM. Mg-acetate, 1.0 mM. ADP, 10 mM. AMP
c). 0.6 mM. NADP, 35 mM. creatine phosphate, 50 ug. HK
d). 25 ug. G-6-PDH, 9.0 mM. glutathione

2. Spectrophotometer

A. Gilford 222, 340 nm.
This instrument is used when analyzing four samples at a time. Temperature control is by means of a 30°C. circulating water bath. Rate of reaction is determined from the strip chart recorder set for 0.200 A full scale and run at one inch per minute.

B. Gilford 300 N, 340 nm.
This instrument is used when analyzing one sample at a time. The change in absorbance per minute is read from the Data Lister print-out. The instrument is used with the thermocuvette set at 30°C.

COMMENTS ON PROCEDURE:

1. Monitor the working substrate for several minutes to check for substrate deterioration and/or instrument drift.

2. If the cuvette is being removed from a water bath to a cuvette well for determining absorbance values, one should work quickly to prevent temperature change in the reaction mixture.

3. Sera with low activity will show a longer lag time than sera with high activity.

4. With an initial A reading over 0.800, or ΔA/minute greater than 0.060 (300 mu./ml.), test must be repeated using a lesser volume of sera. Multiply result by dilution factor.

PROCEDURE:

1. Pipette 1.5 ml. of working CPK substrate into a 1.0 cm. square cuvette and equilibrate to 30°C. for 5 minutes.

2. Using a 50 microliter pipette (calibrated "to deliver"), deliver serum into the cuvette with substrate and mix well by inversion. Return to water bath. Start stop watch.

Creatine Phosphokinase (Continued):

3. After a lag time of approximately 5 minutes (longer for low activity), place cuvette into a 30°C. thermostated cuvette well and monitor the reaction until at least 2 consecutive minutes of linearity are recorded. Spectrophotometer set at 340 nm.

4. CALCULATION:

   Calculation is based on ΔA of 1 minute. Light path is 1.0 cm. in diameter.

   ϵ of NADPH at 340 nm. = 6.22×10^3 Liter/Mole x cm.

$$\frac{\Delta A}{\epsilon \times d} \times 10^6 \times \frac{TV}{SV} \times \frac{1}{Time} = \text{I.U./Liter or mU./ml.}$$

$$\Delta A \times \frac{1}{6.22 \times 10^3 \times 1} \times 10^6 \times \frac{1.55}{0.05} \times \frac{1}{1}$$

$$\Delta A \times F = \text{mU./ml.}$$

$$F = 5000$$

NORMAL VALUES: Males 5 - 90 I.U./Liter or mU./ml.
Females 5 - 70 I.U./Liter or mU./ml.

NOTE:

A laboratory must determine its own normal range of values due to variation type of methodology used and reagent concentration and purity.

REFERENCES:

1. Rosalki, S. B.: Journal of Laboratory and Clinical Medicine, Vol. 4, April, 1967.

2. Bergmeyer: METHODS OF ENZYMATIC ANALYSIS, Academic Press, 1965.

CLINICAL INTERPRETATION:

Creatine phosphokinase exists in different isoenzymes which vary in their tissue distribution. Creatine phosphokinase is found primarily in the brain, heart muscle, skeletal muscle, and thyroid. It

Creatine Phosphokinase (Continued):

is absent in the lung, RBC, liver, and kidney. The greatest amount of enzyme is present in skeletal muscle which is followed next by heart muscle and the brain. The normal serum values are: Males up to 90 International Units, females up to 70 International Units.

Causes for an elevated CPK are:

1. Acute myocardial infarction
2. Muscular dystrophy
3. Hypothyroidism
4. Infarction of the cerebral cortex of the brain
5. Schizophrenia
6. Severe exercise
7. Injections of drugs into skeletal muscle
8. Postpartum period

Decreased CPK levels occur in:

1. Storage
2. Oxalates
3. Inhibitors such as drugs in serum

CPK is an unstable enzyme. When glutathione is added to the serum, 90% of activity persists after four days of storage at 4°C. If no glutathione is added, there will be a 50% loss of activity after 4 to 6 hours of storage at room temperature or after 24 hours of storage at 4°C. If the thiol-stimulated reaction systems for CPK are utilized, there is an extension of time span for CPK beyond the usual time. Persistence of activity with such stimulated reactions may extend for 8 or 9 days.

CPK elevation due to recent electrical defibrillation has been observed. 50% of patients may show an elevation due to powerful intercostal muscle contractions. Prolonged arrhythmias may on occasion produce an elevation of serum enzymes especially with a cardiac rate of over 160.

An increase in the serum CPK occurs within 12 hours after acute myocardial infarction. The peak value is reached within 24 to 36 hours. An elevation of CPK usually returns to normal within the first week after the onset of the myocardial infarction. CPK determination can be helpful in other cardiac or pulmonary conditions. GOT elevation in the serum may be present in congestive heart failure with liver congestion; however, since CPK is not present in the liver, there will be no elevation of CPK when there is congestive heart failure with liver damage. CPK will only be elevated when myocardial damage has occurred. Furthermore, if one wishes to distinguish between acute myocardial infarction and pulmonary infarction, it is important to request LDH and CPK. LDH and CPK are increased in myocardial infarction, but only LDH is increased in pulmonary infarction.

One of the early valuable reasons for determination of CPK was in the clinical appraisal of patients suffering from muscular dystrophy. A markedly elevated CPK may be present in a majority of patients with muscular dystrophy and there may be a 50 fold rise in CPK in these patients. In addition, CPK is present in increased amounts in the serum in patients with active muscle necrosis. Thus, patients with active acute myositis, trauma to the skeletal muscle, surgery involving skeletal muscle, injections of drugs into the skeletal muscle, severe muscular exercise, and acute muscular necrosis in alcoholism, all give elevation of serum CPK. Serum CPK elevation usually is extremely helpful in detecting the preclinical muscular dystrophy state. Asymptomatic carriers usually show a rise in the serum CPK. Thus, in such patients it is extremely important to advise them not to exercise before a serum CPK determination is obtained. CPK is usually normal in atrophy of skeletal muscle and in myasthenia gravis.

A majority of patients with hypothyroidism have been shown to have an increase serum CPK. The exact etiology of this increase in not known. Some investigators feel that there may be a release of CPK from the damaged thyroid tissue. Others feel that there is an induction of CPK activity by excess TSH activity in patients with hypothyroidism. Thus, the assay of CPK activity in the serum is an extremely valuable test in screening for hypothyroidism.

CPK plays a role in the confirmation of disease of the brain. CPK activity is elevated during the acute phase of cerebral disease and it usually returns to normal within two weeks. It is usually elevated in patients with infarction of the cerebral cortex. Furthermore, patients with necrosis within a brain tumor have been shown to have an elevated CPK. It has also been reported to be elevated in patients with schizophrenia. There is a lack of correlation between elevation of serum CPK and degree of central nervous system damage. There is no elevation of the cerebral spinal fluid CPK in transient cerebral ischemia without damage. Demylinating diseases do not cause increased activity. The cerebellum has very little CPK activity and thus cerebellar lesions do not produce a rise in cerebral spinal fluid CPK activity.

The level of CPK in the umbilical cord blood is higher than that found in the serum during infancy and adult life. Higher values

Creatine Phosphokinase (Continued):

are also present in the umbilical cord in infants born to mothers with obstetric difficulties or birth trauma. The elevated CPK levels arising in these conditions usually return to normal values within 30 days after birth. The high levels of CPK in the umbilical cord blood have been reported to be present as a result of intense muscular activity of the mother associated with labor.

A minimal rise in CPK in maternal blood occurs within the first day after delivery. The CPK returns to normal within one week after delivery. The elevation in CPK within this one week postpartum period probably results from involution of the uterus.

CREATININE
(Jaffe´ Reaction)

PRINCIPLE: The most frequently used analytical procedure for Creatinine is based on the Jaffe´ Reaction, i.e. the reaction between alkaline pictrate and creatinine which results in the formation of a red-colored complex. Critical conditions for this reaction include the following: Time of reaction; temperature of reaction, temperature of the creatinine alkaline picrate at the time it is read; degree of agitation during reaction time, concentration of NaOH, and wavelength. This method does not correct for the effect of non-creatinine substances. However, the effects of such chromogens have been minimized by the use of dialysis and the selection of a short reaction time.

SPECIMEN: 0.2 ml. of serum, plasma, or urine.

REAGENTS AND EQUIPMENT:

1. <u>Sodium Chloride</u>
   a). Chemical Composition:

NaCl	18.0 gm.
Distilled water, C.S.	1000.0 ml.
Brij.-35	1.0 ml.

   b). Preparation:
      1). Place approximately 500 ml. of distilled water in a 1 liter volumetric flask.
      2). Add 18.0 gm. of sodium chloride; shake until dissolved.
      3). Dilute to 1000 ml.
      4). Add 1.0 ml. of Brij.-35, and mix.

2. <u>Sodium Hydroxide</u>, 0.6 N $\pm$ 0.05 N.
   a). Chemical Composition:

NaOH	24 gm.
Distilled water, dilute to	1000 ml.

   b). Preparation:
      1). Place approximately 800 ml. of distilled water in a 1 liter volumetric flask.
      2). Add 24 gm. of sodium hydroxide; shake until dissolved.
      3). Dilute to 1000 ml.

3. <u>Picric Acid</u> (saturated)
   a). Chemical Composition:

Picric acid (reagent grade)	13 gm.
Distilled water	1000 ml.

Creatinine (Continued):

b). Preparation:
 1). Place 13 gm. of picric acid in a liter volumetric flask.
 2). Add distilled water to the 1000 ml. mark.
 3). Allow the excess picric acid to remain in contact with the water, and shake occasionally.
 4). Filter; store in a polyethylene bottle.

4. Brij. Water
 a). Chemical Composition:
 Distilled water, q.s.
 Brij.-35

 b). Preparation:
 1). Place 1000 ml. of distilled water in a liter container.
 2). Add 1.0 ml. of Brij.-35 solution, and mix.

5. Sampler
 Technicon Sampler II. Specimens are run at 60 per hour with a 9:1 sample-wash ratio.

6. Pump
 Technicon Pump II

7. Dialyzer - Heating Bath
 24 inch dialysis plates (Technicon) are used. The dialysate is mixed with alkaline picrate and then incubated at 37°C. for exactly 7½ minutes before being read in a spectrophotometer.

8. Spectrophotometer, Gilford 300 N 505 nm.
 The spectrophotometer is equipped with a modified 10 mm. debubbler flow through cuvette (Gilford #3019). The original debubbler was replaced with a glass connector, Technicon A_4, and the lines entering and leaving the cuvette are 0.015 mm. I.D. with a flow rate of 0.075 - 0.128 ml./minute.

9. Recorder
 Heath Kit Recorder, model EU - 203.

10. Data Lister, Gilford 4008
 The data lister is equipped with a 3 second timer board for close monitoring of steady-state plateaus.

COMMENTS ON PROCEDURE:

1. The chemistry of the Jaffé Reaction is linear beyond 40 mg%.

Creatinine (Continued):

2. Urine specimens are normally run on a 1:25 dilution using distilled water.

3. The time from sample pick-up to sample peak is exactly 10 minutes

4. A reference serum is used for a set point.

5. All reagents are filtered.

PROCEDURE:

1. Prior to beginning the daily work, a solution of 20% Purex is flushed in to the flowcell and allowed to remain 25 minutes.

2. Water is pumped through the system for 10 minutes before connecting reagent lines and establishing a reagent baseline.

3. The Gilford 300 N is zeroed with reagents running before sampling begins. When the first reference serum reaches the flowcell, the assayed value of reference serum is set. This is checked with a second reference serum followed by controls. A reference serum is run in duplicate following every ten unknowns throughout the run, the first standard to check drift (acceptable ± 0.1 mg%.) and the second to correct value if necessary. Therefore, the last specimen must be followed by a reference serum to check its validity.

NORMAL VALUES: Normal Adult Serum 0.5 - 1.5 mg%.

REFERENCES:

1. Bonses and Taussky: Jour. Biol. Chem., 158:581, 1945.

2. Chasson, et. al.: Amer. Jour. of Clin. Path., 35:83, 1961.

3. Henry, R. B.: CLINICAL CHEMISTRY: PRINCIPLES AND TECHNIQUES Hoeber, 1964.

4. Roscoe: Jour. of Clin. Path., 6:201, 1953.

5. Owen, et. al.: Biochem. Journal, 58:426, 1954.

Creatinine (Continued):

CLINICAL INTERPRETATION:

Serum creatinine is assesed to determine glomerular filtration. Serum creatinine is a better reflection of the degree of uremia than blood urea concentrations. The relationship of blood urea to creatinine is important in a number of clinical circumstances. The BUN:Creatinine ratio is 10:1. Pre-renal azotemia causes a ratio of greater than 10:1 while intrinsic renal disease causes a BUN:Creatinine ratio of 10:1. Thus in dehydration or gastrointestinal hemorrhage BUN rises with no or only slight increase in serum creatinine. A marked elevation of both creatinine and BUN occurs in acute or chronic renal disease with preservation of the 10:1 ratio.

Urea levels are dependent on protein intake while creatinine concentration is unaffected by diet.

The measurement of serum creatinine is superior to the determination of BUN for the diagnosis, prognosis, and management of renal failure. Creatinine clearance, as determined by measurement of serum and urine creatinine over a specific period of time, is the best method to determine renal glomerular function. Low clearance suggests primary renal disease but may also occur in shock and renal vascular disease.

Technicon ®
AutoAnalyzer ®
Methodology

CREATININE

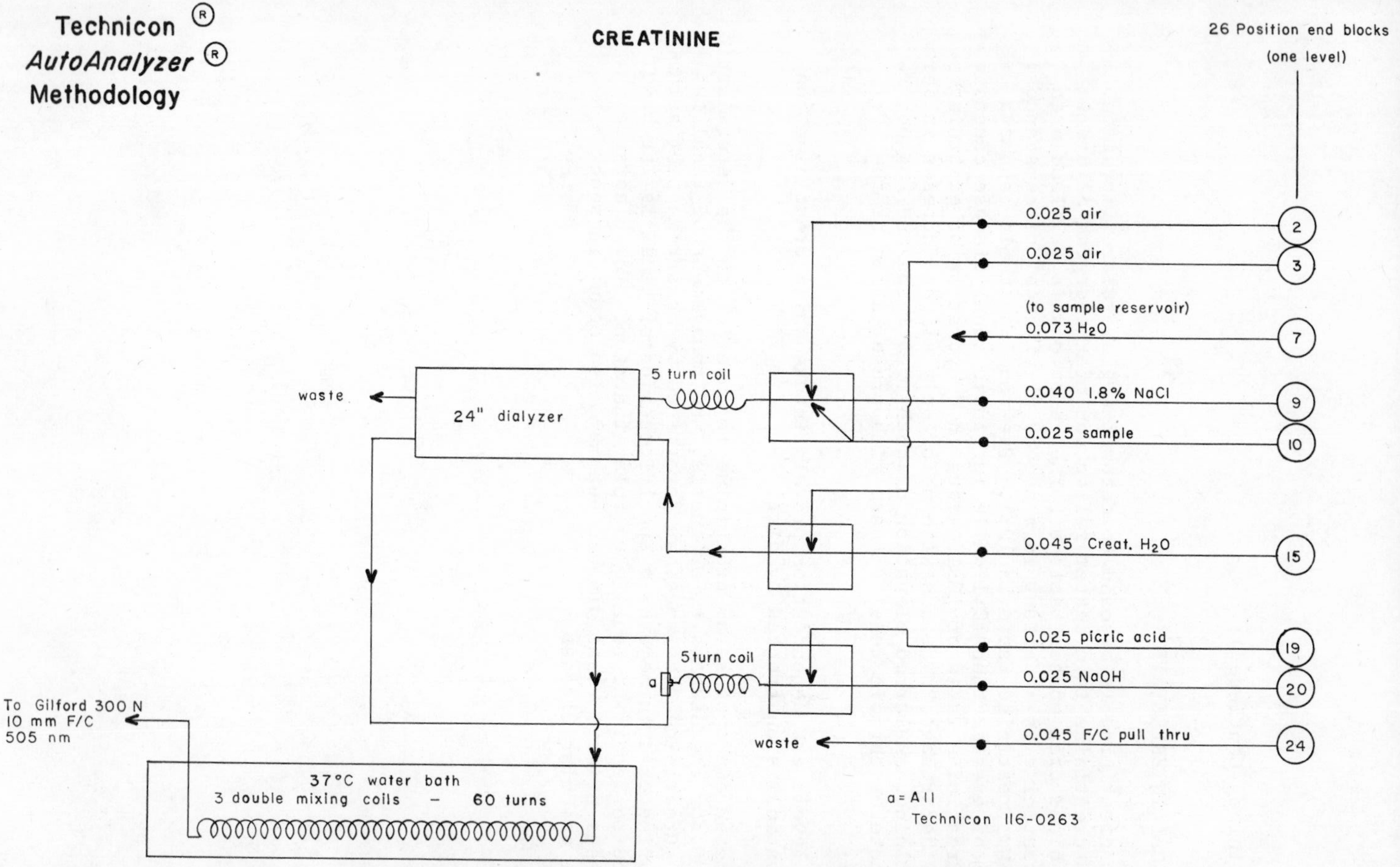

(Modified by Stanford Laboratory – Permission granted by Technicon)

Technicon Instruments Corporation
Tarrytown, New York

(Methodology uses Sampler II, pump with air bar, specimen and reagent injection blocks.)

CRYOGLOBULIN TEST

PRINCIPLE: Cryoglobulin is an abnormal globulin characterized by spontaneous but reversible gelation or precipitation on cooling. It occurs most often in multiple myeloma and cryoglobulinemia. It may also be related to abnormalities in the immune process.

SPECIMEN: 2.0 ml. serum.

PROCEDURE:

1. Using two 12 x 75 mm. tubes, pipette 1.0 ml. of serum in each tube.

2. Appropriately label tubes with patient's name, date and hospital number.

3. Cork both tubes, and place one in the refrigerator and keep the other tube at room temperature.

4. Read both tubes 24 hours later.

5. To read: Compare Room Temperature Control tube with refrigerated tube for discrete white particles or gel formation.

NORMAL VALUES: Negative Test.

CONFIRMATION OF POSITIVE TEST:

Incubate both tubes at 37°C. for 30 minutes. If the precipitate or gel disappears, the test was a true positive.

REFERENCE: Henry, J. B.: CLINICAL CHEMISTRY: PRINCIPLES AND TECHNIQUES, Hoeber, pg. 241 - 242, 1965.

CLINICAL INTERPRETATION:

Cryoglobulins are globulins that precipitate irreversibly on cooling. Most cases of cryoglobulinemia are associated with multiple myeloma or Waldenstrom's Macroglobulinemia. It may also be found in other conditions such as rheumatoid arthritis, other collagen diseases, sarcoidosis and cirrhosis. The globulin may be a 7S Gamma G, a 19S Gamma M or a mixture of both. Symptoms of the condition are purpura, petechiae, and Raynaud's Phenomenon with painful blanching of the extremities on exposure to cold.

Cryoglobulin Test (Continued):

Many cryoglobulins, especially the mixed immunoglobulins may only precipitate if exposed to a temperature of 0 to 4°C. for 12 to 24 hours. Cryoprecipitation depends on the formation of weak, non-covalent bonds between Gamma globulin molecules at low temperatures. It can be prevented by pH changes, ion concentration or addition of urea. Abnormal globulins in rheumatoid Factor type cryoglobulinemia resemble Classic RF. They are Gamma M immunoglobulins with specificity for Fc fragment of a Gamma G Globulin.

FIBRINOGEN
(Semiquantitative)

PRINCIPLE: Thrombin and calcium are added to diluted plasma to convert fibrinogen to fibrin. The time required for the clot to form is inversely proportional to the amount of fibrinogen present.

SPECIMEN: Citrated plasma: 9 parts blood to 1 part 3.8% sodium citrate (EDTA plasma may be used - 7.0 ml. blood to 9.0 mg. of EDTA).

REAGENTS AND EQUIPMENT:

1. <u>Thrombin Stock Solution</u>
   Dilute vials of 5000 units with 0.025 M. Calcium chloride to 1000 units/ml. Aliquot 0.1 ml. amounts in 10 ml. plastic tubes. Store in freezer.

2. <u>Thrombin Working Solution</u>
   10 units/ml. Thaw and dilute as needed 1 tube of stock solution with 9.9 ml. of 0.025 M. Calcium chloride.

3. <u>Imidazole Buffer</u>
   Ingredients include: 6.8 gm. imidazole, 16.7 gm. sodium chloride, 372 ml. 0.1 M. HCl. Dilute to 2 liters and check pH. pH should be 7.35 $\pm$ 0.5.

4. <u>Fibrometer and 0.4 ml. Probe</u>

5. <u>Control - CNP</u>
   Dilute as directed on vial. Stable several days in refrigerator.

SOURCES OF ERROR:

Thrombin is stable in plastic, but rapidly absorbs out on glass. It must be kept in plastic tubes.

PROCEDURE:

1. Prewarm 0.9 ml. of imidazole buffer to 37°C.

2. Prewarm 0.2 ml. aliquots of thrombin in fibrometer cups.

3. Add 0.1 ml. of plasma to the prewarmed buffer. Mix well.

Fibrinogen (Continued):

4. Add 0.2 ml. of diluted plasma to the thrombin, starting the fibrometer.

5. Read value from Curve Chart.

6. If the clotting time is greater than 18 seconds, repeat, using 0.2 ml. plasma and 0.8 ml. buffer. Read value from curve, and divide by 2. If clotting time is less than 9.5 seconds, repeat using 0.1 ml. plasma and 1.9 ml. buffer. Read value from chart, and multiply by 2.

7. Control is run in same manner as patient's plasma. The time in seconds should fall within 2 S.D. of a previously determined time for that Lot No. of CNP.

CALCULATION:

Except in the presence of a rare antithrombin, fibrinogen may be read from a curve previously prepared from a plasma of known fibrinogen value. This gives a variation of 10% and should be reported as ± 10% of the value obtained.

The Curve is made with the following dilutions of plasma in imidazole buffer:

80%	0.40 ml. plasma	0.10 ml. buffer
70%	0.35 ml. plasma	0.15 ml. buffer
60%	0.30 ml. plasma	0.20 ml. buffer
50%	0.25 ml. plasma	0.25 ml. buffer
40%	0.25 ml. of 80% dil.	0.25 ml. buffer
30%	0.25 ml. of 60% dil.	0.25 ml. buffer
20%	0.25 ml. of 40% dil.	0.25 ml. buffer

The dilutions are treated as whole plasma in the test. Clotting times are plotted against mg%. on Log 2 Cycle Graph Paper.

REFERENCES:

1. Clauss, A.: "Gerinnungsphysiologische Schnellmethode zur Bestimmung des Fibrinogens", Acta Haemat., 17:237 - 246, 1957.

2. Morse, Edward: "Automated Fibrinogen Determination", Am. J. Clin. Path., 55:671 - 676, 1971.

Fibrinogen (Continued):

The following is typical of the Calibration Curves used in our Laboratory and is given here only as an example of the scope of the range used. Each laboratory must set up their own curves each and every time reagents and Lot Numbers are changed.

FIBRINOGEN
(EDTA or Sodium Citrate)

Seconds	Mg%.	Seconds	Mg%.	Seconds	Mg%.
9.2	400 ± 10%	12.2	264 ± 10%	15.2	191 ± 10%
9.3	394 ± 10%	12.3	261 ± 10%	15.3	189 ± 10%
9.4	388 ± 10%	12.4	258 ± 10%	15.4	187 ± 10%
9.5	382 ± 10%	12.5	255 ± 10%	15.5	185 ± 10%
9.6	376 ± 10%	12.6	252 ± 10%	15.6	184 ± 10%
9.7	370 ± 10%	12.7	249 ± 10%	15.7	183 ± 10%
9.8	365 ± 10%	12.8	246 ± 10%	15.8	182 ± 10%
9.9	360 ± 10%	12.9	243 ± 10%	15.9	181 ± 10%
10.0	355 ± 10%	13.0	240 ± 10%	16.0	180 ± 10%
10.1	350 ± 10%	13.1	237 ± 10%	16.1	178 ± 10%
10.2	345 ± 10%	13.2	234 ± 10%	16.2	176 ± 10%
10.3	340 ± 10%	13.3	231 ± 10%	16.3	174 ± 10%
10.4	335 ± 10%	13.4	228 ± 10%	16.4	172 ± 10%
10.5	330 ± 10%	13.5	225 ± 10%	16.5	170 ± 10%
10.6	326 ± 10%	13.6	223 ± 10%	16.6	169 ± 10%
10.7	322 ± 10%	13.7	221 ± 10%	16.7	168 ± 10%
10.8	318 ± 10%	13.8	219 ± 10%	16.8	167 ± 10%
10.9	314 ± 10%	13.9	217 ± 10%	16.9	166 ± 10%
11.0	310 ± 10%	14.0	215 ± 10%	17.0	165 ± 10%
11.1	306 ± 10%	14.1	213 ± 10%	17.1	163 ± 10%
11.2	302 ± 10%	14.2	211 ± 10%	17.2	161 ± 10%
11.3	298 ± 10%	14.3	209 ± 10%	17.3	159 ± 10%
11.4	294 ± 10%	14.4	207 ± 10%	17.4	157 ± 10%
11.5	290 ± 10%	14.5	205 ± 10%	17.5	155 ± 10%
11.6	286 ± 10%	14.6	203 ± 10%	17.6	154 ± 10%
11.7	282 ± 10%	14.7	201 ± 10%	17.7	153 ± 10%
11.8	278 ± 10%	14.8	199 ± 10%	17.8	152 ± 10%
11.9	274 ± 10%	14.9	197 ± 10%	17.9	151 ± 10%
12.0	270 ± 10%	15.0	195 ± 10%	18.0	150 ± 10%
12.1	267 ± 10%	15.1	193 ± 10%		

Fibrinogen (Continued):

CLINICAL INTERPRETATION:

If the patient's thrombin time is significantly longer than the Control, a deficiency of fibrinogen or the presence of a substance inhibitory to thrombin-fibrinogen reaction is present. The amount of plasma fibrinogen may be determined from a curve previously prepared from a plasma of known fibrinogen value. A variation of approximately 10% is present and should be reported as $\pm$ 10% of the value obtained. Many conditions may result in a low fibrinogen. Low fibrinogen occurs as a result of a production defect which may b be hereditary or due to severe liver disease. Intravascular conversion of plasma to serum is known as disseminated intravascular coagulation or consumptive coagulopathy. These conditions are:

1. Sanarelli-Schwartzman reaction
2. Purpura fulminans
3. Hemolytic-uremic syndrome (Gasser's syndrome)
4. Kasabach-Merritt syndrome
5. Septic shock
6. Incompatible hemolytic blood transfusion disease
7. Abruptio placentae
8. Dead fetus syndrome
9. Amniotic fluid embolism
10. Acute promyelocytic leukemia
11. Malaria
12. Septic abortion
13. Carcinoma
14. Meningococcemia
15. Waterhouse-Friderichsen syndrome
16. Prostate gland surgery

Other conditions which will predominantly activate the fibrinolytic system are:

1. Preoperative anxiety
2. Pulmonary surgery
3. Neurosurgery
4. Open heart surgery

Conditions which simultaneous or cause coequal activation of clotting system and fibrinolytic system are:

1. Cirrhosis of liver
2. Hepatitis

Fibrinogen (Continued):

An increase in plasma fibrinogen occurs in the following conditions:

1. Acute infection
2. Collagen diseases. The plasma fibrinogen has been used to assess activity in both rheumatic carditis and rheumatoid arthritis.
3. Nephrosis. A very high concentration may be found.
4. Hepatitis (in the absence of severe liver damage)
5. Post-X-ray therapy (indicates tissue damage)
6. Burns (response to tissue injury)
7. Myeloma (some cases)
8. Scurvy

SERUM FOLIC ACID ESTIMATION USING LACTOBACILLUS CASEI

PRINCIPLE: Serum Folic Acid is determined utilizing the following microbiological procedure. The determination of serum folic acid is essential when a possible folic acid deficient macrocytic anemia is present.

SPECIMEN:

1. Serum samples arriving at the laboratory should be labelled and then placed in the deep freeze without delay, that is 30 minutes of receipt. They should not be unfrozen until time of assay.

2. If sample consists of clotted blood, the clot should be "rimmed" with a sterile, acid-washed rod. (A previously unused, sterile wooden applicator stick is also satisfactory.)

3. The tube is then centrifuged for 5 minutes at "3/4 speed", and the supernatant serum aspirated (using previously unused, capillary pipette, or an acid washed standard pipette).

4. This serum is placed in a sterile, acid-washed glass container (or disposable container), is labelled, and promptly placed in deep freeze.
   NOTE: Cap or plug for container must also be sterile, and should either be acid-washed, or washed twelve times in tap water, and rinsed three times in distilled water.

5. Record if serum sample is hemolyzed.

REAGENTS USED IN ASSAY:

1. Ascorbic Acid
   For each assay make up a fresh solution containing 180 mg. in 4.0 ml. distilled water or 270 mg. in 6.0 ml. of distilled water.

2. Pteroylglutamic Acid (Folic Acid)
   Make up fresh every 6 months and store in refrigerator.

   Stock Solution
   100 mg. (0.1 gm.) P.G.A.
   100 ml. sterile, distilled water
   Add 5.0 ml. of 0.1 N. NaOH.

   Be sure all crystals are thoroughly dissolved before proceeding to the Working Solution.

Serum Folic Acid Assay (Continued):

Working Solution
Take 10 ml. of Stock Solution and add to 90 ml. of distilled water. Readjust pH to 7.0 with 0.06 N. H_2SO_4 (approximately 0.8 ml.).

0.06 N. H_2SO_4

a). 1.0 ml. concentrated H_2SO_4 (18 M. or 36 N.) + 35 ml. distilled water = 1.0 N. H_2SO_4
b). 1.0 ml. of 1.0 N. H_2SO_4 + 9.0 ml. distilled water = 0.1 N. H_2SO_4
c). 6.0 ml. of 0.1 N. H_2SO_4 + 4.0 ml. distilled water = 10 ml. of 0.06 N. H_2SO_4

This gives a Working Solution containing 10 mg./100 ml. (0.1 mg./ml.) of P.G.A. or 100,000 ng./ml. This solution should be made up in a 100 ml. flask that has been previously acid washed and sterilized. It should be stored in the refrigerator at 4°C. for weekly use.

3. 0.1 M. Phosphate buffer for Folic Acid
Make up as necessary and store in the refrigerator.

Stock Solution
0.2 M. $NaH_2PO_4 \cdot H_2O$ (Solution "A")
Dissolve 13.8 gm. of the above in a 500 ml. volumetric flask and dilute to volume.
0.2 M. Na_2HPO_4 (Solution "B")
Dissolve 14.2 gm. of the above in a 500 ml. volumetric flask and dilute to volume.

Working Solution
Take 425 ml. of Solution A and 75 ml. of Solution B and mix well in a liter volumetric flask then dilute to volume with distilled water. This resulting solution is then adjusted with a pH meter to a pH of about 6.1. The final solution is 1 liter of 0.1 M. Phosphate Buffer at pH 6.1.

4. 0.25 M. Phosphate Buffer containing 75 mg%. Ascorbic Acid
Combine 150 ml. distilled water
50 ml. Phosphate Buffer 0.1 M.
150 mg. Ascorbic Acid

5. Folic Acid Medium (Difco #0822-15)
Make up desired quantity each time.

Serum Folic Acid (Continued):

NOTE: All equipment (including stirrers) used in the preparation of these reagents must be acid-washed, or very thoroughly washed (12 times in tap water, and then rinsed three times in distilled water).

GENERAL COMMENTS: This is a microbiological assay and is dependent upon meticulous attention to detail for results to be meaningful.

1. Trace contamination with folic acid will completely nullify the significance of the assay, and thus ALL equipment must be acid-washed or extremely thoroughly cleaned, washed 12 times in tap water and rinsed 3 times with distilled water.

2. Bacterial contamination is a problem that is dealt with by autoclaving. However, delays between the various steps of the assay will reduce the reliability of this technique. If any delay becomes necessary, it should always be timed to occur after an autoclaving and before the solutions are handled again.

3. Handling of the initial serum specimens should be minimal, and once deep frozen they should not be thawed out until time of assay.

4. Certain phases of the assay procedures are dependent upon <u>very accurate</u> pipette measurement. These are marked with an asterisk "*".

PREPARATION OF BACTERIA FOR FOLIC ACID ASSAYS:

A. <u>Initial Preparation and Subsequent Maintenance of Culture</u>

1. The initial basic cultures are prepared by rehydrating the lyophilized organisms, <u>Lactobacillus casei</u>.

   These organisms are placed in tubes containing the appropria Difco Culture Broth and are then incubated at $37^{o}C.$ for 48 hours.

   These basic cultures are used for the first subcultures, and then are stored in the refrigerator for possible use if difficulties arise with subsequent cultures.

2. Fresh agar stabs must be prepared every 2 weeks for <u>Lactobacillus casei</u>.

Using overnight broth culture, plunge flamed stab wire into culture and then down into 10 ml. of agar in one stab. Set up 2 agar cultures.

Incubate at 37°C. for at least 48 hours, and then store in refrigerator.

NOTE: Retain previously used stab for 2 further weeks.

B. Subcultures

1. Flame wire loop and place it down side of agar culture tube to cool it. Then place wire loop into the middle of the bacterial growth.

2. Put wire loop into 10 ml. of inoculum broth in a 40 ml. test tube and agitate. Set up 2 tubes.

3. Plug the broth tube with cottonwool and incubate at 37°C. overnight.

C. Harvesting Bacteria

1. Centrifuge the broth culture for 15 minutes at 2,000 rpm in the cold.

2. Pour off the broth leaving a pellet of bacteria.

3. Add sterile, distilled water to the 35 ml. mark and mix the bacterial pellet into this using a sterile, cottonwool plugged pipette.

4. Centrifuge for 15 minutes at 2,000 rpm in the cold and decant the water.

5. Add 5.0 ml. of sterile distilled water using a sterile pipette and resuspend the bacteria.

6. Take 0.6 ml. of this suspension and add it to 20 ml. of sterile distilled water. Thoroughly mix this dilute suspension by means of a sterile, cottonwool plugged pipette.

Serum Folic Acid (Continued):

7. Add *one drop of this dilute suspension into each assay tube (except the blank for each specimen) using a sterile cottonwool plugged Pasteur pipette.

D. Media and Equipment

1. Every 2 weeks prepare fresh "Difco" broth and agar for Lactobacillus casei (dark brown). These should be autoclaved in 40 ml. test tubes (150 x 20 mm.) in 10 ml. amounts using cottonwool plugs.

   Special medium for the actual assay of folic acid is also provided by Difco, but these should be made up freshly for each assay.

   Instructions on how to make up these preparations will be found on the label of the various media bottles. The particular Difco Code Numbers are as follows:

   Lactobacillus casei

Broth	0901-15
Agar	0900-15
Medium	0822-15

2. All equipment used for bacterial preparation must be sterile, and when broth or agar stabs are inoculated, the mouths of the tubes should be flamed.

3. The wire loops and wire stabs used for the culture should be labelled and kept in a glass jar.

4. Always have spare vials of lyophilized Lactobacillus casei, subspecies rhamnous, stored in the refrigerator. (Can be obtained from: The American Type Culture Company, 12301 Parklawn Drive, Rockville, Maryland 20852)

5. Checking Bacteria
   Every 4 weeks send spare broth culture tube (from sub-culture routine) to a Bacteriology Laboratory for a check on culture purity.

Serum Folic Acid (Continued):

AUTOCLAVE ROUTINE:

1. Check "Shut Out" Valve — open (counter-clockwise).

2. Check "By-Pass" Valve — closed (clockwise).

3. Liquid load selection (Drying time = 0).

4. Time selection adjusted.

5. Set temperature timing device.

6. Set pressure in "jacket" of autoclave (on left hand pressure dial) for the level required for particular run.

A. First Autoclave (15 lbs./square inch) - 118°C./3 Minutes

   1. Set indicator at 240°F. (115°C.) - timing for 3 minutes.

   2. Set pressure in jacket at 25 lbs. pressure (Turn pressure reducing valve clockwise).

   3. Close door - bringing up pressure inside chamber to 14 lbs. per square inch in 1 minute - then adjust pressure to 15 lbs. per square inch in both chambers (use pressure-reducing valve).

   4. When 3 minutes are completed and cooling starts, slowly reduce pressure in chamber by very gradually turning "By Pass" Valve counterclockwise ¼ to ½ turn.

Pressure down	1 minute 15 lbs.	-	11 lbs.
in 5 minutes	1 minute 11 lbs.	-	7 lbs.
by gradual	1 minute 7 lbs.	-	4 lbs.
reduction	1 minute 4 lbs.	-	2½ lbs.
	1 minute 2½ lbs.	-	1½ lbs.

Machine now turns off.

B. Second Autoclave (10 lbs./square inch) - 115°C./6 minutes

   1. Set indicator at 230°F. (110°C.) - timing for 6 minutes.

   2. Set pressure in jacket at 15 lbs. pressure (use Pressure-reducing Valve).

Serum Folic Acid (Continued):

3. Close door - bring pressure inside chamber to 9 lbs./square inch in 1 minute - then adjust to 10 lbs./square inch in both chambers (using Pressure-reducing Valve).

4. When 6 minutes are completed and cooling starts, slowly reduce pressure in chamber by very gradually turning "By Pass" Valve <u>counterclockwise ¼ to ½ turn</u>.

Pressure down	1 minute	10 lbs.	- 7 lbs.
in 5 minutes	1 minute	7 lbs.	- 5 lbs.
by gradual	1 minute	5 lbs.	- 3½ lbs.
reduction	1 minute	3½ lbs.	- 2½ lbs.
	1 minute	2½ lbs.	- 1½ lbs.

Machine now turns off.

PROCEDURE:

1. Prepare growth of <u>Lactobacillus</u> <u>casei</u> in broth the night before using (See "Bacterial Preparation").

2. Using 15 ml. graduated centrifuge tubes and take,

	0.2 ml. Ascorbic Acid Solution
Total = 5.0 ml.	2.2 ml. 0.1 M. Phosphate Buffer
	2.1 ml. Distilled water
	0.5 ml. Serum* (add last)

Plug tubes with cotton.

3. Autoclave at 15 lbs. pressure for 3 minutes (See Autoclave Protocol).

4. Cool tubes in water bath.

5. While tubes are cooling, prepare Folic Acid Standard Curve.

Solution "1"	1.0 ml. Standard Solution (bulb pipette) in 99 ml. distilled water (1000 ng./ml.)
Solution "2"	1.0 ml. Solution "1" (bulb pipette) in 99 ml. distilled water (10 ng./ml.)
Solution "3"	2.5 ml. of Solution "2" in 97.5 ml. distilled water (0.25 ng./ml.)
Solution "4"	5.0 ml. of Solution "2" in 45 ml. distilled water (1.0 ng./ml.)

Serum Folic Acid (Continued):

6. Using only Solutions "3" and "4" (all measurements in ml.):

Row Number	1	2	3	4	5	6	7	8	9
*Solution "3"	0	0.1	0.2	0.4	0.8	0	0	0	0
*Solution "4"	0	0	0	0	0	0.4	0.6	0.8	1.0
Distilled water	1.0	0.9	0.8	0.6	0.2	0.6	0.4	0.2	0
Total P.G.A. (ng./ml.)	0	0.025	0.05	0.1	0.2	0.4	0.6	0.8	1.0

This gives a total volume of 1.0 ml. in each tube. The tubes must be set up in quadruplicate. Cover temporarily with parafilm.

7. Now take autoclaved (and cooled) tubes from Step No. 4 and centrifuge at 3/4 speed for 10 minutes. Pour off the supernatant into clean tubes.

8. Using supernatant fluid, prepare unknowns for assay:

   1st Row: 0.25 ml.* supernatant + 0.75 ml. of distilled water.

   2nd Row: 0.50 ml.* supernatant + 0.50 ml. of distilled water.

   NOTE: Both rows should be done in quadruplicate, thus giving a total of 8 assays for each original serum specimen. (Careful marking of tubes with a felt tip marking pen is vital, e.g., Dixon's "Redimark".)

9. Add 1.0 ml. of 0.25 M. buffer, 75 mg%. ascorbic acid to each tube, both standards and unknowns.

10. Add 3.0 ml. of folic medium to each assay tube, both standards and unknowns. Plug and autoclave at 10 lbs. for 6 minutes.

11. After tubes have cooled, place one* drop of Lactobacillus casei in 3 of the 4 tubes of each serum dilution, the fourth tube being set aside as a blank. (This must be done with a sterile Pasteur pipette, taking care to avoid contamination and only inoculating with one drop).

12. Plug all tubes and incubate for 18 hours at 37°C. (including blanks).

Serum Folic Acid (Continued):

13. If growth is excessive it may be necessary to add 2.0 ml. of distilled water to each tube with an automatic syringe before reading.

14. Shake all tubes well before transferring fluid to cuvettes. Read at 600 nm. in a spectrophotometer. Read each dilution against its own blank which must be used to zero the instrument.

15. Plot Standard Curve on arithmetic graph paper (semilog) using spectrophotometric growth densities against the known amount of folic acid per tube.

16. CALCULATIONS:

Calculate the amount of Folic Acid in the unknown serum samples by reading directly from the Standard Curve (utilizing the mean value of the 3 Spectrophotometric readings on the 3 tubes of each dilution). Each serum sample will have 3 readings at dilutions of 1 in 40 (Row No.1) and 3 readings at dilutions of 1 in 20 (Row No. 2).

EXAMPLE

	Mean of 3 Photometer Readings	Folic Acid (ng./ml.)	Dilution	Actual Folic Acid Value (ng./ml.)	Final Average
Row "1"	32	0.170	1:40	6.80	6.87
Row "2"	59	0.347	1:20	6.94	

NORMAL VALUES: 3 - 27 mgm./ml.

REFERENCES:

1. Waters, A. H. and Molin, D. L.: "Studies on the Folic Acid Activity of Human Serum", J. Clin. Path., 14:335, 1961.

2. Herbert, V., Fisher, R. and Koontz, B. J.: "The Assay and Nature of Folic Acid Activity in Human Serum", J. Clin. Invest., 40:81, 1961.

3. Baker, H., Herbert, V., Frank, O., Pasher, I., Hutner, S. H., Wasserman, L. R., and Sobotka, H.: "A Microbiological Method for Detecting Folic Acid Deficiency in Man", Clin. Chem., 5:275, 1959.

CLINICAL INTERPRETATION:

Refer to Interpretation section under Serum Vitamin B_{12} Assay Procedur on page 414.

SERUM GAMMA-GLUTAMYL-TRANSPEPTIDASE

PRINCIPLE: Serum γ-glutamyl-transpeptidase hydrolyzes the substrate γ-glutamyl-p-nitroanilide to yield p-nitroaniline and γ-glutamyl compounds.

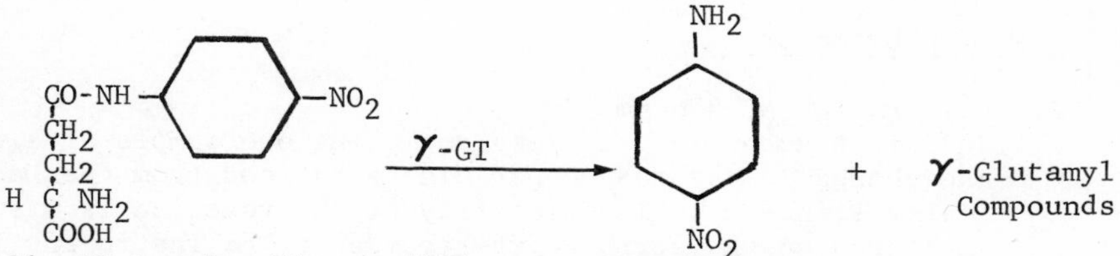

L-γ - Glutamyl-p-Nitroanilide p-Nitroaniline

Since the cleavage product p-nitroaniline has a maximum absorption at 405 nm. (substrate is transparent at this wavelength), its rate of formation can be utilized to determine the activity of the enzyme.

γ-glutamyl transpeptidase activity in human serum has value in differential diagnosis of liver diseases. This enzyme is particularly elevated in obstructive jaundice, hepatic carcinoma and chronic alcoholism.

SPECIMEN: 0.1 ml. of serum is required. Serum should be separated from cells as soon as possible. Only minimum hemolysis allowed. A slight inhibition of the enzyme occurs when oxalate, citrate, and fluoride are present in the usual concentrations. Heparin and EDTA do not inhibit. Serum is stable for over a week at 4°C.

REAGENTS AND EQUIPMENT:

BioChemica Test Combination C-System Kit, 15794 TMBG

1. Substrate (Bottle No. 1)
   4 mM γ-glutamyl-p-nitranilide; 40 mM glycilglycine.

2. Buffer (Bottle No. 2)
   185 mM Tris buffer, pH 8.25.

3. Working Substrate
   Warm the buffer (Solution No. 2) for a few minutes at 50 - 60°C. Add 3.0 ml. of the warmed buffer to each bottle (Bottle No. 1) of substrate. Combine all substrate into a 125 ml. Erlenmeyer flask and connect to the microdilutor.

Serum Gamma-Glutamyl-Transpeptidase (Continued):

4. Ultramicro Dilutor
   The sample syringe is set to take up 100 microliters of sample, and the flush syringe is set to deliver 1.5 ml. of substrate.

5. Spectrophotometer

   A. Gilford 300 N, 405 nm.
      This instrument is used when analyzing one sample at a time. The change in absorbance per minute is read from the Data Lister Print-out. The linearity of the reaction is visually monitored on the strip chart recorder. The instrument is used with the thermocuvette set at 30°C.

   B. Gilford 222, 405 nm.
      This instrument is used when analyzing four samples at a time. Temperature control is maintained by means of a 30°C. circulating water bath. Rate of reaction is determined from the strip chart recorder set for 0.200 A full scale and run at one inch per minute.

COMMENTS ON PROCEDURE:

1. Working substrate is stable for 24 hours at 20 - 25°C.

2. Dilute all specimens when the change in rate is greater than 0.150 ΔA/minute.

3. Substrate exhaustion is rate-limiting; the reaction will be linear for 10 minutes.

4. Reaction mixture and cuvette chamber must be at temperature before making a measurement.

PROCEDURE:

1. With the ultramicro dilutor, take up 100 microliters of serum and flush with 1.5 ml. of working substrate. For the Gilford 222, use Pyrex cuvettes 10 mm in width and a 10 mm. light path. Mix well.

2. Dilute up three more specimens in the same manner.

3. Place 4 specimens in the cuvette chamber of the Gilford 222.

Serum Gamma-Glutamyl-Transpeptidase (Continued):

4. Set the baseline of Specimen No. 1. with the slit, and No.'s 2,3 and 4 with the Off-Set knobs. Switch to "Auto" and scan all 4 channels.

5. Record 2 - 3 minutes of linear reaction time at 405 nm. Determine the mean ΔA/minute from recorder chart.

6. CALCULATIONS:

$$\frac{\Delta A}{\epsilon \times d} \times 10^6 \times \frac{TV}{SV} \times \frac{1}{Time} = \text{I.U./Liter or mU/ml.}$$

ϵ of p-nitroaniline at 405 nm. = 9.9×10^3 Liters/Mole x cm.

d = Diameter of light path = 1.0 cm.
TV = Total Volume
SV = Sample Volume
Time = 1 Minute

10^6 converts Moles/Liter into Micromoles/ml.

$$\frac{\Delta A}{1} = \frac{1}{9.9 \times 10^3 \times 1} \times 10^6 \times \frac{1.6}{0.1} \times \frac{1}{1} = \text{mU/ml.}$$

$\Delta A \times F$ = mU/ml.
F = 1616

NORMAL VALUES:

Normal values should be established by each laboratory running the test. A normal typical value for 25°C. would be 0 - 11 I.U.

REFERENCES:

1. Szasz, G.: Clin. Chem., 15:124, 1969.

2. Zein, M., Discombe, G.: The Lancet, pg. 748 - 750, Oct., 1970.

CLINICAL INTERPRETATION:

This enzyme is usually elevated in patients with hepatobiliary tract disease. Elevated levels are usually associated with metastatic cancer to the liver and common bile duct obstruction due to neoplasms.

Serum Gamma-Glutamyl-Transpeptidase (Continued):

Icteric or non-icteric patients may have high values if there is a hepatobiliary intra or extra-hepatic obstruction. The increased levels are partly due to interference with excretion. Various other liver diseases may be associated with increased values. Recently high values have been found in post-myocardial infarction patients, and in patients with neurologic disease in which highly vascular or necrotic changes in vascular endothelium have been demonstrated.

GASTRIC ANALYSIS
(The Histalog Test)

PRINCIPLE: A measured amount of gastric juice is titrated with 0.1 N. NaOH utilizing the pH Meter. The gastric juice is collected after Histalog stimulation.

REAGENTS AND EQUIPMENT:

1. Sodium Hydroxide, 0.1 N.

2. Nasogastric tube

3. 20 ml. Glass Syringe

4. Centrifuge

5. pH Meter

6. Histalog, 1.7 mg./Kg. Body Weight

COMMENTS ON PROCEDURE:

1. Patient is fasted at least 12 hours before test.

2. A nasogastric tube is passed by the Technologist (pass tube to above 1 inch from the second to last black mark on the tube).

3. If any doubt as to position of tube, get it checked in X-ray.

PROCEDURE:

1. Ask patient to lie on left side, semi propped up.

2. Empty stomach and discard aspirate.

3. Make two 30 minute collections (samples 1 and 2) — viz: basal 1 hour collection. Aspirate by hand every 2 to 3 minutes, using a 20 ml. glass syringe. Do not aspirate too hard (an indication of this blood staining). If tube appears blocked, clear with 15 ml. of air, and then aspirate again.

4. Doctor gives I. M. histalog injection — using 1.7 mg./Kg. body weight.

Gastric Analysis, Histolog Test (Continued):

5. Make 5 x 15 minute collections (samples 3, 4, 5, 6, and 7) of gastric juice.

6. Remove nasogastric tube.

7. A careful check on the patient throughout the test should be maintained. If any serious change in patient's condition develops, doctor must be notified at once.

CALCULATIONS:

A. Measure and record volume and pH of aspiration samples (1) - (7). Also note bile, mucus and blood.

B. Take aliquots and centrifuge at "3/4 speed" for 15 minutes to get clear supernatant.

C. Take 5.0 ml. aliquots from each sample (less if necessary, but record volume used). Titrate with 0.1 N. NaOH to pH 7.4 (using pH meter).

D. Record volume of 0.1 N. NaOH used for each aliquot, and estimate mEq./L. of acid and total mEq. of acid in each sample.

<u>Example</u>:

1). 5.0 ml. aliquot from 38 ml. sample used 2.5 ml. 0.1 N. NaOH

2). Thus 5.0 ml. aliquot was equivalent to 0.25 mEq. NaOH

3). Thus 1000 ml. = $\frac{0.25}{5} \times 1000 = 50$ mEq. (viz: mEq./L.

4). Total sample of 38 ml. = $\frac{50}{1000} \times 38 = 1.9$ mEq.

(Total acid in sample).

These figures represent acid in mEq.

NORMAL VALUES: See Clinical Interpretation.

REFERENCES:

1. Zaterka, S. and Neves, D. P.: "Maximal Gastric Secretion in Human Subjects after Histalog Stimulation", <u>Gastroenterology</u>, 47:251, 1964.

Gastric Analysis, Histolog Test (Continued):

2. Ward, S., Gillespie, I. E., Passaro, E. P., and Grossman, M. I.: "Comparison of Histalog and Histamine as Stimulants for Maximal Gastric Secretion in Human Subjects and in Dogs", Gastroenterology, 44:620, 1963.

3. "Effect of Large Doses of Histamine on Gastric Secretion of HCl: An Augmented Histamine Test", Brit. Med. Journal, 2:77, 1953.

CLINICAL INTERPRETATION:

The basal acid secretion for normal males is 1.3 to 4.0 mEq./L. per hour. Lower values occur in females and with aging. Low values occur in gastric cancer and benign gastric ulcer higher values are present in duodenal ulcer. Extremely high acid output is present in patients with the Zollinger-Ellison Syndrome.

When the Histamine Test is utilized, the maximum rate of acid secretion is attained in 15 minutes and is maintained for 30 minutes. Basal levels are achieved in 60 minutes, the maximum acid output representing the sum of the acid outputs for the four 15 minute post histamine samples. This is the generally accepted expression of gastric acid secretion. The range for maximal acid output in normal males is 4.9 to 38.9 mEq. per hour.

A maximal acid output of greater than 210 mEq. per hour is found in about 40 per cent of males with duodenal ulcer. Zollinger-Ellison Syndrome patients have a high ratio of basal to maximal acid output. Ratios greater than 60 per cent are indicative of this disorder.

Anacidity in the Histamine Test is found in adults with gastric cancer and pernicious anemia with atrophic gastritis. It is also found in conditions such as hypochromic anemia, rheumatoid arthritis, steatorrhea, aplastic anemia, and myxedema.

Surgeons utilize gastric acid analysis in determining the surgical procedure to be performed. Elevated maximal acid output indicates the need for gastric resection. Elevated basal secretion with slightly elevated maximal secretion is an indication for vagotomy.

GASTRIC ANALYSIS

PRINCIPLE: A measured amount of gastric sample is titrated with 0.1 N. NaOH until the indicator Topfer's reagent (dimethylaminoazobenzene) is bright yellow (pH 4.0). This is the end point for "free hydrochloric acid". To measure the amount of "combined acid", weakly ionized protein salts, organic acids and free acidic groups of protein, phenolphthalein which changes from colorless to red at pH 8.5 is used as the indicator and the titration continued to a pink color.

REAGENTS:

1. <u>Sodium Hydroxide</u>, 0.1 N.

2. <u>Topfer's Reagent</u>, 0.5%
   In a 100 ml. volumetric flask weigh out 0.5 gm. p-dimethylaminoazobenzene dilute to volume with 95% ethyl alcohol.

3. <u>Phenolphthalein</u>, 1.0%
   Weigh out 1.0 gm. of phenolphthalein into a 100 ml. volumetric flask. Dilute to volume with 95% ethyl alcohol.

PROCEDURE:

1. <u>Macroscopic Examination</u>

   A. Record volume of each specimen.
      Normal: 20 - 50 ml.

   B. Note color
      1). Normally colorless
      2). Green from bile
      3). Faint red from blood
      4). Coffee-colored from altered blood remaining in stomach for some time.

   C. Odor
      1). Normally odorless or sour
      2). Fecal due to bowel obstruction or gastrocolic fistula
      3). Pungent odor caused by cancer or catarrhal gastritis

   D. Appearance, normally three layers
      1). Top - Mucus
      2). Mid - Opalescent fluid
      3). Lower - Sediment residual food

Gastric Analysis (Continued):

2. Chemical Determination

A. To 5.0 ml. gastric specimen in a porcelain evaporating dish add 2 drops of Topfer's reagent
   1). Lemon yellow color indicates NO free hydrochloric acid
   2). If red or orange color, titrate with 0.1 N. NaOH until lemon-yellow endpoint. Note volume of NaOH used. If less than 5.0 ml. of gastric sample is used, titrate with 0.01 N. NaOH.

B. To the above sample (after free HCl determination) add 2 drops phenolphthalein. Continue titration with 0.1 N. NaOH until pink color develops. Note volume of NaOH used.

CALCULATIONS:

$$\text{ml. 0.1 N. NaOH used} \times \frac{100}{\text{Vol. gastric sample used in titration}} = \text{Degrees acidity}$$

If 5.0 ml. gastric sample used:

ml. NaOH used to yellow endpoint x 20 = Degrees (or units) FREE HCl.

ml. NaOH used in both titrations x 20 = Degrees (or units) TOTAL ACID.

NORMAL VALUES:

Since a broad range of normal values exist for free hydrochloric acid in gastric juice, the primary importance of quantitating free HCl is the demonstration of achlorhydria. The absence of free HCl after gastric stimulation and the presence of stomach lesions is strongly suggestive of stomach cancer.

REFERENCES:

1. Ham, T.: SYLLABUS OF LABORATORY EXAMINATION IN CLINICAL DIAGNOSIS, Harvard Univ. Press, 1958.

2. Hepler, O.: MANUAL OF CLINICAL LABORATORY METHODS, 3rd. Ed., 1958.

CLINICAL INTERPRETATION:

Refer to Gastric Analysis - The Histolog Test, page 183.

GLUCOSE OXIDASE
(GOD - Perid Method)

PRINCIPLE: Glucose is oxidized by the specific enzyme glucose oxidase to gluconolactone which, in aqueous solution, is converted to gluconic acid.

$$\text{Glucose} + O_2 + H_2O \xrightarrow{\text{GOD*}} \text{Gluconic acid} + H_2O_2$$

In the presence of peroxidase, the hydrogen peroxide produced in the above reaction oxidizes with the formation of a green color.

$$H_2O_2 + \underset{\text{(reduced-colorless)}}{\overset{\text{ABTS**}}{\text{Hydrogen donor dye}}} \xrightarrow{\text{POD***}} \underset{\text{(oxidized-green)}}{\text{Hydrogen donor dye}} + 2H_2O$$

The intensity of the dye is proportional to the glucose concentration

SPECIMEN: 10 microliters of serum. The specimen must be spun down or chilled to 4^oC. within 15 minutes after collection.

REAGENTS AND EQUIPMENT:

Reagents available as Biochemica Test combination for Blood Sugar GOD - Perid Method, 15756 TEAP, 3 x 175 tests, Boehringer Mannheim Corporation.

1. Stock Glucose Standard, 10 mg./ml.
   Place 1000 mg. of dessicated glucose (dextrose) into a 100 ml. volumetric flask. Dissolve and bring to volume with saturated benzoic acid.

   Glucose oxidase is specific for beta-D glucose; solutions of glucose freshly prepared from the dry chemical should be allowed to stand for at least two hours to insure that mutarotation has reached a state of equilibrium. No such precautions are necessary when preparing dilutions from a Stock Standard solution

2. Working Glucose Standards

   A. 50 mg%. glucose: Quantitatively transfer 5.0 ml. of stock glucose standard into a 100 ml. volumetric flask, and dilute with saturated benzoic acid.

Glucose Oxidase (Continued):

B. 100 mg%. glucose: Quantitatively transfer 10 ml. of stock glucose standard into a 100 ml. volumetric flask, and dilute with saturated benzoic acid.

C. 300 mg%. glucose: Quantitatively transfer 30 ml. of stock glucose standard into a 100 ml. volumetric flask, and dilute with saturated benzoic acid.

3. Working Substrate
One bottle containing 100 mM. phosphate buffer, pH 7.0; 20 micrograms POD/ml., 180 micrograms GOD/ml., 1.0 mg. ABTS/ml. is dissolved with deionized water and brought to 1000 ml. in a volumetric flask. Keep in a dark bottle. Stable for 6 weeks at 4°C.

4. Ultramicro Dilutor
Sample syringe set to take up 10 microliters of specimen, and flush syringe set to deliver 5.0 ml. of substrate. The reagent is stored in the dilutor with a dark bottle. The dilutor is stored in a light-tight cupboard.

5. Gilford 300 N
Equipped with either regular or microaspiration cuvette.

COMMENTS ON PROCEDURE:

1. It is essential that all glassware and reagents used in the procedure be chemically clean. Chemical contamination may interfere with the enzyme system.

2. The Working Substrate must be kept light-tight for reproducible results within a series; the reagent darkens progressively upon exposure to light.

3. Stability of the Working Substrate has been proven to be good up to 10 days at room temperature.

4. The determination is carried out at room temperature; it is not critical that the temperature be exact as long as the temperature is stable during the analytical run. Temperatures of greater than 38°C. may inactivate the enzyme. The analysis may also be performed at other temperatures, but the reaction time is affected.

Glucose Oxidase (Continued):

5. The green color which is measured attains its maximum absorption after 20 minutes at room temperature.

6. Adult serum separated within 30 minutes of drawing of the blood sample has a glucose level identical with that of heparinized plasma analyzed as soon as possible. Once the serum has been removed from the cells, the glucose concentration remains constant for several days at 25 - 30°C. without the addition of a preservative. However, when serum is left in contact with the cells, disappearance of glucose commences within a few minutes. The specimen must be spun down or chilled to 4°C. within 15 minutes after collection. The rate of glycolysis is higher in newborn infants than in adults. So it is important that the cells are separated immediately. (See Table I)

TABLE I

RECOVERY OF GLUCOSE FROM BLOOD SPECIMEN NOT SEPARATED FROM CELLS

Time Specimen Left Standing	1 Hour	2 Hours	3 Hours	4 Hours
Chilled	100%	100%	100%	100%
Room Temperature	90%	84%	84%	63%

7. Hemoglobin and Bilirubin do not seem to interfere, although elevated uric acid values can lower the results. (See Table II)

TABLE II

RELATIONSHIP OF CONCENTRATION OF URIC ACID IN SERA WITH AMOUNT OF GLUCOSE RECOVERED

Concentration Uric Acid	34 Mg%.	19 Mg%.	11 Mg%.	7.6 Mg%.	4.0 Mg% Reference
% Glucose Recovered	51%	77%	83%	90%	100%

PROCEDURE:

1. Dispense 5.0 ml. of substrate into a new 16 x 100 test tube; label as "Reagent Blank".

2. At ½ minute intervals, take up 10 microliters of Standards, Controls, and Specimen; dispense into new tubes with 5.0 ml. of substrate. After each dilution, rapidly vortex the tube.

Glucose Oxidase (Continued):

3. Allow to stand 20 minutes, and read samples at ½ minute intervals.

4. Determine the absorbance of the Reagent Blank in the Gilford 300 N at 650 nm. Record absorbance of Reagent Blank. Then zero the machine with the Reagent Blan, and read samples.

5. Calculate glucose concentration of the Controls and samples from the linear graph paper.

6. Plot quality controls.

7. Be sure to leave enough substrate at room temperature for the next analysis.

NORMAL VALUES:

1. Newborn
   6 - 59 mg%. (BMC) increasing to 13 - 75 mg%. by the sixth day.
   20 - 80 mg%. (O'Brien & Ibbott)
   30 - 100 mg%. (Cornblath & Schwartz)

2. Normal Fasting Levels
   50 - 95 mg% (Boehringer-Mannheim Corp.)

REFERENCES: (Method)

1. Werner, W., Rey, H., Rey, G., and Wiebinger, H.: Z. Anal. Chem., 252:224, 1970.

2. Schmidt, F. H.: Internist, 4:554, 1963.

ABBREVIATIONS USED:

*GOD (Glucose Oxidase
**ABTS (2,2' Azinodiethylbenzthiazoline-Sulfonic Acid)
***POD (Peroxidase)

CLINICAL INTERPRETATION:

Diabetes mellitus is undoubtedly the most common disease responsible for hyperglycemia. One must always insist that the blood sample be taken fasting to rule out post-prandial hyperglycemia. It is also

Glucose Oxidase (Continued):

surprising how many times there is failure to note whether or not the patient is receiving glucose intravenously when the blood sample is drawn.

Diminished glucose tolerance occurs in other diseases. Decreased ability to store glycogen occurs in chronic liver disease and hyperglycemia is present. Chronic liver disease is thus a common cause for hyperglycemia. Pregnancy results in hyperglycemia which is transient during the course of the pregnancy. Uremia is characterized by presence of serum uremic toxins which are anti-insulin with resultant hyperglycemia. Certain endocrine abnormalities are frequently associated with hyperglycemia. These are hyperthyroidism, acromegaly with excess growth hormone, pheochromocytoma, and Cushing's syndrome. Brain trauma results in prominent hyperglycemia. This is known as the diabetic pique. Brain trauma, neoplasms or rupture of a cerebral vascular aneurysm may stimulate a hypothalamic-adrenal medullary response resulting in release of norepinephrine and glycogenolysis with hyperglycemia.

The most common cause for hypoglycemia is excessive use of insulin. Insulin treatment in diabetes mellitus may result in severe hypoglycemia. Malabsorption may result in hypoglycemia due to poor absorption of carbohydrate. Severe hepatic disease is associated with fasting hypoglycemia as is severe infection. Insulin-producing pancreatic islet cell neoplasms and bulky mesenchymal neoplasms cause severe hypoglycemia by insulin production or glucose consumption. Finally endocrine disease may induce hypoglycemia. It may be seen in hypopituitary or hypoadrenal states.

GLUCOSE-6-PHOSPHATE DEHYDROGENASE
(U. V. Method on Red Cell Hemolysate)

PRINCIPLE: The enzyme G-6-PDH catalyzes the reaction which takes place when glucose-6-phosphate is converted to 6-phosphogluconic acid. In carbohydrate metabolism, this enzyme introduces the reaction which starts the pentose phosphate (hexose monophosphate) pathway. For this reason, an active and adequate concentration of G-6-PDH must be present in the red cell under certain conditions of stress. Low levels of G-6-PDH in the red cell have been associated for some time with hemolytic episodes in individuals following exposure to agents such as primaquine (and some other aromatic, heterocyclic structured medications) and fava beans.

G-6-PDH catalyzes the hydrogen transport reaction indicated beneath. The rate of formation of $NADPH_2$, which absorbs strongly at 340 nm. and 366 nM., is utilized as a measure of enzyme activity.

$$\text{Glucose-6-Phosphate} + \text{NADP} \xrightleftharpoons{\text{G-6-PDH}} \text{6-Phosphogluconate} + \text{NADPH}_2$$

SPECIMEN: Collect whole blood in sodium citrate or sodium heparin anticoagulant. This may be stored at refrigerator temperatures for up to four days. (See specimen preparation under "Comments on Procedure.")

REAGENTS AND EQUIPMENT:

Available as Biochemica Test Combination TC-W 15993 (20 Determinations).

1. Triethanolamine Buffer, 0.05 M., pH 7.6
   Dissolve the contents of Bottle #1 in 100 ml. of redistilled water. Also contains 0.005 M. EDTA. Stable for one year at room temperature.

2. NADP, 0.01 M.
   Dissolve the contents of Bottle #2 in 2.0 ml. of redistilled water. Stable for four weeks at approximately 4^oC.

3. Glucose-6-Phosphate, 0.031 M.
   Dissolve the contents of bottle #3 in 1.5 ml. of redistilled water. Stable for four weeks at approximately 4^oC.

4. Digitonin, approximately 0.02%
   Use the solution in bottle #4 undiluted. Stable for one year at room temperature.

Glucose-6-Phosphate Dehydrogenase (Continued):

COMMENTS ON PROCEDURE:

1. Specimen Preparation
   a. Wash 0.2 ml. of whole blood three times with 2.0 ml. physiological saline. Centrifuge after each washing for ten minutes at approximately 3000 rpm.
   b. Do a red count on 20 lambda of the packed cell button.
   c. Suspend the washed button in 0.5 ml. digitonin and allow to stand at 4°C. for 15 minutes. Recentrifuge for clear hemolysate.

2. Derivation of Factor (at 340 nm.):

$$\frac{0.001}{6.22} \times 10^6 \times \frac{1}{1} \times \frac{3.25}{0.10} \times \frac{6}{1} \times \frac{5}{1} = 15661$$

3. Refer to original methodology for G-6-PDH determination on serum.

PROCEDURE:

1. Into a 1.0 cm. square glass cuvette, pipette the following solutions: 3.0 ml. of triethanolamine buffer, 0.1 ml. of NADP, and 0.1 ml. of hemolysate.

2. Mix the contents of the cuvette well by inversion and incubate in the water bath at 25°C. for approximately five minutes.

3. Pipette 0.05 ml. of glucose-6-phosphate into the cuvette, mix, and determine the absorbance at 340 nm. or 366 nm. Immediately start stopwatch and record absorbance change in the temperature controlled cuvette at 1, 2, and 3 minutes. (Or scan for three minutes on a recorder.)

4. Determine the mean absorbance change per minute (Δ E/min.) against an air reference. Absorbance differences greater than 0.060/min. at 340 nm. or 0.030/min. at 366 nm. require a dilution of 1:10 with physiological saline.

5. CALCULATION:

$\Delta E_{340\ nm.}$/min. x 15661 = milli-units (mU)/#RBC in 1.0 ml. blood.

$\Delta E_{366\ nm.}$/min. x 29600 = milli-units (mU)/#RBC in 1.0 ml. blood.

Glucose-6-Phosphate Dehydrogenase (Continued):

NORMAL VALUES: 120 - 240 mU./10^9 erythrocytes.

REFERENCES:

1. Kornberg, A., Colowick and Kaplan, METHODS IN ENZYMOLOGY, Vol. I, 322, Academic, 1955.

2. Bergemeyer, METHODS OF ENZYMATIC ANALYSIS, Academic, 1965.

3. Bishop, J. Lab. and Clin. Med., 68:149, 1966.

4. Batsakis and Brierre, INTERPRETIVE ENZYMOLOGY, C. Thomas, 1967.

CLINICAL INTERPRETATION:

During the last few years a number of hemolytic anemias have been investigated which are caused by inherited errors in the metabolism of glucose. The erythrocyte derives all its energy from the metabolism of glucose by the Embden-Meyerhof cycle along with the pentose monophosphate shunt. Red cells do not have all the tricarboxylic acid cycle components or the cytochrome oxidase system, but the mature red cell does possess a functioning Krebs cycle. Thus, over 90% of glucose consumption occurs through the Embden-Meyerhof anaerobic pathway. Glucose must be phosphorylated before it can be metabolized by the erythrocytes. A direct oxidative pathway of glucose metabolism is also present within the red cell, this is known as the pentose monophosphate shunt. Usually less than 10% of glucose undergoes direct oxidation by this shunt. The primary function of the anaerobic cycle is to provide sufficient adenosine triphosphate for the necessary metabolism of the red cell. It has been demonstrated that many of the recognized abnormalities of red cells are associated with enzyme deficiencies which occur with increasing age of the red cells. Thus, it might be valuable to consider these conditions as metabolic abnormalities which are imposed on the normal aging process.

Young red cells usually possess higher levels of glycolytic enzymes. The older red cells usually have a lesser amount of these essential enzymes. Young red cells such as reticulocytes utilize more glucose and have more glycolytic enzymes. The aging red cells have a very rapid decrease in activity in G-6-PD activity. Along with this decrease in this important enzyme, there is a decrease in ATP as well as DPN or NAD. One group of investigators has demonstrated

Glucose-6-Phosphate Dehydrogenase (Continued):

that G-6-PD activity in reticulocytes is ten times greater than the activity in the older red cells.

Deficiency of glucose-6-phosphate dehydrogenase occurs on a world-wide basis. It may exist as a number of genetic variants related to different changes in enzyme structure. It is a common cause for neonatal jaundice and kernicterus in highly affected groups such as in the black race; it has been more commonly recognized as the cause for hemolytic anemia in patients who have taken various drugs.

The underlying mechanism and one laboratory test is as follows: Glucose-6-phosphate dehydrogenase is an essential enzyme in the pentose monophosphate shunt which produces TPNH. In the absence of sufficient TPNH, there is a marked reduction in available reduced glutathione. Although the role of reduced glutathione in preserving red cell integrity is not well understood, it appears to be an important one. The glutathione stability test was originally proposed by Beutler. Reduced glutathione sensitive cells are extremely sensitive to oxidation in the presence of drugs such as acetyl-phenylhydrazine. It has been demonstrated that this test is not as reliable in detecting heterozygous females since in one series 30 to 50% false negative results occurred. The dye reduction test is based on the inability of sensitive red cells to reduce the dye, brilliant cresyl blue. It is also unsatisfactory for the detection of the female heterozygote with the disease. In this test, decolorization should be complete within one hundred minutes; partial decolorization at one hundred minutes indicates partial enzyme deficiency. Because of the qualitative nature of the test, intermediate results are sometimes obtained and are difficult to interpret; those samples showing no decolorization at 3 to 6 hours have virtually no enzyme activity. The direct estimation of glucose-6-phosphate dehydrogenase activity is the most informative method, in particular, with respect to the detection of the heterozygous patient. Normal values are 150 to 215 units per 100 ml. of red cells. Less than ten units per 100 ml. red cells is indicative of a glucose-6-phosphate dehydrogenase deficiency.

A deficiency of G-6-PD was recognized many years ago to be present throughout the world. (It was recognized in the black races and in the Sephardic Jews.) A hemolytic anemia may arise in patients with this enzyme deficiency; these patients usually become anemic when they are exposed to certain drugs or fava beans. Patients with G-6-PD deficiency may have a hemolytic crisis when drugs of the following types are taken:

Glucose-6-Phosphate Dehydrogenase (Continued):

1. Analgesics or antipyretics such as aspirin
2. Sulfa drugs such as Sulfapyridine or Sulfasoxazole (Gantrisin)
3. Antimalarias such as Primaquine, Pentaquine (Atabrine), or Quinine
4. Nitrofurantoin or Nitrofurazone
5. Other drugs such as Chloromycetin, PAS, Quinidine

Recently, conditions such as viral hepatitis and diabetic acidosis have been shown to induce hemolytic anemia in patients with G-6-PD deficiency. It is thought that certain oxidants increase in the blood of patients with viral hepatitis or diabetic acidosis, and this leads to an acute hemolytic crisis.

The deficiency is more severe in the Middle East as in the Sephardic Jew than in the black races. It is unwise to use donor blood from patients who have severe G-6-PD deficiency for blood transfusion purposes from affected individuals from the Middle East. However, individuals in the United States who have G-6-PD deficiency of the minimal type may serve as blood donors.

Patients who have an acute hemolytic anemia progress through various phases: the first phase in the acute hemolytic reaction may last two weeks, a 7 to 10 day recovery phase follows, and this is followed by an equilibrium phase in which the anemia disappears. The exact mechanism for the hemolysis is not clear; however, it has been suggested that because of the G-6-PD deficiency, there is a defect in the sodium pump of the red cell membrane and associated with this defect, there is a defective generation of ATP because of glutathione deficiency. The potassium loss of G-6-PD deficient red cells has been reported to be greater than that of normal red cells following incubation of these with drugs such as Primaquine. However, no significant differences in potassium content of G-6-PD deficient or normal red cells after incubation with acetyl phenyl hydrazine has been demonstrated. These experiments emphasize that there are differences in the type of injury induced by different drugs, and that no single explanation suffices to explain the mechanism of the hemolysis. More and more drugs are being implicated in the causation of hemolysis in patients who have G-6-PD deficiency.

Usually spontaneous hemolysis does not occur in the newborn with this enzyme deficiency. The administration of an offending drug to the mother with a G-6-PD deficient fetus prior to delivery is considered hazardous.

GLUCOSE - UREA NITROGEN
(Technicon AutoAnalyzer II)

PRINCIPLE: <u>Glucose</u> - Cupric neocuproine chelate in an alkaline medium is reduced by glucose producing a colored cuprous neocuproine complex. After heating to $85^{o}C. \pm 3^{o}$, the percentage of transmitted light is measured at 460 nm. in a 15 mm. flowcell. An electronic log-linear converter permits read-out directly in mg%.

<u>Urea Nitrogen</u> - Urea in a weak acid solution forms a colored complex with diacetyl-monoxime. Addition of thiosemicarbazide and ferric ion plus heating to $90^{o}C. \pm 3^{o}$ intensify the colored product. The percentage light transmitted is measured at 520 nm. in a 15 mm. flowcell. Log-linear conversion gives direct read-out in mg%.

SPECIMEN: 0.2 ml. of serum, plasma or other clear fluid. Urine urea nitrogen is usually a 1:25 dilution. See "Comments on Procedure for urine glucose specifications.

REAGENTS:

A. Glucose:

1. <u>Neocuproine Copper Reagent</u> (Technicon T01 - 0176)
   a). Chemical Composition:

Copper sulfate pentahydrate ($CuSO_4 \cdot 5H_2O$		0.2 gm.
Neocuproine hydrochloride		0.4 gm.
Distilled water	q.s.	1000 ml.

   b). Preparation:

   Add 0.2 gm. of copper sulfate to approximately 500 ml. distilled water in a liter volumetric flask and dissolv Add 0.4 gm. of neocuproine hydrochloride and dissolve. Dilute to volume and mix.

2. <u>Sodium Carbonate</u>, 0.25 M. (Technicon T01 - 0315)
   a). Chemical Composition:

Sodium carbonate (Na_2CO_3)		26.5 mg.
Distilled water	q.s.	1000 ml.
Brij.-35 (30% Solution)		1.0 ml.

Glucose-Urea Nitrogen (Continued):

b). Preparation:

Place approximately 800 ml. distilled water in a liter volumetric flask and add 26.5 gm. sodium carbonate. When dissolved completely, dilute to volume. Add 1.0 ml. Brij.-35 and mix.

3. Sodium Chloride, 0.9% with Brij.-35 (Technicon T01 - 0381)
   a). Chemical Composition:

Sodium chloride (NaCl)	9.0 gm.
Distilled water	q.s. 1000 ml.
Brij.-35 (30% Solution)	1.0 ml.

   b). Preparation:

Place approximately 800 ml. distilled water in a liter volumetric flask. Add 9.0 gm. of sodium chloride and mix until dissolved. Dilute to volume and add 1.0 ml. Brij.-35 and mix again.

B. Urea Nitrogen:

1. Diacetyl Monoxime, 2.5%
   a). Chemical Composition:

Diacetyl monoxime (2,3-butanedione-2 oxime)	25 gm.
Distilled water	q.s. 1000 ml.

   b). Preparation:

Place 25 gm. of diacetyl monoxime in a liter flask. Add approximately 500 ml. of distilled water and mix until completely dissolved. Dilute to volume and filter. Transfer filtered solution to a sealable amber bottle.

2. Thiosemicarbazide, 0.5%
   a). Chemical Composition:

Thiosemicarbazide	5.0 gm.
Distilled water	q.s. 1000 ml.

Glucose-Urea Nitrogen (Continued):

b). Preparation:

Place approximately 500 ml. distilled water in a liter beaker. Add 5.0 gm. thiosemicarbazide and stir on magnetic mixer until completely dissolved. Transfer with rinsing to a liter volumetric flask and dilute to volume with distilled water. Mix well. Transfer contents to a sealable amber bottle.

3. BUN Color Working Solution (Technicon T01 - 0401
Place approximately 200 ml. distilled water into a liter volumetric flask. Add 67 ml. of 2.5% diacetyl monoxime and 67 ml. of 0.5% thiosemicarbazide. Mix and dilute to volume with distilled water. Add 1.0 ml. of Brij.-35 (30% Solution), mix, and transfer to a sealable bottle. This reagent is stable at room temperature for 6 months.

4. Ferric Chloride - Phosphoric Acid
a). Chemical Composition:

Ferric chloride hexahydrate ($FeCl_3 \cdot 6H_2O$)		15 gm.
Phosphoric Acid (85%)		300 ml.
Distilled water	q.s.	450 ml.

b). Preparation:

Place 90 - 100 ml. of distilled water in a 500 ml. graduated mixing flask. Add 15 gm. of ferric chloride hexahydrate and dissolve. While slowly swirling the flask, add 300 ml. of phosphoric acid. Dilute to 450 ml. total volume with distilled water, mix and transfer to a sealable amber bottle.

5. Sulfuric Acid, 20%
Place approximately 500 ml. distilled water in a liter Erlenmeyer flask. Add, with mixing, 200 ml. of concentrated sulfuric acid. Cool the solution, transfer to a liter volumetric flask and dilute to volume with distilled water.

6. BUN Acid Working Solution (Technicon T01 - 0162)
Place approximately 500 ml. of 20% sulfuric acid in a liter volumetric flask. Add 1.0 ml. of ferric chloride-phosphoric acid reagent and mix. Dilute to volume with 20% sulfuric acid and mix.

Glucose-Urea Nitrogen (Continued):

7. BUN Water
Place 1000 ml. distilled water in a liter container. Add 1.0 ml. Brij.-35 (30% Solution) and mix.

EQUIPMENT: Technicon AutoAnalyzer II System

1. Sampler II with a 60 specimen/hour cam having a 9:1 sample to wash ratio.

2. Proportioning Pump III.

3. AA II Colorimeter equipped with 460 nm. and 520 nm. filters and two 15 mm. flowcells. The Control Panel is located on this unit and it encloses the colorimeter electronics.

4. Recorder with dual - pen.

5. AutoAnalyzer manifold assemblies for BUN and Glucose Methodologies.

6. Voltage Stabilizer 115 Volts, 15 amp. The stabilizer provides a regulated 6.3 volts for the lamp of the colorimeter and a voltage at line level.

COMMENTS ON PROCEDURE:

1. The glucose method reduces the interference of ascorbic acid and sulfhydryl groups by treating the dialysate with sodium carbonate before the addition of the color reagent.

2. The glucose values obtained with this method are comparable to the Technicon Ferricyanide Procedure.

3. Serum values over 500 mg%. glucose should be diluted with 0.9% saline and rerun.

4. The Urea Nitrogen Method is linear to concentrations of 200 mg%. Samples of greater value are repeated on dilution with 0.9% saline.

5. All specimens should be well centrifuged and clear supernatants used to avoid introducing clots or debris into the system.

6. Noisy records may indicate worn pump tubes, poor bubble pattern, improper gain, dirty slide wire and/or power supply problems.

Glucose-Urea Nitrogen (Continued):

7. If a single test is ordered, the volume of specimen required may be decreased by disconnecting the sample line of the test not required and sealing of the PT-2 outlet with a short piece of tubing with a tight knot tied in it (use 0.015 ID).

8. It is adviseable to clean the system periodically (i.e. weekly) to prevent build-up of protein or reagent deposits. First disconnect the heating baths and allow them to cool. Them pump 1.0 N. NaOH or 10% Purex through the system for 15 - 20 minutes followed by a 15 minute flush with distilled water. The dialyzer membranes must be changed after this wash procedure. Remember to reconnect the heating baths.

9. The life of the manifold pump tubing is approximately 180 hours of action.

10. Urine Glucose Preparation
Record volume of urine and check urine for glucose using TesTape Using approximately 5.0 ml. of urine, add 1.0 gm. of Fuller's Earth (Lloyd's Reagent). Mix well and allow to stand at least five minutes. Centrifuge and run glucose on supernatant. It may be necessary to make urine dilutions using the treated urine.

If the TesTape indicates:

1+ use an undiluted specimen
2+ use a 1:10 dilution
3+ use a 1:10 dilution
4+ use a 1:25 dilution

Multiply the results by dilution factor and give results in mg%.

PROCEDURE:

1. Turn the power "ON" to the colorimeter, recorder and heating baths. Allow a warm-up period of 5 - 10 minutes. Check the levels of all reagents and fill if needed and empty the waste receptacles.

2. Position the manifold and begin pumping with distilled water for 10 minutes. Check the flow system for good bubble pattern and leaky connections.

Glucose-Urea Nitrogen (Continued):

3. Turn the valve to reagent position and begin pumping reagents. Once reagents have stabilized in the colorimeter (10 - 15 minutes), balance the energy of the reference, urea nitrogen and glucose phototubes to the same level (85 Units on the recorder paper). In "Operate" and "Direct" modes adjust the baseline to zero on the recorder paper and leave the chart drive "On".

4. Prepare the sample wheel beginning with three cups of assayed Reference Standard. Follow these with a normal and an elevated Control and two more cups of assayed Standard. Then ten unknowns, two assayed Reference Standards, ten unknowns and two assayed Reference Standards, etc.. Always end the run with an assayed Standard to varify the standard set-points and validity of the last group of unknowns.

5. Start the sample pick-up. When the first Reference Standard reaches the flowcell (approximately 9 - 10 minutes after sample pick-up), the standard assayed values are adjusted to the corresponding values on the chart paper using the Standard Calibration Knob. Coarse adjustment may be made with the first Standard, fine adjustment achieved with the second, and the third Standard should serve as a set-point check. The values of the Controls are read directly from the chart paper as are the Standards and specimens. The paired Standards between groups of unknowns should serve as follows: Allow the first one to check for drift (BUN: $\pm$ 2.0 mg%. and CHO $\pm$ 5.0 mg%.), and use the second to reset the set-point if necessary. If there is excessive drift, check for possible causes and repeat all samples.

6. Upon completion of the run, the system should be flushed with distilled water for 15 - 20 minutes.

NORMAL VALUES:	Adults:	BUN	10 - 20 mg%.
	Adults:	CHO	65 - 110 mg%.

REFERENCES:

1. Technicon: Operation Manual for the AutoAnalyzer, TAO - 0211 - 00, September, 1970.

2. Technicon: Illustrated Parts Breakdown and Telephone Manual for the Colorimeter, TAO - 0212 - 00, 1970.

Glucose-Urea Nitrogen (Continued):

3. Technicon: Parts Catalog for AA II Manifold Assemblies, TMO - 0170 - 00, 1970.

4. Marsh, W. H., et. al.: "Automated and Manual Direct Methods for the Determination of Blood Urea", Clin. Chem., 11:624 - 627, 1965.

5. Brown, M. E.: "Ultramicro Sugar Determination using 2, 9-Dimethyl-1, 10 Phenanthroline Hydrochloride (Neocuproine)", Diabetes, 10:60 - 62, 1961.

6. Bitner, D. and McCleary, M.: "The Cupric Phenanthroline Chelate in the Determination of Monosaccharides in Whole Blood", Amer. J. Clin. Path., 11:423, 1963.

CLINICAL INTERPRETATION:

Diabetes mellitus is undoubtedly the most common disease responsible for hyperglycemia. One must always insist that the blood sample be taken fasting to rule out post-prandial hyperglycemia. It is also surprising how many times there is failure to note whether or not the patient is receiving glucose intravenously when the blood sample is drawn.

Diminished glucose tolerance occurs in other diseases. Decreased ability to store glycogen occurs in chronic liver disease and hyperglycemia is present. Chronic liver disease is thus a common cause for hyperglycemia. Pregnancy results in hyperglycemia which is transient during the course of the pregnancy. Uremia is characterized by presence of serum uremic toxins which are anti-insulin with resultant hyperglycemia. Certain endocrine abnormalities are frequently associated with hyperglycemia. These are hyperthyroidism, acromegaly with excess growth hormone, pheochromocytoma, and Cushing's syndrome. Brain trauma results in prominent hyperglycemia. This is known as the diabetic pique. Brain trauma, neoplasms or rupture of a cerebral vascular aneurysm may stimulate a hypothalamic adrenal medullary response resulting in release of norepinephrine and glycogenolysis with hyperglycemia.

The most common cause for hypoglycemia is excessive use of insulin. Insulin treatment in diabetes mellitus may result in severe hypoglycemia. Malabsorption may result in hypoglycemia due to poor absorption of carbohydrate. Severe hepatic disease is associated with fasting hypoglycemia as is severe infection. Insulin-producing

Glucose-Urea Nitrogen (Continued):

pancreatic islet cell neoplasms and bulky mesenchymal neoplasms cause severe hypoglycemia by insulin production or glucose consumption. Finally endocrine disease may induce hypoglycemia. It may be seen in hypopituitary or hypoadrenal states.

The blood urea nitrogen is related to the glomerular filtration rate. BUN is not as good as serum creatinine to evaluate glomerular filtration. Approximately 50 to 75 per cent of total nephrons must be non-functional before the BUN is elevated. Urea is a potent osmotic diuretic. With increasing renal impairment, an osmotic diuresis occurs in the remaining functional nephrons. Thus, the filtered urea escapes reabsorption in the damaged nephrons. With greater damage to the nephrons, the blood urea clearance approximates the glomerular filtration rate.

Blood urea level is a net expression of diet, protein metabolism and glomerular function. Ingestion of large amounts of readily absorbable protein produces a marked rise in BUN. Blood has much protein. Thus, gastrointestinal hemorrhage causes an absorption of nitrogen products from the blood in the gastrointestinal tract. The rise in BUN may also relate to a lower glomerular filtration rate. Moderate exercise will not change the BUN.

The usual cause for an elevated BUN is acute or chronic renal disease due to glomerulonephritis, nephrosclerosis, or pyeloniphritis. Hemolytic anemia, hypotension due to hemorrhage, severe injury, burns, dehydration, heart failure or myocardial infarction may compromise glomerular filtration and cause a high BUN. Chemical injury or urinary tract obstruction may elevate the BUN.

Since BUN is synthesized by the liver, a low BUN suggests chronic hepatic disease. Other causes are excessive infusion of intravenous glucose with a resultant overhydration and low BUN. Pregnancy results in physiologic hydremia and BUN is thus low in pregnancy.

Technicon®
AutoAnalyzer®
Methodology

GLUCOSE
Technicon AA II
(Simplified Flow Diagram)

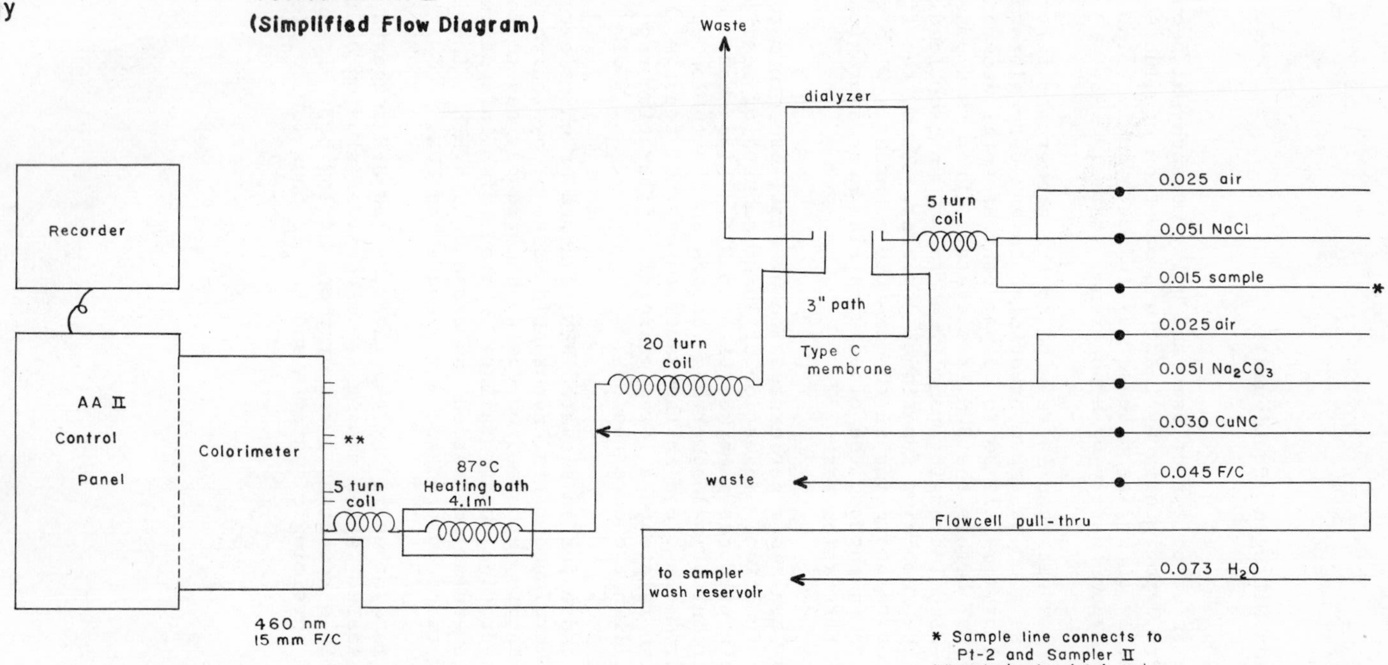

* Sample line connects to Pt-2 and Sampler II
** Colorimeter inlet and outlet for BUN

Technicon Instruments Corporation
Tarrytown, New York

Technicon®
AutoAnalyzer®
Methodology

Urea Nitrogen

Technicon AA II
Simplified flow diagram

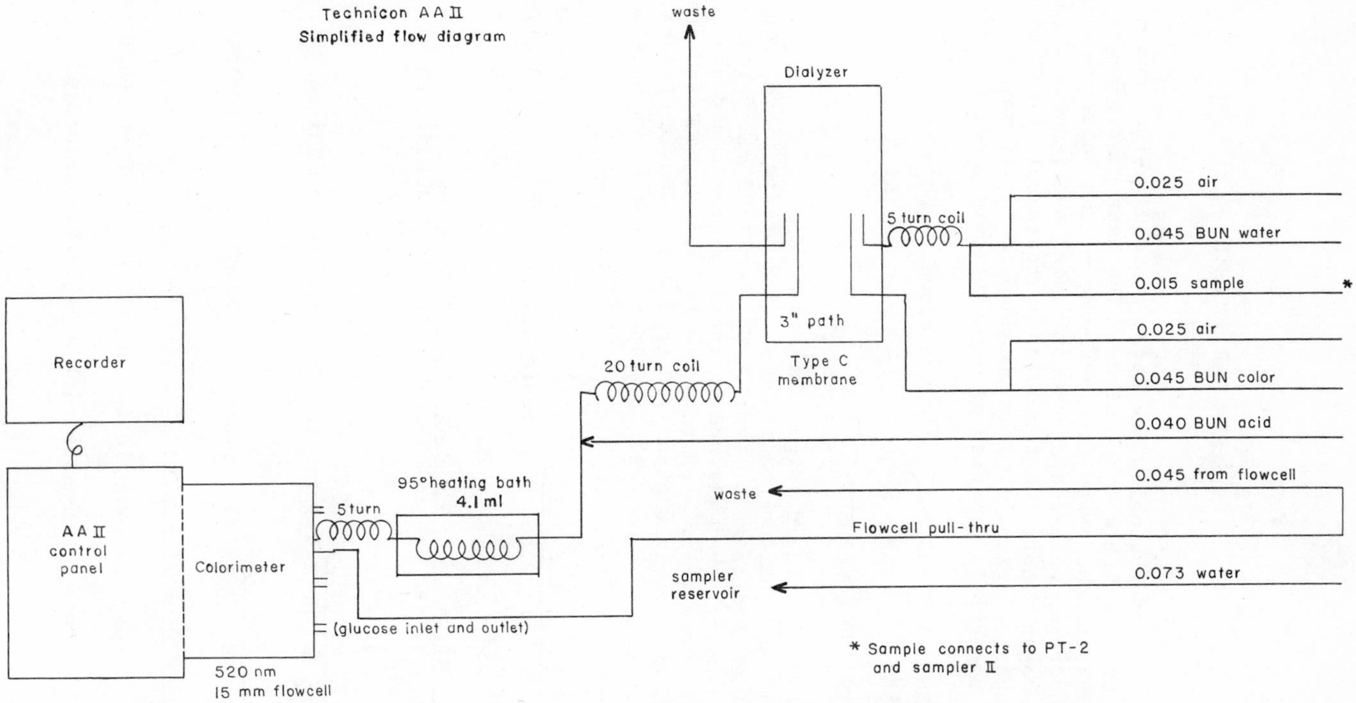

* Sample connects to PT-2 and sampler II

Technicon Instruments Corporation
Tarrytown, New York

SERUM HAPTOGLOBINS

PRINCIPLE: Haptoglobins are serum proteins which combine with hemoglobin or hemoglobin derivatives. The haptoglobin-hemoglobin complex enhances the peroxidase activity of the latter substance. Using hydrogen peroxide as the oxidizing substance and guaiacol as the hydrogen donor, the formation of tetraguaiacol is measured spectrophotometrically. The standardization is such that the haptoglobin content is measured as methemoglobin binding capacity.

SPECIMEN: 5.0 - 8.0 ml. clotted blood, separated, and refrigerated until run.

REAGENTS AND EQUIPMENT:

1. Guaiacol, liquid (Sigma Chemical)
   1.66 ml. guaiacol in a 500 ml. volumetric flask, add 50 ml. 1.0 M. acetic acid, swirl, add about 100 ml. deionized water. Mix well. When all the guaiacol is mixed, add about 250 ml. deionized water. pH to 4.0 using 1.0 M. NaOH (pH starts about 2.9 and requires about 7.0 to 8.0 ml. NaOH). Dilute to volume with deionized water. Refrigerate in a brown bottle. Good for 2 to 3 weeks.

2. 1.0 M. Acetic Acid
   54 ml. of 99 to 100% acetic acid diluted to 1000 ml. with deionized water.

3. Normal Serum for Standard Curve
   Lyophilized special chemistry control serum. Should be fairly fresh.

4. 0.15 M. NaCl, 0.85%
   8.5 gm. NaCl diluted to 1000 ml. with deionized water.

5. 0.1% $K_3Fe(CN)_6$
   0.1 gm. $K_3Fe(CN)_6$ diluted to 100 ml. with deionized water.

6. Methemoglobin Solution

   A. Assay the hemoglobin concentration of any hemolysate.

   B. Dilute hemolysate to 1.0 gm%. with deionized water.

Serum Haptoglobins (Continued):

C. In a 50 ml. volumetric flask place 2.5 ml. of the 1.0 gm%. hemolysate and 1.0 ml. of the 0.1% $K_3Fe(CN)_6$. Let sit for 10 minutes. Dilute to volume with deionized water.

D. Methemoglobin is not stable and must be made up each week.

7. 0.05 M. Hydrogen Peroxide, 30%
0.575 ml. H_2O_2 diluted to 100 ml. with deionized water. MUST BE USED WITHIN 30 MINUTES.

PROCEDURE:

1. Standard Curve. Run each week.

A. Pipette 5.0 ml. guaiacol into each of 9 tubes numbered Blank through 8.

B. Place in covered wath bath at room temperature.

Tube	Blank	1	2	3	4	5	6	7	8
Normal Serum	0	0	0.1	0.2	0.3	0.4	0.5	0.6	0.7
Saline	1.0	1.0	0.9	0.8	0.7	0.6	0.5	0.4	0.3
Met. Hgb.	0	1.0	1.0	1.0	1.0	1.0	1.0	1.0	1.0

Mix well and let sit 10 - 20 minutes.

C. 0.1 ml. of each of the above tubes is pipetted into the corresponding guaiacol tube.

D. 1.0 ml. of the Working Solution of H_2O_2, which has been brought to room temperature, is immediately added, tubes mixed, and incubated covered for 8 minutes.

E. Read Standard Curve within 4 minutes at 470 nm. on the Beckman DU.

2. Unknowns. Always run a control sample.

A. Pipette 5.0 ml. guaiacol into each of two tubes for each serum. Place in water bath.

B. In a test tube combine 2.0 ml. 0.15 M. NaCl and 0.5 ml. serum. Mix well.

Serum Haptoglobins (Continued):

C. Label two tubes for each serum: One Blank and One Test.

	Blank	Test
Dilute Serum	1.0	1.0
Met. Hgb.	0	1.0
Water	1.0	0

Mix well.

D. 0.1 ml. of each of the above samples is pipetted into the corresponding guaicol.

E. Immediately add:
1.0 ml. Working H_2O_2. Mix. Let sit covered for 8 minutes. Read O.D. against water within 4 minutes.

3. CALCULATIONS:

A. Standard Curve

1). Substrate tube 1, Reagent Blank, from each of the following tubes.

2). Plot O.D. vs. tube number using linear graph paper.

3). Draw slope, usually using the first four points.

4). Draw straight line using points 6, 7, and 8.

5). The point at which these lines intersect is 50 mg%. haptoglobin binding capacity.

6). Calculate the value of each square by dividing 50 mg%. by the total number of squares. This is the factor for this curve and these reagents only.

B. Unknowns

1). Subtract the O.D. of the blank from the O.D. of the tes

2). From the graph, obtain the number of squares for each O.D.

3). Multiply the number of squares by the factor obtained in Step No. 6 above. This is the mg%. value of the dilute sample.

Serum Haptoglobins (Continued):

4). Multiply the mg%. value by 5 (dilution) to obtain the haptoglobin binding capacity in mg%.

NOTES AND PRECAUTIONS:

1. Methemoglobin solution is not stable and must be prepared each week. In case a stat haptoglobin is requested, assay it using the methemoglobin and its curve, then repeat the sample when a new curve is prepared.

2. Reconstitute lyophilized serum every 1 to 2 weeks.

3. Working H_2O_2 is not stable for more than 30 minutes. Stock as long as it is well covered, refrigerated and kept in a brown bottle. It appears to be stable until it is gone.

4. Guaiacol is light sensitive and tubes should be placed in a covered water bath as soon as they are pipetted. The stock solution is good for 2 to 3 weeks. Thereafter it becomes increasingly more sensitive.

5. The brownish tetraguaiacol color fades on standing. Samples must be read within 4 to 5 minutes.

6. The normal control can be any serum previously assayed and found to be within normal limits. This value should be within ± 10% in each successive run.

7. If curve reads lower than usual, make up new methemoglobin.

NORMAL VALUES: 53 - 150 mg%.

REFERENCE: Owen, J.A.: "A Simple Method for the Determination of Serum Haptoglobins", Am. J. Clin. Path., 13:163, 1960.

CLINICAL INTERPRETATION:

Males have higher levels of serum haptoglobin than females. Premature infants do not have haptoglobins and levels are low in the cord blood of most term infants. The amount of haptoglobin rises slowly throughout infancy and reaches adult levels by the age of 6 months.

Serum Haptoglobins (Continued):

Haptoglobin levels are decreased in hemolytic anemia. The disappearance of haptoglobin is accentuated by its combination with free hemoglobin. Normally the amount of plasma haptoglobin is sufficient to combine with 3.0 gm. of hemoglobin. Serum haptoglobin thus combines with free hemoglobin to protect the renal tubule from damage. The measurement of haptoglobin is thus a valuable test to ascertain the presence of hemolytic anemia. Hemoglobinemia and hemoglobinuria are usually transient following hemolysis, but depressed haptoglobin levels persist for a longer period.

Other causes for low haptoglobin are loss into the urine due to nephrosis, decreased synthesis in liver disease, and hereditary deficiency frequently associated with Glucose-6-Phosphate Dehydrogenase deficiency.

An increase in haptoglobin occurs in inflammatory states, both acute and chronic. In acute infections, haptoglobin reaches a maximum in two weeks and returns to normal during the healing phase. Chronic infections such as tuberculosis may cause the haptoglobin level to be 300 mg%. Immunologic disease such as rheumatoid arthritis and glomerulonephritis, and metastatic carcinoma cause prominent elevations of plasma haptoglobins.

ALPHA-HYDROXYBUTYRATE DEHYDROGENASE
(α-HBD)

PRINCIPLE: Advantage is taken of the clinical differentiation of the serum lactic dehydrogenase (LDH) isoenzymes. It has been shown that the faster moving isoenzymes reduce α-ketobutyrate while the slower moving ones show negligible activity. Damage to heart muscle, altered erythropoiesis and extreme damage to other tissue could lead to increased HBD activity.

The HBD measurement in relation to total LDH becomes of great value in the measurement of the "heart" fractions of LDH isoenzymes.

This test has the advantage of being performed quite rapidly and simply.

The following reaction takes place:

$$\alpha\text{-oxobutyrate} + NADH_2 \underset{}{\overset{HBD}{\rightleftharpoons}} \alpha\text{-hydroxybutyrate} + NAD$$

SPECIMEN: A pipetable 100 microliters (0.10 ml.) of serum, free from hemolysis, is required. This specimen is stable for several days at room temperature or $4^{o}C$.

REAGENTS AND EQUIPMENT:

Biochemica Kit TC-HD, Cat. No. 15953 THAD for 60 determinations.

Bottle No. 1 0.05 M. Phosphate Buffer, pH 7.4
Bottle No. 2 0.004 M. NADH
Bottle No. 3 0.10 M. alpha-oxobutyrate

1. Dissolve contents of Bottle No. 1 in 200 ml. deionized water.

2. Dissolve contents of Bottle No. 2 in 2.5 ml. deionized water.

   Mix these reagents together in a brown glass bottle. This makes Solution 1. (Stable four weeks at $4^{o}C$.)

3. Dissolve the contents of Bottle No. 3 in 6.5 ml. of deionized water. This makes Solution 2. (Stable three months at $4^{o}C$.)

   Do <u>not</u> mix the two solutions.

Alpha-hydroxybutyrate Dehydrogenase (Continued):

4. DBG or Gilford 300 N Spectrophotometer
An instrument with temperature controlled cuvette well and recorder should be used.

5. Eppendorf Pipette
Use a 100 microliter Eppendorf pipette for dispensing serum sample.

PROCEDURE:

Place into a 10 mm. square cuvette:

1. 3.1 ml. of Solution 1.

2. 0.10 ml. (100 microliters) of serum.

Mix and place into a 30°C. water bath for approximately 15 minutes.

3. Add 0.10 ml. (100 microliters) of Solution 2. Mix. Place cuvette in cuvette well and record A. change for two minutes of linearity.

4. CALCULATIONS:

Based on $\Delta A/1$ minute and a light path of 1.0 cm.

(ϵ NADH at 340 nm. = 6.22 Liter/Mole x cm.)

$$\frac{\Delta A}{\epsilon \times d} \times 10^6 \times \frac{TV}{SV} \times \frac{1}{Time} = \text{I.U./L. or mU/ml.}$$

$$\Delta A \times \frac{1}{6.22 \times 10^3 \times 1} \times 10^6 \times \frac{3.3}{0.1} \times \frac{1}{1} = \text{mU/ml.}$$

$$\Delta A \times F = \text{mU/ml.}$$

$$F = 5305$$

NORMAL VALUES: Up to 168 mU/ml. at 30°C.

Alpha-hydroxybutyrate Dehydrogenase (Continued):

REFERENCES:

1. Rosalki, S. B. and Wilkinson, J. H.: Nature, 188:1110, 1960.

2. Rosalki, S. B.: British Heart Journal, Vol. XXV, No. 6, 795, 1963.

3. Preston, J. A., Batsakis, J. G. and Briere, R. O.: "Serum Alpha-Hydroxybutyrate Dehydrogenase, A Clinical and Laboratory Evaluation in Patients with Myocardial Infarct.", Am. J. Clin. Path., 41:237, 1964.

CLINICAL INTERPRETATION:

The value of determining alpha-hydroxybutyrate dehydrogenase is that this enzyme corresponds to LDH-one which is from a myocardial source. Myocardial damage results in liberation of this enzyme and elevation in the serum. Hemolytic anemia and metastatic malignant melanoma will result in an elevated serum alpha-hydroxybutyrate dehydrogenase. The enzyme is present in the erythrocyte and the malignant melanoma cells.

When myocardial damage is present, LDH-one or alpha-hydroxybutyrate dehydrogenase will be increased in the serum. At times the total LDH will be normal, but the former isoenzymes will always be increased. It is more difficult to perform an electrophoresis to ascertain the level of LDH-one. In addition the determination of heat stable LDH requires more time than the determination of alpha-hydroxybutyrate dehydrogenase. Heat stable LDH corresponds to LDH-one and alpha-hydroxybutyrate dehydrogenase. It is determined by ascertaining the amount of LDH remaining after total serum LDH is heated in a water bath at $60^{o}C$. for 30 minutes. Heat stable LDH originates from myocardial cell mitochondria and the erythrocyte. Heat labile LDH is derived from the liver and skeletal muscle. Thus, a rapid method to ascertain the presence of myocardial necrosis is the determination of alpha-hydroxybutyrate dehydrogenase. It is more rapid and simpler than the determination of LDH-1 by electrophoresis or heat stability.

SEROTONIN 5-HYDROXYINDOLE ACETIC ACID

PRINCIPLE: In the malignant carcinoid syndrome, the tumor cells liberate amounts of serotonin into the blood stream. The serotonin is partially metabolized and excreted chiefly as 5-hydroxyindole acetic acid (5HIAA). 5HIAA reacts with 1-nitroso-2-naphthol in the presence of a reducing agent such as nitrous acid to produce a purple complex. This complex is further characterized by being extractable with ethylene dichloride. False positive results have been reported due to the ingestion of bananas or acetanilid. False negatives may be caused by administration of phenothiazine derivatives.

SPECIMEN: Single voided urine specimen.

REAGENTS:

1. <u>1-Nitroso-2-Naphthol</u>, 0.1% in 95% ethanol.

2. <u>Sodium Nitrite</u>, 2.5%
   Solution must be refrigerated and made up fresh every two weeks.

3. <u>Sulfuric Acid</u>, 2 N.

4. <u>Nitrous Acid</u>
   Freshly prepared by adding 0.2 ml. of 2.5% sodium nitrite to 5.0 ml. 2.0 N. sulfuric acid.

5. <u>Ethylene Dichloride</u>

PROCEDURE:

1. <u>Run a positive control with each test</u>. To make positive control Weigh out 1.5 mg. 5-Hydroxyindol-3-Acetic Acid (J. T. Baker Co.- Kept in dessicator) and dissolve this in 10 ml. deionized water. Aliquot 0.2 ml. in 10 ml. tubes and freeze. This gives a 15% control.

2. Pipette 0.2 ml. of urine into a 10 ml. test tube.

3. Add 0.8 ml. deionized water.

4. Add 0.5 ml. 1-nitroso-2-naphthol and mix.

5. Add 0.5 ml. nitrous acid and let stand for 10 minutes.

Serotonin 5HIAA (Continued):

6. Add 5.0 ml. ethylene dichloride and shake vigorously. If turbid, centrifuge before reading.

NORMAL VALUES:

A definite layering occurs at the top of the column of fluid. Result is positive when top layer is purple, deep purple, or almost black, indicating excretion of 5-hydroxyindole acetic acid in excess of 4.0 mg./100 ml. or more. Negative urines are colorless to yellowish, usually representing normal 5-HIAA excretion (2.0 to 9.0 mg. per 24 hours).

REFERENCE: Page and Culver: SYLLABUS OF LABORATORY EXAMINATIONS IN CLINICAL DIAGNOSIS, Harvard, pg. 325, 1966.

CLINICAL INTERPRETATION:

Carcinoid tumors are benign or malignant. They may arise in the bronchus, small intestine, colon or appendix. Malignant carcinoids usually arise in the colon or small intestine. They produce excessive serotonin. 5-hydroxyindole acetic acid is a metabolite of serotonin which appears in large amounts in the urine in metastatic carcinoid, especially to the liver. Extensive liver metastases compromises the hepatic function of metabolizing serotonin and its metabolites. Normal excretion of 5-hydroxyindole acetic acid in 24 hours is 1.0 to 5.0 mg. A random specimen of urine is usually sufficient for screening purposes; if a 24 hour collection is made, it should be acidified with HCl or acetic acid. Patients with malignant metastatic carcinoid may excrete up to 350 mg. of 5-hydroxyindol acetic acid per day. False positive tests occur in patients with liver disease, psychiatric disorders on drugs, excessive consumption of bananas, and sprue.

17-HYDROXYCORTICOID METHOD
(Porter-Silber)

PRINCIPLE: In the Porter-Silber method, after hydrolyzing the conjugates with B-glucuronidase to liberate the steroids, the urine is extracted with methylene chloride, the aqueous phase removed by aspiration, and the organic layer washed with dilute sodium hydroxide to remove chromogens. An aliquot of the methylene chloride is shaken with phenylhydrazine-sulfuric acid-ethanol reagent, the methylene chloride removed by aspiration, and the yellow color produced measured in a spectrophotometer.

SPECIMEN: 24 hour urine specimens collected under refrigeration. Measure total volume and freeze approximately 100 ml. until determination made.

REAGENTS:

1. Sodium Acetate, 0.2 N., anhydrous.
   In a one liter volumetric flask, dissolve 16.408 gm. of sodium acetate and bring to volume. Store refrigerated.

2. Acetic acid, 0.2 N.
   In a 1000 ml. volumetric flask, measure 11.4 ml. of glacial acetic acid and bring to volume.

3. Beta-glucuronidase, 5000 u./ml.
   Warner-Chilcott division of General Diagnostics Ketodase.

4. Penicillin, 1,000,000 units.
   Add 5.0 ml. sterile saline to bottle containing Penicillin G, crystalline potassium, U.S.P. Lilly, 1,000,000 units.

5. Methylene Chloride.
   Use Merck spectrophotometric grade.

6. NaOH, 0.1 N.
   Into a 1000 ml. volumetric flask, dissolve 4.0 gm. of sodium hydroxide in about 600 ml. of deionized water, and bring to volume.

7. H_2SO_4, 22 N.
   Add 620 ml. concentrated H_2SO_4 to about 380 ml. distilled water. When cool, dilute to volume. Use Pyrex volumetric flask and pack in ice.

17-Hydroxycorticoid Method (Continued):

8. Ethanol, 95%
   3 times distilled; first time with 2,4-dinitrophenylhydrazine.

9. Phenylhydrazine HCl
   Recrystallized four times from 95% ethanol. Start with 25 gm. and dissolve in 1 liter of aocohol with heat. Raise heat to not over 65°C. Add more alcohol gradually until completely dissolved. Filter through a Buchner funnel while still hot. Let cool at room temperature and place in cold room overnight. Filter, redissolve precipitate in about 250 ml. ethanol; heat and add alcohol until dissolved. Filter as above and repeat twice. Dry and keep over $CaCl_2$. Keep in dark at all times.

10. Stock Standard Hydrocortisone, in alcohol
    Stock standard of 50 mg. in 100 ml. 3 times distilled ethanol. Make working standard of 20, 40, and 60 micrograms/ml. in 3 times distilled ethanol. Determine absolute amount by ultraviolet absorption in quartz cuvettes.

11. Sodium Bisulfite ($NaHSO_3$)
    In a 1000 ml. volumetric flask, dissolve 250 gm. of $NaHSO_3$ in about 600 ml. of deionized water, and bring to volume.

PROCEDURE:

DAY ONE:

1. Mix thoroughly specimens of urine and centrifuge approximately 30 ml.

2. Test each urine for sugar phenothiazine and acetone, and record.

3. Put 8.0 ml. aliquot ot each centrifuged urine into 10 ml. disposable beaker (if result is expected to be high, as after ACTH, take 4.0 ml. of urine and 4.0 ml. of water or 2.0 ml. urine and 6.0 ml. of water).

4. Add 0.2 ml. of 0.2 N. Na Ac to each.

5. Adjust pH to 4.8 - 4.9 with 0.2 N. HAc (if urine is very alkaline, use 0.1 ml. of 10% HAc or 5 N. NaOH if very acid, to avoid increasing volume above 10 ml.).

6. Adjust volume to 10 ml. with water.

17-Hydroxycorticoid Method (Continued):

7. Place 5.0 ml. aliquot of each in glass-stoppered 50 ml. centrifuge tube for hydrolysis.

8. Prepare 2 "Reagent Blank" tubes and 3 "Standard" tubes:

	Vol. of Std./ml.	H_2O ml.	0.2 N. Na. Acetate ml.	HAc 0.2 N. ml.
Standard I	1.0	3.4	0.2	0.4
Standard II	1.0	3.4	0.2	0.4
Standard III	1.0	3.4	0.2	0.4
Reagent Blank (in duplicate)	---	4.4	0.2	0.4

9. Add about 5,000 (1.0 ml.) Units glucuronidase solution to each tube, including Reagent Blanks and Standards. Add 2 drops of penicillin solution aseptically to each tube.

10. Stopper and place in water bath at 37°C. overnight.

DAY TWO:

1. Prepare Porter-Silber Reagent and keep in dark:
   A. Blank Reagent: 2 parts 22 N. H_2SO_4 + 1 part EtOH. Let cool to room temperature (P. S. B. Reagent.)
   B. Color Reaction Reagent: Immediately before use, dissolve 44 mg. phenylhydrazine HCl in 100 ml. of the above solution

2. Remove specimens from water bath. Add 1.0 ml. $NaHSO_3$ if sugar is present. Cool. Add 25 ml. methylene chloride to each tube, stopper, and shake vigorously for exactly 15 seconds.

3. Centrifuge and remove aqueous phase by suction.

4. Add 2.0 ml. 0.1 N. NaOH to each tube, stopper, and shake vigorously for exactly 10 seconds.

5. Centrifuge. (Prepare color reagent at this time.) Remove aqueous layer by suction.

6. Take two 10 ml. aliquots of methylene chloride phase from each tube, 1 for Porter-Silber Blank reagent (P. S. B.) and 1 for color reaction reagent. (Ø)

17-Hydroxycorticoid Method (Continued):

7. Add 3.0 ml. color reaction reagent (Solution B) quantitatively to each tube marked Ø, and to both Standards.

8. Add 3.0 ml. blank reagent (Solution A) quantitatively to each tube marked P. S. B. Cap tubes with stoppers.

9. Shake all vigorously for 15 seconds.

10. Centrifuge, remove and discard most of the methylene chloride layer by suction.

11. Place in water bath at 60°C. for 30 minutes. Remove from water bath and cool.

12. Using the Beckman DU spectrophotometer, read at 370 nm., 410 nm., and 450 nm. wavelengths.

    A. Phenylhydrazine treated "Reagent Blank" (marked RBØ) against H_2SO_4·EtOH treated "Reagent Blank" (marked RB/PSB).

    B. Standards and unknowns against phenylhydrazine treated Blank (marked RBØ).

    C. Unknown blanks against H_2SO_4·EtOH treated "Reagent Blank" (RB/PSB).

CALCULATIONS:

Subtract each blank reading from corresponding color reaction reading. Add the corresponding 370 + 450 readings. Double the reading at 410. Subtract the 370 + 450 total from the 2 X 410 reading. Using the density obtained above, determine the micrograms of 17-OH in the aliquot of urine used from the standard curve. Divide this by the number of ml. of urine in the aliquot (½ the volume of the original aliquot taken in Step No. 3 of procedure) to determine micrograms/ml. or mg./L. Multiply by number of liters of 24 hour specimen to determine mg./24 hours.

NORMAL VALUES:

Male: 6 - 12 mg./24 hours
Female: 4 - 10 mg./24 hours

0 - 2 Years: 2 - 4 mg./24 hours
2 - 6 Years: 3 - 6 mg./24 hours
6 - 10 Years: 6 - 8 mg./24 hours
10 - 14 Years: 8 - 10 mg./24 hours

17-Hydroxycorticoid Method (Continued):

Newborn up to 24 months: 0.5 mg./24 hours.

REFERENCES:

1. Silber and Porter: J. Biol. Chem., Vol. 210:2, Oct., 1954.

2. Ibbott and Obrien: PEDIATRIC MICRO AND ULTRAMICRO- BIOCHEMICAL TECHNIQUES, Hoeber, Ed. 3 (Reference for normal values of children.

CLINICAL INTERPRETATION:

The urinary 17-hydroxycorticoids measure cortisol, cortisone, 11-desoxycortisol and their metabolites or any corticosterone with 20, 21 alpha ketol and 17-hydroxyl groups. They permit quantitation of the overall 24 hour urinary excretion of cortisol and its derivatives and afford an indirect measure of adrenal cortical secretory activity. The measurement of 17-hydroxycorticosteroids has enjoyed wide use since they are a more sensitive and significant measure of overall adrenal cortical secretory activity than the ketosteroids. If the metabolites are included in the measurement, the amounts of steroid measured are ten times larger than the unconjugated, biologically active free corticoid fraction. One shortcoming of the method is that it does not indicate the plasma level of cortisol. In some conditions, blood cortisol is normal but urinary 17-hydroxycorticoids are elevated, such as in obesity and hyperthyroidism.

SERUM IRON AND TOTAL IRON-BINDING CAPACITY

PRINCIPLE: Serum iron is bound to the protein transferrin in the ferric form. It can be directly measured without protein precipitation, by complexing the reduced form (Fe^{+2}) with magnesium bathophenantroline sulfonate. The three important steps for the reaction are: 1). Dissociation of ferric iron from transferrin, 2). Reduction of ferric iron to ferrous iron, and 3). Complexing with iron complexing agent. The simplicity and sensitivity of the method is worth mentioning, since the reaction occurs in one tube, without the problem of incomplete recovery due to coprecipitation or lack of filtrate clarity. The entire reaction is also irreversible, insuring complete recovery.

$$\text{(TRANSFERRIN)}\ (Fe^{+3})_2 \rightleftharpoons \text{TRANSFERRIN} + 2Fe^{+3} \longrightarrow Fe^{+2}$$

$$\text{(red)}\quad Fe{<}^{\beta}_{\beta}{}^{\beta} \longleftarrow 3\beta\ \text{(BATHOPHENANTROLINE)}$$

The total iron-binding capacity (TIBC) of serum is the sum of the serum iron plus the ferric iron that combines with transferrin following the addition of ferric iron to the serum. The excess ferric iron is removed by adsorption with magnesium carbonate.

SPECIMEN: UNHEMOLYZED, pipettable 2.0 ml. of serum is needed to run both serum iron and total iron-binding capacity. A diurnal variation has been found, which in general is higher in the morning and lower in the afternoon. It is therefore recommended that fasting blood specimens be drawn for the determination. The serum is stable in the refrigerator for one week, and for several weeks in the freezer.

EQUIPMENT AND REAGENTS:

1. Iron Blank Reagent
   0.1 M. Hydroxylamine (reducing agent) in acetate buffer at pH 4.0 Available from American Monitor, Kit # R5350 (for 60 determinations). Stable indefinitely at room temperature.

2. Iron Color Reagent
   Magnesium bathophenantroline sulfonate. Available from American Monitor, Kit # 5350 (for 60 determinations). Stable indefinitely at room temperature.

Serum Iron (Continued):

3. Iron Binding Reagent
For use in the determination or iron-binding capacity of serum. 500 micrograms of Ferric Chloride per 100 ml. in 1/200th M. HCl with Redox Preservative. American Monitor Kit # 5350.

4. Iron Adsorbent
$MgCO_3$ powder, supplied in capsules, one capsule for each TIBC Test. American Monitor Kit # 5350.

5. Iron Working Standard
200 micrograms per 100 ml. Dissolved in HCl with preservative. American Monitor Kit # 5350.

6. Calibration Curve
Stock Iron Standard, 20 mg%. (20,000 ug%.). Transfer 1.404 gm. ferrous ammonium sulfate hexahydrate (MW. 392), $Fe(NH_4)_2(SO_4)_2 \cdot 6H_2O$ to a one liter volumetric flask. Dissolve in about 800 ml. of deionized water. Add 0.5 ml. of concentrated sulfuric acid, and dilute to the mark. This solution is stable. Working standards of 100, 200, 300, 400, and 500 ug%. can be made by diluting 0.5, 1.0, 1.5, 2.0, and 2.5 ml. of the stock standard to 100 ml. with acidified (HCl) deionized water respectively. pH should be around 2.4.

7. Gilford Spectrophotometer 300 N.

NOTE: The lot number of color and of blank reagent should be the same. Do not mix lots.

GLASSWARE: Glass pipettes should be rinsed with dilute HCl (10 ml. concentrated HCl, q.s. to about 500 ml.) and rinsed several times with deionized water.

PROCEDURE:

Serum Iron

1. Using disposable culture tubes, 75 x 12 mm., set up a "Standard Blank" and two tubes for "Standard".

2. For each serum, set up a "Test" and "Serum Blank".

3. Controls are Hyland Normal and Hyland Abnormal.

Serum Iron (Continued):

4. Pipette 0.5 ml. of water to "Standard Blank". Pipette 0.5 ml. of Std. or Stds. (Calibration Curve) to "Standard". Add 0.5 ml. serum to both "Test" and "Serum Blank".

5. Add 1.6 ml. of Blank Reagent to "Standard Blank" and "Serum Blank". Add 1.6 ml. of Color Reagent to "Standards" and "Test". Cover all tubes with parafilm, and mix well.

6. Incubate all tubes at $37^{o}C$. for 5 minutes.

7. Set the machine at 0.000 absorbance with water and read absorbance of "Blanks", "Standards", and "Tests" at 535 nm. Final product should be read as soon as possible.

CALCULATIONS: $\frac{\text{O.D. Test}}{\text{O.D. Standard}} \times 200 = \text{ug Fe/100 ml.}$

TIBC

1. Pipette 1.0 ml. of serum into a 12 ml. glass stoppered centrifuge tube. Controls are Hyland Normal and Hyland Abnormal.

2. Add 2.0 ml. of Iron-Binding Reagent. Mix well, and let stand for 5 minutes.

3. Add $MgCO_3$ (one capsule). Stopper, and mix by inversion for 2 minutes.

4. Centrifuge for at least 20 minutes.

5. Run 0.5 ml. of supernatant in place of serum for test and blank in the above procedure.

CALCULATIONS: $\frac{\text{O.D. Test}}{\text{O.D. Standard}} \times 600 = \text{TIBC in ug./100 ml.}$

$$\text{\% Saturation} = \frac{\text{Fe}}{\text{TIBC}} \times 100$$

Serum Iron (Continued):

NORMAL VALUES:

Serum Iron
Men — 80 - 160 ug./100 ml.
Women — 60 - 135 ug./100 ml.

TIBC
250 - 350 ug./100 ml.

PROCEDURE FOR ULTRAMICRO SERUM IRON AND TOTAL IRON-BINDING CAPACITY

Serum Iron

1. Using disposable 3.0 ml. A.A. cups, set up a "Standard Blank" and 2 cups for "Standard".

2. For each serum, set up a "Test" and a "Serum Blank". Run the same controls.

3. Using a pre-calibrated Hamilton syringe, deliver 0.5 ml. of Blank Reagent to the "Standard Blank" and to all the Serum Blanks. Rinse well with deionized water, then with the color reagent, and dispense the same amount (0.5 ml.) of Color Reagent to the cups marked "Standards" and "Tests".

4. Pipette 100 ul. of water to the "Standard Blank". Pipette 100 ul. of Standard or Standards (Calibration Curve) to Standards. 100 ul. of serum is added to both Test and Blank.

5. Incubate the same way, and read in the Gilford Spectrophotometer 300 N with the micro cuvette.

TIBC

1. Pipette 0.5 ml. of serum into a 12 ml. glass stoppered centrifuge tube. Use the same controls.

2. Add 1.0 ml. of Iron-Binding reagent. Mix well, and let stand for 5 minutes.

3. Add $MgCO_3$ (one-half capsule). Stopper, and mix by inversion for 2 minutes.

Serum Iron (Continued):

4. Centrifuge for 20 minutes.

5. Run 0.1 ml. of supernatant in place of serum for test and blank in the above procedure.

CALCULATIONS: The same.

REFERENCES:

1. Goodwin, J. F., Murphy, B., and Guilemette, M.: Direct Measurement of Serum Iron and Binding Capacity. Clin. Chem. 12:2, 1966.

2. Levy, A. C., and Vitacca, P.: Direct Determination and Binding Capacity of Serum Iron. Clin. Chem. 7:241, 1961.

3. Ramsay, W. N. M.: Plasma Iron. Advances in Clin. Chem. 1:1, 1958.

CLINICAL INTERPRETATION:

The most common cause for a low serum iron is blood loss which may be gross or occult, especially from the urogenital or gastrointestinal tract. Malnutrition or malabsorption may contribute to low serum iron. Gastrointestinal lesions which may cause iron deficiency anemia are esophageal varices, peptic ulcer, gastric carcinoma or carcinoma of the colon. Other causes for decreased serum iron are chronic infection and azotemia.

Elevated serum iron is found in hemolytic anemia and acute hepatitis. Iron is released from the red blood cells and necrotic liver. Hemochromatosis and siderochrestic anemia may be associated with an elevated serum iron.

An increased iron binding capacity or unsaturated capacity is found in iron deficiency anemia, while a decreased iron binding or saturated capacity is found in loss due to nephrosis, hemochromatosis, azotemia, hereditary transferrin deficiency, hemolytic anemia and siderochrestic anemia.

ISOCITRATE DEHYDROGENASE
(U. V. Methodology)

PRINCIPLE: Mammalian tissue contains two isocitrate dehydrogenase enzymes; the one which is of clinical interest is that enzyme linked to NADP. The enzyme shows molecular heterogeneity, with four distinct isoenzymes being separated following electrophoresis. The activity of I.C.D. in man is found in greatest concentration in the liver, then heart, tumors and muscle. Abnormal I.C.D. activity is best considered as a relatively sensitive indicator of hepatocellar damage. Elevations of serum I.C.D. may also be found with myocardial infarction, but not consistently so, due to the rapid denaturation or "clearing" of the heart isoenzyme fraction in the body. Red cells and platelets also display significant I.C.D. activity.

The spectrophotometric determination is based upon the reaction indicated beneath. The reduction of NADP to $NADPH_2$ is followed spectrophotometrically at 340 nm. or 366 nm. $NADPH_2$ is strongly absorbant at the preceding wavelengths and its increase is proportional to I.C.D. concentration.

$$\text{Isocitrate} + \text{NADP} \underset{Mn^{++}}{\overset{\text{I.C.D.}}{\rightleftharpoons}} \text{Oxalosuccinate} + NADPH_2$$

Mn^{++} is an activator and is added to the reaction mixture along with the substrate.

SPECIMEN: 0.5 ml. of <u>unhemolyzed</u> serum. Do not use plasma specimens; anticoagulants interfere with enzyme activity.

REAGENTS AND EQUIPMENT:

Reagents for U. V. Method are available as Biochemica Test Combination TC-ID #15933 TIAA. One kit is sufficient for 25 determinations.

1. <u>Triethanolamine Buffer</u>, 0.1 M., pH 7.5 and <u>DL-Isocitrate</u>, 0.0046 M.
   Dissolve contents of bottle #1 in 75 ml. redistilled water. Stable for 3 months at approximately 4°C. Bottle also contains 0.052 M. NaCl.

Isocitrate Dehydrogenase (Continued):

2. NADP, 0.0091 M. and $MnSO_4$, 0.12 M.
   Dissolve contents of Bottle No. 2 in 3.0 ml. of redistilled water. Stable for four weeks at approximately 4°C.

3. Water Bath, at 30°C.

4. Spectrophotometer, with temperature controlled cuvette well (30°C.)

PROCEDURE:

1. Into a glass cuvette of 1.0 cm. light path, pipette 2.5 ml. of buffer and isocitrate reagent.

2. Deliver 0.5 ml. of serum specimen into the cuvette containing reagent. Mix the contents and place the cuvette in a 30°C. water bath for 5 minutes.

3. Deliver 0.1 ml. of NADP-$MnSO_4$ reagent into the cuvette and mix well by inversion.

4. Place cuvette in thermostated cuvette well and start recorder.

5. Monitor the change in absorbance for at least two consecutive minutes of linearity.

6. CALCULATIONS:

   Calculations based on mean A change of 1 minute.

   $NADPH_2$ at 340 nm. = 6.22×10^3 Liters/mole x cm.

$$\frac{\Delta A}{\epsilon \times d} \times 10^6 \times \frac{TV}{SV} \times \frac{1}{Time} = \text{I.U./L. or mU/ml.}$$

$$\frac{\Delta A}{1} \times \frac{1}{6.22 \times 10^3 \times 1} \times 10^6 \times \frac{3.1}{0.5} \times \frac{1}{1} = \text{mU/ml.}$$

$$\Delta A \times F = \text{mU/ml.}$$
$$F = 997$$

NOTE: Sera with absorbance changes per minute greater than 0.100 at 340 nm. should be diluted 1:10 with physiological saline and repeated on 0.5 ml. of this dilution. Multiply final answer x 10.

Isocitrate Dehydrogenase (Continued):

NORMAL RANGE: Up to 11 mU/ml.

REFERENCES:

1. Batsakis and Briere: INTERPRETIVE ENZYMOLOGY, C. Thomas, 1967.

2. Ochoa, in Colowick & Kaplan: METHODS IN ENZYMOLOGY, Vol. 1, pg. 699, Academic Press

3. Wolfson: Proc. Soc. Exp. Biol. & Med., 92:231, 1957.

4. Wolfson, Ann: N.Y. Acad. Sci., 75:260, 1958.

5. Henry, J. B.: WORKSHOP ON CLINICAL ENZYMOLOGY; PRE- WORKSHOP MANUAL, ASCP, 1964.

6. Clin. Chem., 6:208, 1960 (Colorimetric Determination)

7. J. Lab. & Clin. Med., 62:148, 1963 (Colorimetric Determination)

8. Sterkel, R. & Wolfson: J. Lab. Clin. Med., 52:176, 1958.

9. Cohen: Ann. of Int. Med., 55:604, 1961.

CLINICAL INTERPRETATION:

Isocitric dehydrogenase is an ubiquitous enzyme which is present in the mitochondria of the cell. It occurs in two isoenzyme forms: a fast and a slow electrophoretic fraction. The liver contains the fast component, while the slow component is present in heart muscle. The slow portion is not as heat stable as the fast component. The enzyme is present in the liver, heart, skeletal muscle, and certain neoplasms. Activity has also been demonstrated in platelets and in erythrocytes; very little activity is demonstrable in normal plasma.

Heart isocitric dehydrogenase activity is heat labile; it is inactivated at 56°C. Isocitric dehydrogenase derived from a liver source, however, is heat stable at 56°C. After an acute myocardial infarct, the heat labile heart isoenzyme persists for a very short time in the blood; but with liver damage there is persistence of activity of isocitric dehydrogenase in the serum. An elevation of isocitric dehydrogenase is a sensitive indicator of parenchymal

Isocitrate Dehydrogenase (Continued):

hepatic disease, but the enzyme activity cannot be used to differentiate various liver diseases. A normal isocitric dehydrogenase activity is usually found after an acute myocardial infarct; however, if heart failure occurs as a complication, there may be an elevation due to the intense hepatic congestion with necrosis of centrilobular cells. Various lesions such as viral hepatitis, metastatic carcinoma, hepatoma, and severe congestive heart failure have caused an elevation of isocitric dehydrogenase. The enzyme activity is elevated early in the course of infectious hepatitis, it persists for approximately three weeks and then returns to normal with recovery from the illness. When isocitric dehydrogenase levels remain elevated, one may assume that there is persistence of the viral hepatitis.

It has been observed that isocitric dehydrogenase is elevated in patients with megaloblastic anemia, presumably the enzyme is produced in large amounts by the megaloblasts. This is the same situation that exists in pernicious anemia with serum elevations of LDH. Both isocitric dehydrogenase and LDH are mitochondrial enzymes, and staining the megaloblasts by the tetrazolium-formazan technique demonstrates a large amount of isocitric dehydrogenase and LDH activity in the megaloblasts. With proliferation of these cells, a large amount of enzyme activity will be contributed to the serum. In addition, intra-bone marrow destruction of megaloblasts results in elevated LDH. There are rare reports of elevations of isocitric dehydrogenase in patients who have carcinoma of the pancreas, carcinoma of the prostate, infarction of the placenta, and myeloid leukemia. Thus, this enzyme increase is from necrosis of the pancreas associated with carcinoma of this organ, from the placenta, from carcinoma of the prostate, and from myeloid leukemic cells which contribute it to the serum. Isocitric dehydrogenase activity is increased in the cerebrospinal fluid when there is a primary or metastatic carcinoma to the central nervous system, and in patients who have had cerebral infarction or an acute bacterial meningitis.

17 - KETOSTEROIDS

PRINCIPLE: The 17 - ketosteroids are steroid compounds which have a ketone group on Carbon 17. The chief steroids measured in urine are androsterone, dehydroisoandrosterone, etiocholanolone, and dehydroisandrosterone. These are often referred to as "Neutral" 17 - ketosteroids.

The 17 - ketosteroids are excreted as sulfate and glucuronide conjugates which are hydrolyzed with concentrated HCl at 80°C. The free steroids are extracted from the hydrolyzed urine with ether, and the extract is washed with 10% NaOH to remove phenolic compounds (estrogens). The extract then is washed with distilled water to remove the excess base.

An aliquot of the ether extract is evaporated to dryness. Color development is based on the Zimmerman color reaction (a reaction between a CH_2CO grouping and m-dinitrobenzene in the presence of a strong alkali). The reaction involves the active methylene group in the 16 position adjacent to the keto group in position 17.

A reagent blank is run (in triplicate) to correct for any color produced by KOH and m-dinitrobenzene, and a urine blank is run to correct for any color present in the specimen. Tests, standards, and blanks are read at 520 nm. against a 75% alcohol blank.

SPECIMEN: 20 ml. of a well mixed, 24 hour urine collection. No preservative is required. Urine specimens containing preservatives for other purposes may be used. Store <u>duplicate</u> aliquots frozen.

REAGENTS:

1. <u>HCl</u>, Concentrated, reagent grade.

2. <u>Ether</u>, anhydrous, J. T. Baker

3. <u>Ethyl Alcohol</u>, absolute, reagent grade (Gold Shield).

4. <u>NaOH</u>, 10%
   An aqueous solution of reagent grade NaOH. Titration is not necessary.

5. <u>Ethyl Alcohol</u>, 75%
   This is used as a diluent and is made from 95% EtOH.

17-Ketosteroids (Continued):

6. KOH, 5.0 N.
Titrate with 1.0 N. HCl using methyl red as an indicator. Store in a polyethylene bottle in the refrigerator. (Test for purity by adding 10.4 ml. 75% EtOH to 0.2 ml. of alkali solution. The mixture should give no turbidity.)

7. m-dinitrobenzene (DNB), 2.0%
Purchased from Sigma Chemical Company, Grade V. Dissolve 0.5 gm. in 25 ml. of absolute alcohol (Gold Shield). Keep in a brown bottle and at constant temperature. Discard after 2 weeks.

8. Stock Standard Dehydroisoandrosterone.
28.62 mg. of dehydroisoandrosterone acetate is dissolved in Gold Shield alcohol and stored in the deep freeze. 28.62 mg./ 25 ml. (If dehydroisoandrosterone is used = 25 mg./25 ml. EtOH.)

9. Working Standard
Make Working Standards containing 20, 40, 60, and 80 micrograms per 0.5 ml. in Gold Shield alcohol. Refrigerate, stable for 6 months.

PROCEDURE:

1. Hydrolysis:
Add 6.0 ml. of concentrated HCl to each 20 ml. aliquot of urine in round bottom tubes, stopper, mix, and place in 80°C. water bath for 10 minutes. Cool in tap water.

2. Extraction:

   A. In 50 ml. graduated cylinder, measure 30 ml. aliquot of anhydrous ether. Transfer to separatory funnel. Add urine aliquot and rinse tube with 4.0 ml. deionized water. Extract with a 30 second shake. (With high protein concentration gel formation occurs. Add an additional 30 ml. ether and mix. Sometimes this must be centrifuged to separate.)

   B. Discard urine. Wash ether extract with 10 ml. 10% NaOH and shake 10 seconds.

   C. Discard NaOH, wash extract with 10 ml. distilled water, shake for 10 seconds and discard water.

   D. Remove extract into test tube and pipette from it two 5.0 ml. aliquots into 2 tubes, labelled T and B.

17-Ketosteroids (Continued):

E. Evaporate extract in a 45 - 50°C. water bath with air.

F. Pipette 0.5 ml. of each Standard and evaporate to dryness. Do in duplicate.

3. Color Development:

Reagent	Alcohol Blank	Reagent Blank	Standard	Urine Test	Urine Blank
Standards					
ABs. EtOH	0.8	0.4	0.4	0.4	0.8
2.0% DNB		0.4	0.4	0.4	
5.0 N. KOH	0.6	0.6	0.6	0.6	0.6

The Reagent Blanks and Standards are run in duplicate. Shake all tubes gently and place in the water bath at 18 - 20°C. Cover and allow to stand 90 minutes in the dark. Add 10 ml. of 75% ethyl alcohol, mix and read at 520 nm. in a 19 x 105 cuvette.

4. CALCULATIONS:

A. Use the optical density chart to convert % T. readings to O. D.

B. Average the 3 reagent blanks and the 3 standards.

C. Standard O. D. = average of 3 Standards - average of 3 Blanks.

D. Unknown O. D. = O. D. reading of the test - O. D. Urine Blank - average of blanks.

E. Concentration of Unknown =

$$\frac{\text{O.D. Unk.}}{\text{O.D. Std.}} \times \frac{\text{Conc. Std.}}{1} \times \frac{1}{\text{Vol. Urine used}} \times \frac{\text{Total ether extract}}{\text{ether}}$$

$$\text{Conc. Unk.} = \text{O.D. Unk.} \times \frac{30}{1} \times \frac{1}{20} \times \frac{30}{5}$$

$$\frac{9}{\text{O.D. Std.}} = K$$

K x O.D. Unk. = mg./L.

Mg./L. x Vol./24 hours = mg./24 hours.

17-Ketosteroids (Continued):

NORMAL VALUES: Male: 10 - 22 mg./24 hours
Female: 5 - 18 mg./24 hours

REFERENCES:

1. Drekter, et. al.: J. Clin. Endocrin., 7:795, 1947.

2. Sunderman and Sunderman: LIPIDS AND THE STEROID HORMONES IN CLINICAL MEDICINE, p. 158, 1960.

CLINICAL INTERPRETATION:

The urinary determination of 17-ketosteroids must be thought of primarily in its approximate correlation with the weakly androgenic secretions of the adrenal cortex and not as a measure of overall adrenal cortical activity. As a measure of androgenic activity, the 17-ketosteroids are only a rough estimate, since the individual compounds have widely differing androgenic potencies. Fractionation of the 17-ketosteroids may permit separation of the 11-hydroxy derivatives of adrenal origin which may reveal an excess production in the adrenal rather than in the testes or a masculinizing tumor of the ovary, if a prominent rise in dehydroepiandrosterone should suggest the presence of a carcinoma of the adrenal cortex.

LACTATE DEHYDROGENASE
(Wacker Method)

PRINCIPLE: LDH catalyzes the following reaction:

$$\text{Lactate} + NAD^+ \rightleftharpoons \text{Pyruvate} + NADH + H^+$$

The reduction of NAD^+ proceeds at the same rate as the oxidation of of lactate and in equimolar amounts. The rate at which NADH is formed can be determined by the increase in absorbance at 340 nm. This is the forward reaction according to Wacker utilizing lactate as the substrate at an alkaline pH.

SPECIMEN: 0.050 ml. (50 microliters) of hemolysis-free serum (or body fluid) is required. Serum should be separated from cells soon after clotting takes place. Specimens for LDH may be stored at room temperature or 4°C. for several days.

REAGENTS AND EQUIPMENT:

1. Working Substrate
   Biochemica Test Combination LDH-L 10 Test reagents are used.

   Contents of 10 Test System LDH-L vial:
   a). 150 mg. NAD/vial (7.10 mM after reconstitution)
   b). Stabilized 0.05 M. pyrophosphate buffer, pH 8.6; 0.045 M. L-Lactate.

2. Spectrophotometer

   A. Gilford 222, 340 nm.
   This instrument is used when analyzing four samples at a time. Temperature control is by means of a 30°C. circulatin water bath. Rate of reaction is determined from the strip chart recorder set for 0.200 A. full scale and run at one inch per minute.

   B. Gilford 300 N, 340 nm.
   This instrument is used when analyzing one sample at a time. The change in absorbance per minute is read from the Data Lister print-out. The instrument is used with the thermocuvette set at 30°C.

3. Bailey Microdilutor
   The sample syringe is set to pick up 0.050 ml. (50 microliters), and the reagent syringe is set to dispense 1.5 ml. of substrate.

LDH (Continued):

COMMENTS ON PROCEDURE:

1. Substrate exhaustion is rate limiting. The reaction will be linear for only a few minutes. DO NOT read after 5 minutes.

2. The reaction mixture and cuvette chamber must be at temperature before taking a reading.

3. During the run, the substrate and cuvettes are kept at 30°C. in the water bath.

4. Dilute all specimens when ΔA is greater than 0.080/ minute.

PROCEDURE:

1. With a microdilutor, take up 50 microliters of specimen, and flush with 1.5 ml. of pre-incubated substrate. Mix well and return to the water bath.

2. Quickly dilute up 3 more specimens in the same manner.

3. When 4 specimens have been diluted, immediately dry cuvettes and place in Gilford 222 (normally takes about 45 seconds).

4. Quickly set the baseline of Specimen No. 1 with the slit, and No.'s 2, 3, and 4 with their Off-Set knobs. Switch to "AUTO" and scan all 4 channels.

5. Record 2 to 3 minutes of linear reaction time at 340 nm. Determine ΔA/minute from recorder chart.

6. CALCULATIONS:

$$I.U. = \frac{\Delta A}{\epsilon \times d} \times 10^6 \times \frac{TV}{SV} \times \frac{1}{Time}$$

ϵ = Molar Extinction Coefficient

ϵ of NADH at 340 nm. = 6.22×10^3 Liter/Mole x cm.

d = Diameter of light path (1.0 cm.)

TV = Total Volume: 1.55 ml.

SV = Sample Volume: 0.05 ml.

LDH (Continued):

T = Time in minutes.

10^6 converts Moles/Liter (or mMoles/ml.) into Micromoles/L. (or Millimicromoles/ml.)

1 I.U./Liter = 1 mU/ml.

THEREFORE:

$$\frac{\Delta A}{6.22 \times 10^3 \times 1} \times 10^6 \times \frac{1.55}{0.05} \times \frac{1}{1} = \text{mU/ml.}$$

When all conditions of the assay remain constant, a factor may be derived:

$$\frac{\Delta A}{1} \times \frac{1}{6.22 \times 10^3 \times 1} \times 10^6 \times \frac{1.55}{0.50} \times \frac{1}{1} = \text{mU/ml.}$$

$\Delta A \times F$ = mU/ml. at 30°C.

F = 4984

WE report results as mU/ml. at 37°C.

I. U. at 37°C. = mU/ml. at 30°C. x Temperature Conversion Factor
Temperature Conversion Factor = 2 (For this Laboratory)

F = (9968 Actual) We round off to 10,000.

NORMAL VALUES: 85 - 200 mU/ml. (For this Laboratory)

REFERENCES:

1. Wacker, W. E. C., Ulmer, D. D., and Valu, B. L.: New Eng. J. Med., 255:449, 1956.

2. Amador, E. L. D., Dorfman, And Wacker, W. E. C.: Clin. Chem., 9:391, 1963.

3. Gay, R. J., McComb, R. B., and Bowers, C. N.,Jr.: Clin. Chem., 14:740, 1968.

LDH (Continued):

CLINICAL INTERPRETATION:

Lactic dehydrogenase acts in the glycolytic cycle to catalyze the conversion of lactic and pyruvic acid. This enzyme also may catalyze the reduction of other keto acids, and it is widely distributed in the body. The isoenzymes of lactic dehydrogenase are composed of two basic units. Each of the isoenzymes contain four of the units in one of five different combinations. These five different forms of the enzyme differ in their mobility in an electrophoretic field, with the pattern of lactic dehydrogenase related to the metabolic activity of the tissue. The isoenzyme which exhibits optimal activity where there are high levels of lactic acid predominates in those cells where this intermediate tends to be present. The isoenzyme which acts on pyruvic acid is usually present in greater amounts in those tissues which have a richer supply of oxygen. Lactic dehydrogenase isoenzymes are tetramers formed from various combinations of two types of subunits.

The stabilities of LDH isoenzymes must be considered before electrophoresis. LDH_5 is heat labile in contrast to LDH_1. Thermal stability during storage is important because with storage LDH_5 disappears. LDH_1 concentration remains constant at 4°C., -20°C., and 25°C. for one month. At -20°C., there is a rapid loss of LDH_4 and LDH_5 (two days) and after 8 to 10 days LDH_2 and LDH_3 are severely decreased at 25°C. There is no decrease in LDH_2, LDH_3, LDH_4, and LDH_5 for ten days. After more storage LDH_2 and LDH_3 decrease with a marked decrease in LDH_4 and LDH_5.

The best way to store sera is at room temperature avoiding excessive heat and bacterial contamination.

Lactic dehydrogenase is an intracellular enzyme. Usually an increase in the serum level of the enzyme is present where there is cellular death and leakage of enzyme from the cell. In addition, when neoplastic cells proliferate, the serum lactic dehydrogenase will be elevated. Strenuous exercise may increase the serum lactic dehydrogenase from skeletal muscle. Furthermore, the enzyme is elevated postpartum due to muscle exertion incurred during labor.

The level of serum lactic dehydrogenase is not influenced by meals. Lactic dehydrogenase is somewhat higher in infants and children. Hemolysis will increase the serum lactic dehydrogenase because a large amount of enzyme is present within red cells. It remains stable in stored serum.

LDH (Continued):

Oxalate inhibits lactic dehydrogenase. Thus, it is advisable to utilize serum rather than plasma in determining lactic dehydrogenase. Inhibitors of lactic dehydrogenase are present in urine, and these should be removed by dialysis if lactic dehydrogenase is determined in the urine.

The causes for an increase in serum lactic dehydrogenase are:

1. Acute myocardial infarction
2. Acute leukemia
3. Pernicious anemia
4. Acute pulmonary infarction
5. Malignant neoplasms
6. Acute renal infarction
7. Hepatic disease
8. Sprue
9. Skeletal muscle necrosis
10. Shock with necrosis of various major organs

The causes for a decrease in serum lactic dehydrogenase are:

1. Clofibrate
2. Oxalate anticoagulant

As previously mentioned, lactic dehydrogenase exists as five isoenzymes. A sixth isoenzyme has been identified in testicular tissue. When the isoenzymes are electrophoretically separated, it has been demonstrated that isoenzyme one travels between albumin and alpha 1 globulin; LDH_2 travels with alpha-one globulin; LDH_3 travels with beta globulin; LDH_4 travels with the fast gamma globulin; and LDH_5 travels with the slow gamma globulin.

The total serum lactic dehydrogenase may be increased in many different conditions as listed above. In order to determine which tissue is diseased, it is recommended that the total lactic dehydrogenase be separated into its isoenzymes. Separation into isoenzymes is best accomplished by electrophoresis. However, various other methods are available to separate these isoenzymes. Lactic dehydrogenase one is heat stable while lactic dehydrogenase five is heat labile. The heat stability-lability test is performed by diluting the serum with buffer at pH 7.4 and incubating the serum for thirty minutes at 65°C. If lactic dehydrogenase persists, this indicates heat stable LDH_1. Lactic dehydrogenase one reacts with the substrate for alpha-hydroxybutyrate dehydrogenase. However, the other LDH

LDH (Continued):

isoenzymes show progressive decreasing activity. Thus, alpha-hydroxybutyrate dehydrogenase is considered similar to LDH_1. Another less used procedure for the chemical fractionation of lactic dehydrogenase is based on the fact that lactic dehydrogenase one reacts best at low substrate concentrations of either lactate or pyruvate and is not inhibited by urea. In contrast, LDH_5 reacts best at high substrate concentrations and is inhibited by urea.

A review of the localization of the various isoenzymes is as follows: Lactic dehydrogenase of heart is one and two, as is reticuloendothelial tissue lactic dehydrogenase and kidney cortex. Lactic dehydrogenase of the lung and placenta is two and three. Lactic dehydrogenase of the pancreas is four and five, as is lactic dehydrogenase of the liver and skeletal muscle. When lactic dehydrogenase four and five are frozen, there is loss of activity. Repeated freezing and thawing of LDH may cause dissociation and recombination of the H and M subunits, and thus, for practical purposes storage at 4°C. is recommended.

Patients who suffer an acute myocardial infarction have an elevation of lactic dehydrogenase within the first 12 hours after the onset of the infarct. A peak is reached within 72 hours, and there is persistence of elevation for seven days. Alpha-hydroxybutyrate dehydrogenase may persist for two weeks, thus if the elevation of GOT or lactic dehydrogenase or CPK is missed and a myocardial infarct is suspected, alpha-hydroxybutyrate dehydrogenase should be determined, and this serum elevation will persist for as long as two weeks. If there is an elevation of lactic dehydrogenase, there is an indication that a myocardial infarct has occurred. Generally coronary insufficiency without infarction does not cause elevation in enzyme activity. The increase in serum lactic dehydrogenase is proportional to the size of the myocardial infarction. It may also increase to a greater extent if the patient has sustained congestive heart failure associated with the myocardial infarct. In patients who suffer an acute pulmonary infarct, an elevation of lactic dehydrogenase two and three occurs. Generally GOT and CPK are normal.

In patients with liver disease, total lactic dehydrogenase will increase with hepatocellular damage. The elevation is usually due to LDH_4 and LDH_5, especially 5. However, there are other causes for an elevation of LDH_5 such as skeletal muscle necrosis or proliferation of malignant neoplasms.

LDH (Continued):

Various hepatic lesions may result in an elevated LDH_5. These include congestive heart failure with necrosis of the centrilobular cells, acute and chronic active hepatitis, and carcinoma metastatic to the liver. Lactic dehydrogenase is increased with necrosis of skeletal muscle. The necrosis may result from trauma or due to inflammatory lesions. Usually LDH_5 is increased.

Patients with infarction of the renal cortex have an increase in serum lactic dehydrogenase. The LDH_1 and 2 fractions are usually elevated. Furthermore, there has been recent interest in the measurement of lactic dehydrogenase in the urine which may result from the presence of necrosis of the renal cortex or due to carcinoma of the kidney. Inhibitors of lactic dehydrogenase must be removed from the urine before determination. Other urinary tract conditions will cause an elevated lactic dehydrogenase in the urine. These include acute cystitis, acute pyelonephritis, and acute glomerulonephritis.

Malignant neoplasms produce lactic dehydrogenase 2,3,4, and 5. Various malignant neoplasms will present with an elevated lactic dehydrogenase in the serum. This is usually due to proliferation of neoplastic cells containing the enzymes. Furthermore, patients with leukemia such as acute or chronic granulocytic leukemia, and acute or chronic myelomonocytic leukemia may present with an elevated lactic dehydrogenase in the serum. Various lymphomas also may present with elevated serum lactic dehydrogenase. The use of LDH as a screening test for leukemia or cancer is not entirely reliable since various series differ in the incidence of elevated lactic dehydrogenase in the serum. The reported incidences vary from 40 to 90%. Various anemias, especially megaloblastic anemias such as pernicious anemia result in an elevated lactic dehydrogenase. The enzyme is produced and released from the megaloblasts. Destruction of the megaloblasts also accounts for the elevation of LDH. By utilizing the tetrazolium formazan technique and staining sections of bone marrow from patients with pernicious anemia, it has been shown that lactic dehydrogenase is present in abundant amounts in the cytoplasm of the megaloblasts. Treatment with vitamin B_{12} causes a rapid decrease in the serum lactic dehydrogenase. Patients with hemolytic anemia may also have an elevated lactic dehydrogenase since there is an abundant amount of lactic dehydrogenase in the red cell.

Only a minimal elevation of lactic dehydrogenase occurs during pregnancy. This is only found during labor and shortly after. The elevation occurs for approximately two days after delivery and then becomes normal. The elevation may be due to the increased muscular activity during labor or it may be due to necrosis of the placenta. Isoenzymes three and four may be elevated. The activity of lactic dehydrogenase may be elevated if there is an abruptio placenta. This elevation is on the basis of necrosis of the placenta. The rise in lactic dehydrogenase is also related to the presence of blood clot associated with the abruptio placenta. The highest lactic dehydrogenase levels have been found in patients who have hypofibrinogenemia induced by various abnormal pregnancy conditions. An elevation of lactic dehydrogenase has been found in patients with choriocarcinoma and hydatidform mole. Lactic dehydrogenase of the umbilical cord blood is greater than the activity in the blood of the normal adult. The activity rises during the first two days of life and returns to normal about the first week after birth. The activity in the umbilical cord blood is higher in jaundiced babies. An exchange transfusion which is utilized in erythroblastosis causes a decline in lactic dehydrogenase associated with a decrease in jaundice.

Lactic dehydrogenase isoenzymes two and three are elevated in the spinal fluid when there are destructive lesions of the central nervous system. In addition, there is an elevation of lactic dehydrogenase in the spinal fluid in Tay-Sachs disease with the level of lactic dehydrogenase in the spinal fluid reaching its highest level in the second year of the disease, then it declines to normal. In contrast, lactic dehydrogenase of the spinal fluid is normal in Niemann-Pick disease. In addition, there is elevation of lactic dehydrogenase in the spinal fluid in patients who infarction of the cerebral cortex, tuberculous leptomeningitis, convulsive disorders, and hemorrhage into the cerebral cortex.

Lactic dehydrogenase is elevated in the gastric juice of patients with pernicious anemia. It is also elevated in patients with carcinoma of the stomach. It is not elevated in the gastric juice in patients with peptic ulceration.

Lactic dehydrogenase is present in increased amounts in synovial fluid in patients who have rheumatoid arthritis. The increased lactic dehydrogenase in this condition in the synovial fluid results from the increased number of cells in the synovial fluid producing lactic dehydrogenase.

Lactic dehydrogenase is present in effusions in elevated amounts which are exudates. Recent reports indicate that determination of lactic dehydrogenase in an effusion is one of the better methods

LDH (Continued):

of differentiating a transudate from an exudate. The increased level of lactic dehydrogenase in exudates reflects the greater cell count of an exudate over that of a transudate. Thus, effusions which are exudates and are caused by a malignant neoplasm metastatic to the mesothelial surface or the presence of an inflammatory or immunological disease will be characterized by increased numbers of cells in the effusion with a resultant increased level of lactic dehydrogenase in the fluid. Approximately 25% of all transudates may give a slight rise in the lactic dehydrogenase which might be secondary to an increase in the lactic dehydrogenase in the serum. The effusion LDH/serum LDH ratio would be greater than one in a malignant effusion.

LDH ISOENZYMES BY ELECTROPHORESIS

PRINCIPLE: The isoenzymes of lactate dehydrogenase are separated by electrophoresis on agarose, thin-gel plates. Following electrophoresis, a reagent film consisting of sodium lactate and the coenzyme, nicotinamide adenine dinucleotide (NAD^+), is spread over the gel surface. According to the following reaction, NAD is reduced to NADH, which is fluorescent.

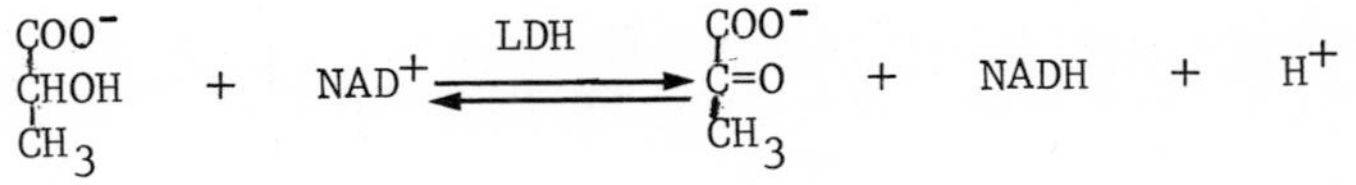

$$\underset{\text{Lactate}}{\begin{matrix} COO^- \\ | \\ CHOH \\ | \\ CH_3 \end{matrix}} + NAD^+ \underset{}{\overset{LDH}{\rightleftharpoons}} \underset{\text{Pyruvate}}{\begin{matrix} COO^- \\ | \\ C=O \\ | \\ CH_3 \end{matrix}} + NADH + H^+$$

The intensity of fluorescence of each of the five fractions is directly proportional to the concentration. The relative fluorescence of the fractions is determined by scanning the gel strips in a specially adapted door for the Turner Fluorometer.

SPECIMEN: 1.0 microliter of unhemolyzed serum. Should be stored at room temperature if unable to perform analysis on day of specimen collection. Analysis should be made within three days of specimen collection. Do not freeze samples.

EQUIPMENT AND SUPPLIES:

   Unless otherwise specified, the following equipment is obtained through Analytical Chemists, Inc., Palo Alto, California. The ACI part numbers are indicated in parentheses.

1. Cassette Electrophoresis Cell and Power Supply (#1-3300)
   1 set required.

2. Quantitative Microliter Sample Dispenser (Elevitch), (#1-4100)
   a modified 10 microliter Hamilton syringe. 1 each required.
   Disposable Sample Tips (#1-4110) (100 tips/vial)

3. Electrophoresis Buffer, Barbital, pH 8.6 (#1-5100) 0.05 M.
   with 0.035% EDTA. 2 sets required.

4. Film Cutter (#1-4300)
   1 each required.

LDH Isoenzymes (Continued):

5. Incubator/Oven (#1-3500)

6. Agarose UNIVERSAL Electrophoresis Film[R] (#1-1000-96)
12 Films (96 Determinations) per package.

7. Fluorometric Lactate Dehydrogenase Isoenzyme Substrate (#1-150(
One package is enough for 96 determinations.

8. Sta-Moist Paper (#1-3550)
100 Sheets/package, for use in incubator.

9. Turner Fluorometer (#111 - G. K. Turner Associates, 2524 Pulgas Ave. Palo Alto, California)

Prepare the fluorometer as follows:
Filter Selection:
Primary: 365 nm.
Secondary: 410 nm.

10. Turner Strip Scanning Door, Automatic (#110-525)
Designed to fit the above fluorometer.

11. Varian Chart Recorder, Model G-22, 10 mv. - Varian Associates Palo Alto, California.

12. Chart Paper for above Recorder #5A (#00-940507-01)
Varian Aerograph 2700 Mitchell Dr., Walnut Creek, California.

13. Tablet of scratch paper, 8½ x 11 inches, lined.

14. Labelling Tape (3 MM white pressure sensitive tape 1 inch width

PROCEDURE:

1. Determine Total LDH value for all specimens. Samples with a Total LDH of 200 I.U. yield optimal tracings. Dilute samples with values greater than 200 I.U. with saline or isoenzyme buffer so that the dilution falls in the optimal range. The dilution need not be exact; drop-counting with a Pasteur pipett is satisfactory. If the value is much lower than 100 I.U., a double application can be made.

LDH Isoenzymes (Continued):

2. Each film has eight (8) numbered strip positions. In this laboratory, we conventionally run all samples in duplicate (reporting averaged results) and reserve positions No. 1 and No. 8 for Control Serum. Assign strip positions to each specimen and record these together with pertinent sample identification information in a log or workbook.

3. The plastic bag enclosing each box of films contains a small amount of EDTA solution which acts as a bacteriostatic and to maintain moisture. Save the tape on the plastic bag and use it to reseal the package. Remove the box of films carefully so as not to spill the EDTA. Handle the individual films carefully, avoiding pressure on the soft (film) side. If necessary, rinse the film with a small amount of distilled water and drain and wipe gently. Reseal the box of unused films into the plastic bag.

4. Arrange the work area for application: have on hand
   a). A clean pad of lined tablet paper
   b). Appropriately diluted samples
   c). Hamilton or Elevitch syringe with a clean disposable tip for each sample.

5. Peel the film from its rigid plastic backing, which may be discarded. Place the film, agarose side up, on the lined paper. The film may be secured in place by taping to the paper at the corners. The numbered edge of the film with the application troughs is the cathode (negative) edge. (See Diagram Below)

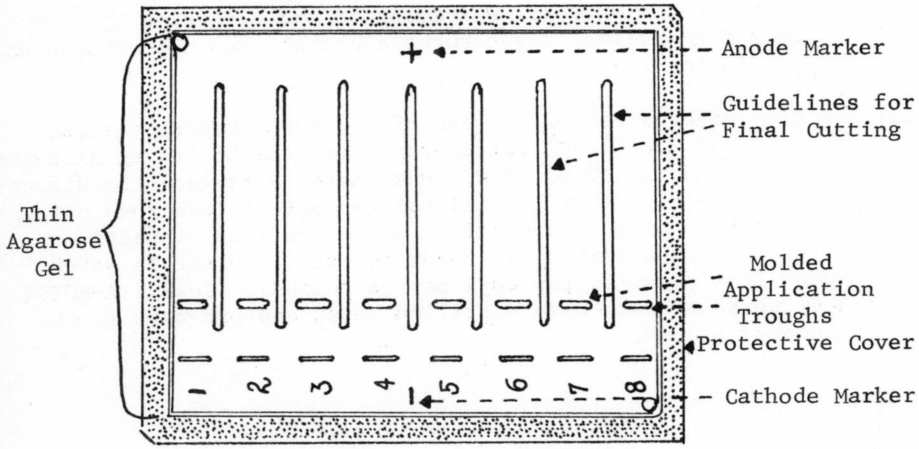

LDH Isoenzymes (Continued):

6. Apply 1.0 microliter of appropriately diluted specimen to each of the assigned troughs. Dispense the aliquot of sample in several consecutive portions, allowing each portion to soak into the gel. This will minimize the application artifact. Fill each position; if there are extra spaces, fill these with repeat samples or additional controls.

7. While waiting for the application to be completely absorbed into the gel, prepare the electrophoresis box. Connect the cassette to the power supply with the banana plugs. Measure 190 ml. of buffer with a graduated cylinder and add to the cassette, tilting the cassette to equilibrate the liquid level between the chambers. Cover with a pair of discarded film backings to protect from evaporation and contamination.

8. When the application troughs look "dry", pick up the film by the protective backing and flex into a "U" or trough shape with the protective cover on the outside and the agarose gel on the inside or concave aspect. Anode and Cathode markers will be opposite each other. Insert the flexed film into the trough holder in the electrophoresis box lid, being certain that the + and - markers on the film correspond to the + and - markings on the lid. (The lid markings are engraved on the outside of the lid. We have found it useful to clearly mark the inside of the lid with a bright wax marker crayon). Make sure that the film is snuggly inserted and that its edges are caught and held in place by the lip or "gutter" of the holder.

9. Invert the lid over the box and position gently to avoid splashing of the buffer. Placement of the lid trips the "ON" switch and the signal light turns a bright red. Electrophorese for 35 minutes.

10. Ten (10) minutes before the end of the run, prepare the substrate. Add 2.0 ml. of refrigerated lactate to frozen dispenser bottle containing 10 mg. NAD. Make sure to replace the dispenser cap firmly so that it will not pop off or leak when using the dispenser as a drop-bottle. Store the light-sensitive substrate in the dark until ready to use. (Substrate saved frozen from previous few days may be reused). Soak STA-MOIST paper to saturation with distilled water and pre-warm in the incubator at 38°C.

LDH Isoenzymes (Continued):

11. At the end of the 35 minute run, remove the lid vertically; again avoid splashing of the buffer. Place on a piece of absorbant toweling to drain. At the end of the run, the anode pH is about 8.2 and the cathode pH about 9.0 in contrast to the initial uniform pH of 8.6. Before removing film from lid, blot excess liquid from edges with a piece of clean tissue, but do not scar the agarose gel. Remove from lid maintaining the "U"-shape and release gradually to avoid spraying of condensation from plastic protector. Blot away any excess liquid, but do not touch the agarose gel.

12. Place the film, agarose side up, on a clean lined paper pad with the cathode side away from the operator. Tape the anode (near) edge to the paper, aligning the film edge with the lines on the paper which serve as a guide during application of the substrate.

13. Turn on the fluorometer and recorder to warm up at this time.

14. Using the dropper bottle in which it has been prepared, distribute 16 - 20 drops of the NAD-Lactate Substrate (about 0.6 ml.) near to and parallel to the anode edge of the agarose. Lay a plastic 5.0 ml. serological pipette along the drops, parallel to the edge and resting lightly on the agarose, so that the drops coalesce into a line along and ahead of the pipette. Check visually to see that the meniscus formed is continuous from side to side. Holding the pipette ends lightly between the fingertips, raise the pipette slightly off the agarose, but not enough to break the meniscus or separate the substrate from the agarose. Very slowly advance the pipette towards the cathode, pushing the substrate forward without touching the gel with the pipette. Watch the lateral edges to check that the meniscus is showing through the pipette. Keep the pipette parallel to the lines on the paper and advance right off the cathode end onto the paper. The process should be slow enough so that only a small amount of substrate is pushed off, and the bulk of it is absorbed by the end of the application.

15. As soon as the substrate is absorbed, place the film, agarose side up, on the wet filter paper on the incubator shelf. The shelf is removable for convenience. "Float" the film onto saturated filter paper laying down first at one edge to avoid trapping air bubbles (insulators). Uniform heating is essential. Plug in shelf for 15 minutes. During this time turn on the oven to pre-heat.

LDH Isoenzymes (Continued):

16. When incubation is complete, remove film and dry the plastic side before transferring it to the removable oven shelf. Lay the film on the shelf so that the anode-cathode axis is at right angles to a line extended from the banana plugs on the shelf. In this position the film can be held in place by elastic bands placed over the film edges and the shelf (the plastic tends to flex). Dry for 12 minutes. Remove to a dark dust-free area until ready to scan.

17. View the film under an ultra-violet light source to locate the extremes of the fluorescing patterns; mark the edges on the protective backing; trim along mark leaving a clearance of about 2.0 cm. from the pattern ends. (Careful trimming and taping will permit four pattern strips to be loaded onto the scanning drum simultaneously). Take a strip of 1-inch white paper tape the same length as the protective backing (4½ inch) and place it so that ¼ inch of its width adheres to the film (the unused application troughs make a handy guide for positioning). Place film and tape on the cutter and split the tape in half lengthwise. Position the trimmed-off strip of tape on the opposite side of the film, again allowing ¼ inch to overlap and press firmly into place. When the film is positioned gel-side up and with the application slots towards the operator, the left-most side of the film is position No. 1. It is advisable to number the positions on the labelling tape before cutting apart.

18. Line up the untaped edge of position No. 1 on the cutter plate and trim away just the protective backing without cutting into the agarose. Roll the cutter blade firmly and allow the tape to hold the film firmly in place. Release the film and tape carefully so that the agarose is not pulled away from the backing. Realign the film so that the interstrip division line is on the edge of the cutter plate and trim off the first pattern strip. Leaving the rest of the film in position on the cutter, take strip position No. 1 and apply it to the rotating drum of the fluorometer.

19. Hold the strip so that the application trough is on the right and position it on the drum so that the right-hand tape margin lines up with the marker "1A", on the vertical face of the drum. The drum width is exactly wide enough to seat the film, so it is important that the sides of the strips be cut parallel. Neither film nor tape must be allowed to project over the metal

rim of the drum channel; the clearance as the drum rotates is extremely narrow and any projections will catch and tear or crumple the film.

20. Cut the next strip, No. 2 and again keeping the application-trough end to the right, position the strip on the drum to the left of Strip No. 1, allowing the tape to overlap somewhat. In this manner four trimmed strips will fit onto the drum which normally will hold only three full-length strips.

21. The Turner Fluorometer should be allowed to warm-up for half an hour; after turning Power to "ON" and activating the "START" toggle switch, check visually for the blue light of the mercury vapor lamp. The Varian recorder is warmed up with the voltage switch (left hand toggle switch) at "STANDBY". The recorder pen is activated by the "RECORD" switch on the fluorometer. Chart speed is controlled by the right-hand toggle switch on the recorder and is left at "LOW". Chart motion is stopped and started by the left-hand toggle switch of the recorder and is used in the "HIGH" position. The scanning drum may be rotated manually in either direction by means of the knurled knob underneath the drum housing. When scanning automatically the drum moves counterclockwise until it hits the automatic stop. While scanning, the red indicator light in front of the drum is ON, when stopped, the light goes OUT. To reactivate scanning, the drum is rotated manually counterclockwise (approximately 3/4 turn of the knurled knob). In the starting position, the notation "OA" will be visible in the window on the top of the drum cover. Rotation is stopped or started by turning the "SAMPLE" toggle switch on the fluorometer to "ON".

22. After loading the strips onto the drum, replace the drum cover. Rotate the drum until the letters "OA" appear in the window. Set all switches on the fluorometer and recorder to "running" positions. The light on the drum should be "ON". Rotate the drum manually, clockwise, until the light goes out and then back again, counterclockwise until the light just comes ON again. This places the drum in the starting position. One complete rotation takes approximately ten minutes.

23. Monitor the recorder during the scan for pattern and sensitivity. The pattern should stay between 0 - 10 on the recorder scale. Generally, if the specimens have been prediluted, no adjustment of the sensitivity will be needed. Label the tracings as they appear.

LDH Isoenzymes (Continued):

24. If a pattern goes off scale, the sensitivity must be readjusted and the tracing repeated. Sensitivity is controlled by the knurled knob on the top left of the drum housing. If the setting of this knob is changed, the baseline adjustment (Blank-Adjust knob on top of fluorometer) must also be reset.

EVALUATION OF TRACINGS:

1. Draw the baseline. Application artefacts and various chemicals in medications can cause irregularities.

2. Draw a perpendicular line from each peak to the baseline; measure the height in mm. and record. The five peaks are equally based isosceles triangles, so that their comparative areas are proportional to their heights.

3. CALCULATIONS:

Calculate the per cent of each fraction by the following formula

$$\frac{\text{Height of fraction in mm.}}{\text{Total of all fractions in mm.}} \times 100 = \text{Fraction \%}$$

Average the duplicates and then multiply each percentage by the total LDH. This gives International Units per fraction. Report both the International Units and the percentage of Total.

4. INTERPRETATION:

Fractions are numbered from one-to-five, beginning at the anode. Fraction one is generally considered LDH of cardiac origin. Fractions four and five are primarily of liver origin.

NORMAL VALUES:

	Fraction 1	Fraction 2	Fraction 3	Fraction 4	Fraction 5
%:	17 - 27%	28 - 38%	19 - 27%	5 - 16%	5 - 16%
LDH:	Up to 54 I.U.	76 I.U.	54 I.U.	32 I.U.	32 I.U.

REFERENCES:

1. Methodology available through Analytical Chemist, Inc.

LDH Isoenzymes (Continued):

2. Elevitch, F.: PROGRESS IN CLINICAL PATHOLOGY, Ed. by Stefanini, Grune and Stratton, 1966.

CLINICAL INTERPRETATION:

Refer to Interpretation section under Lactate Dehydrogenase (Wacker Method) on page 237.

BLOOD LACTIC ACID

PRINCIPLE: At the present time there is no other known lactic acid in the blood other than L (+) lactic acid. It may be noted that in experiments with the enzymatic method, DL-glyceric acid was the only substance whose reaction simulated, to an appreciable degree (12.5% that of lactic acid. It was found, however, that when L (+) lactic acid and DL-glyceric acid were determined in the same test solution the glyceric acid did not influence the lactic acid values. The enzyme, lactic dehydrogenase, is an enzyme of the glycolytic cycle, catalyzing reversibly the conversion of lactate to pyruvate, utilizing B-diphosphopyridine nucleotide (B-DPN). Equilibrium is obtained for reduction of lactate to pyruvate by means of excess DPN, addition of glycine-hydrazine buffer (pH 9.0 - 9.5) with the resultant pyruvate formed as pyruvic acid hydrazine. This is a quantitative conversion and the amount of B-DPNH formed is measured on the Beckman DU at a wavelength of 340 nm. with a slit width of 0.17 nm.

SPECIMEN: Pipette into the proper number of tubes (dependent upon the number of samples to be drawn) 2.0 ml. of cold 3.5% perchloric acid. One may also use 1.0 ml. $HClO_4$ in a test tube and weigh it. If one is to determine pyruvic acid as well as lactic acid use 7% perchloric acid but when the supernatant is used for analysis make a 1:1 mixture of supernatant and perchlorate mixture. If one uses the 7% $HClO_4$ a further factor of 2 must be used in the final calculations for the determination of lactic acid. Be sure supernatant is scrupulously clear of any extraneous precipitated material. The blood can be measured directly from a syringe, but most prefer pipetting the sample. One must remember speed in measuring and precipitating of the proteins as clotting of the sample may occur. <u>The blood sample must be drawn without the use of a tourniquet</u>, or if a tourniquet is applied, release several seconds before commencing to draw the blood as an <u>uncongested vein</u> must be used or results are not reliable. Mix the blood and perchloric acid with a thin glass stirring rod and centrifuge at 3000 rpm for 5 - 10 minutes. A refrigerated centrifuge is preferred unless one is able to complete the precipitation and centrifuging within the minimal time possible as there may be some glycolysis taking place, increasing the results. 0.1 ml. of the supernatant is used for analysis. If the sample is expected to be increased, dilute the supernatant to a convenient dilution using a 1:1 mixture of perchloric acid:water. The filtrates may be stored in the refrigerator if not analyzed at once or kept frozen for longer periods of time.

Blood Lactic Acid (Continued):

REAGENTS AND EQUIPMENT:

1. Preparation of Lactic Acid Standard

A. Stock Standard
This is prepared from 85% commerically available. This is a racemic mixture so as when a standard amount is prepared, only 50% of it appears in the enzyme analysis. Dilute the 85% (approximately 10 N. solution) by taking 15 ml. of it and dilute to 50 ml. Boil gently 10 - 15 minutes, cool, and replace the volume. Titrate this against 0.1 N. NaOH which has been previously standardized against 0.1 N. Potassium acid phthalate. Dilute to exactly 2.5 N. (by chemical methods). Stopper tightly and store in the refrigerator. Stable for several months.

B. Working Standard
1). Take 0.1 ml. of Stock Standard and dilute to 25 ml. with water.

2). Take 1.0 ml. of Solution No. 1 and dilute to 10 ml. with water.

3). OR one can merely take 0.1 ml. of the Stock Standard and dilute this to 250 ml.

Use 0.1 ml. of Solution No. 2 or No. 3 for analysis. This contains 4.5 micrograms of L (+) lactic acid and an optical density of 0.104 units. (This contains 9.0 micrograms by chemical analysis.)

Working Standard must be made up fresh daily.

2. Glycine Buffer, pH 9.0; 0.5 M. Glycine and 0.4 M. Hydrazine
Glycine 37.5 gm./L.
Hydrazine (95%) 12.82 gm./L.

Dissolve glycine in 700 ml. of water in a volumetric flask, add 12.6 ml. of hydrazine in a ventilated exhaust hood with an automatic pipetter, and using a pH meter adjust to pH 9.0 with concentrated HCl or 1.0 N. NaOH and dilute to volume. Keep refrigerated. When using this reagent daily, take only the necessary amount from the stock bottle as it has been found that this may easily become contaminated. Check the pH periodically.

Blood Lactic Acid (Continued):

3. B-Diphosphopyridine Nucleotide (B-DPN), 0.027 M.
This compound varies slightly due to different amounts of water of hydration and different percent purity. However, 0.027 M. is usually between 19 and 20 mg./ml., so it is permissible to use 20 mg./ml. in making this up. The important thing is to use the same batch of B-DPN throughout a single analysis. It is convenient to make up 25 - 30 ml. of this, put in 5.0 ml. portions in small vials and freeze. After thawing once keep in the refrigerator. It will keep for a week if refrigerated.

4. Perchloric Acid, 3.5%
Dilute 5.0 ml. of 70% $HClO_4$ to 100 ml.

5. Perchlorate Mixture
Take equal volumes of 3.5% perchloric acid and water.

6. Lactic Dehydrogenase
Use a concentrated Stock Solution (Sigma Type II), which is an ammonium sulfate suspension of rabbit muscle or beef heart LDH, which is made up for use by taking 0.1 ml. of Stock Solution and adding 2.5 ml. of water. Dilute 1:20 in water of 0.01 M. saline. Keep in dark bottle in refrigerator. Stock Solution should have activity of about 17000 - 25000 Vester units/ml.

7. Reagent Blank
A reagent blank must be run with each sample. (This will usuall give a 2 - 5% error. For greater desired accuracy it may be necessary to read the reagent blank and samples against water.) It is recommended to run a "Control" sample (made up of pooled filtrates) with the unknown sample or a "recovery" which contain the unknown plus an added amount of standard.

COMMENTS ON PROCEDURE:

1. The supernatant or blood filtrate, must be clear or results are not reliable.

2. It is important to maintain a pH of 9.0 - 9.5 of the glycine-hydrazine mixture.

3. Since 85% lactic acid is the standard, it is well to remember this is a racemic lactate solution (L + D) known to be present in equal amounts so therefore only 50% is in actuality L (+) lactate.

Blood Lactic Acid (Continued):

4. The enzymatic determination of lactic acid using lactic dehydrogenase has proven to be specific and quantitatively 100%.

PROCEDURE: (Do not exceed 25 micrograms/cuvette)

Reagent	Blank	Blood Filtrate	Std.	Recovery
Glycine buffer	2.00 ml.	2.00 ml.	2.00 ml.	2.00 ml.
$HClO_4$ mixture	0.10 ml.		0.10 ml.	
Dil. of Std.			0.10 ml.	0.10 ml.
Blood Filtrate		0.10 ml.		0.10 ml.
LDH	0.03 ml.	0.03 ml.	0.03 ml.	0.03 ml.
DPN	0.20 ml.	0.20 ml.	0.20 ml.	0.20 ml.
H_2O	0.67 ml.	0.67 ml.	0.57 ml.	0.57 ml.
TOTAL	3.00 ml.	3.00 ml.	3.00 ml.	3.00 ml.

It is of great importance to observe the strict order and exact volumes as given above. Pipette directly into Beckman 10 mm. pyrex cuvettes. When adding reactants, try to avoid leaving droplets on the side of the cuvette, stir well, and incubate all samples in a water bath at 25°C. for one hour. Theoretically the timing should be made from the addition of the B-DPN, but good results have been obtained after the stirring as the values do not begin to rise until well after the hour is up. (We have followed the reaction for as long as two hours after and have not observed any significant increased values). The important point is to remember to do the procedure the same way all the time.

CALCULATIONS:

1. When using equal volumes of blood and perchlorate and using 0.1 ml. of filtrate for analysis:

$$\frac{\text{ugm./cuvette} \times 1.85}{0.1} \times 0.1 = \text{Mg\%. Lactic acid}$$

2. When using serum as in (1)

$$\frac{\text{ugm./cuvette} \times 2}{0.1} \times 0.1 = \text{Mg\%. Lactic acid}$$

NORMAL VALUES: 5 - 20 mg%.

Blood Lactic Acid (Continued):

REFERENCES:

1. Lundholm, L., Mohme-Lundholm, E., and Svedmyr, N.: "Comparative Investigation of Methods for Determination of Lactic Acid in Blood and in Tissue Extracts", Scandinav. J. Clin. & Lab. Investigation, 15:311 - 316, 1963.

2. Olson, G. F.: "Optimal Conditions for the Enzymatic Determination of L-Lactic Acid", Clin. Chem., 8:1, 1962.

3. Horn, H. D. and Bruns, F. H.: "Quantitative Bestimung von L (+) Milchsäure mit Milchsäuredehydrogenase", Biochem. Biophys. Acta (Amst), 21:378, 1956.

4. Scholz, R., Schmitz, H. and Bücher, T., Lampen, J. O.: "Über die Wirkung von Nystatin auf Bäckerhefe", Biochem. Z., 331:71, 1959.

5. Pfleiderer, G. & Dose, K.: "Eine Enzymatische Bestimmung der L (-) Milchsäure mit Milchsäuredehydrase", Biochem. Z., 326:436, 1955.

6. Heb, B.: "Über eine kinetisch-enzymatische Bestimmung der L (+) Milchsäure im Menschlichen Serum und anderen Biologischen Flüssigkeiten", Biochem. Z., 328:110, 1956.

7. Hohorst, H. J.: "Enzymatische Bestimmung von L (+) Milchsäure", Biochem. Z., 328:509, 1956.

8. Barker, S. B. & Summerson, W. H.: "The Colorimetric Determination of Lactic Acid Biological Material", J. Biol. Chem. 138:535, 1941.

CLINICAL INTERPRETATION:

The small amounts of lactate and pyruvate in the blood are derived from glycolytic metabolism from skeletal muscle and erythrocytes. Anaerobic glycolysis will markedly increase the blood lactate and cause some increase in pyruvate especially with prolonged muscle exercise. Lactate is metabolized by the liver to glycogen and glucose. The main causes for increase in blood lactate and pyruvate is anoxia. Conditions responsible for severe anoxia are shock, pneumonia, and congestive heart failure. Lactic acidosis may also occur in renal failure and leukemia. Large numbers of malignant

Blood Lactic Acid (Continued):

leukemic cells which are rapidly proliferating produce lactic acid.

Beriberi is characterized by a thiamine deficiency. Thiamine is needed for the transformation of pyruvate to acetate. Thus, thiamine deficiency results in an increased pyruvate and lactate.

Diabetic ketoacidosis is associated with increased blood lactate and pyruvate especially after administration of glucose and insulin.

Finally an increase in blood lactate and pyruvate is found in patients with severe liver disease or with acute hepatic necrosis, due to lack of metabolism of lactate and pyruvate by a poorly functioning liver.

LEUCINE AMINOPEPTIDASE (LAP)
(Goldbarg and Rutenburg: Modified by Sigma Chemical Company)

PRINCIPLE: LAP is an enzyme which catalyzes the following reaction

$$\text{1-Leucyl-}\beta\text{-Naphthylamide} + H_2O \xrightarrow{LAP} \text{Leucine} + \beta\text{-Naphthylamine}$$

$$\beta\text{-Naphthylamine} + NaNO_2 + \text{Dye Base} \longrightarrow \text{Blue Dye}$$

The absorbance of the dye produced is proportional to the amount of enzyme present.

SPECIMEN: 0.1 ml. serum added to 4.9 ml. H_2O for a 1:50 dilution. The serum may be stored up to 7 days at 4^oC. without significant loss of activity. Serum bilirubin up to 20 mg%. and serum hemoglobin up to 0.2% do not affect LAP activity.

REAGENTS:

(Sigma Chemical Company)

1. <u>LAP substrate</u>, Stock #251-1
   1-leucyl-β-Napthylamide HCl. Store at 4^oC.

2. <u>Sodium Nitrite</u>, Stock #251-4 (2.0 mg. Tablets)
   Each tablet is dissolved in 2.0 ml. H_2O to make a 0.1% solution. Prepare fresh daily.

3. <u>Ammonium Sulfamate</u>, Stock #251-3
   0.5% Solution. Store at 4^oC.

4. <u>N-(1-Napthyl)-Ethylenediamine-2-HCl</u>, Stock #251-5
   Pre-weighed bottle. 0.05% Alcoholic Solution. To the pre-weighed bottle add 110 ml. Ethyl Alcohol. Store at 4^oC. Discard when solution becomes cloudy.

5. <u>TCA</u>, 25%

6. <u>TCA</u>, 13.3%
   For calibration curve only. Add 9.0 ml. H_2O to 10 ml. of 25% TCA.

7. <u>LAP Calibration Standard</u>, Stock #251-10
   A Standard Solution of β-Napthylamine for calibration.

LAP, Goldbarg and Rutenburg Method (Continued):

PROCEDURE - CALIBRATION CURVE:

1. Preparation of Calibration Curve

Tube No.	Ml. of Stock No. 251-10	Ml. of 13.3% TCA	G-R Units LAP/ml. of 1:50 serum
1	0	1.0	0
2	0.1	0.9	108
3	0.2	0.8	216
4	0.4	0.6	432
5	0.6	0.4	648
6	0.8	0.2	864

2. Add 1.0 ml. of 0.1% sodium nitrite to each tube, mix, and allow to stand for exactly 3 minutes at room temperature.

3. Add 1.0 ml. of ammonium sulfamate to each tube, mix, and allow to stand for exactly 2 minutes at room temperature.

4. Add 2.0 ml. of alcoholic solution Stock No. 251-5 1-Naphthyl - Ethylenediamine to each tube, mix, and allow to stand for 30 minutes at room temperature.

5. Transfer to 1.0 cm. square cuvettes and read A on the Gilford 222, using water as the reference solution at 580 nm. Plot the calibration curve on graph paper.

COMMENTS ON PROCEDURE:

After a 60 minute incubation of serum and substrate at 37°C., hydrolysis is terminated by protein precipitation with TCA. Diazotization of β-naphthylamine is accomplished by the addition of $NaNO_2$. Excess $NaNO_2$ is decomposed by the addition of ammonium sulfamate. The addition of N-(1-Naphthyl) ethylenediamine dihydrochloride produces a blue azo-dye which is proportional to the amount of β-naphthylamine liberated. This is measured spectrophotometrically at 580 nm.

PROCEDURE:

1. Dilute serum 1:50 (0.1 ml. + 4.9 ml. deionized water)
2. Prepare three tubes as follows:

	Reagent Blank	Serum Blank	Test
Deionized Water	1.0 ml.	1.0 ml.	—
Substrate 251-1	1.0 ml.	—	1.0 ml.
Diluted Serum	—	1.0 ml.	1.0 ml.

LAP, Goldbarg and Rutenburg Method (Continued):

3. Incubate at 37°C. for exactly 1 hour. Stop reaction by adding 1.0 ml. 25% TCA. Invert with parafilm. Let stand at room temperature for 10 minutes and then centrifuge.

4. Transfer 1.0 ml. supernatant into another series of test tubes. Add the following reagents at exact time increments:

   A. 1.0 ml. $NaNO_2$. Mix by lateral tapping. Wait 3 minutes.

   B. Add 1.0 ml. ammonium sulfamate. Mix. Wait 2 minutes.

   C. Add 2.0 ml. N-(1-Naphthyl)-Ethylenediamine Dihydrochloride. Mix by inversion with parafilm. Let color develop for 30 minutes.

5. Read A against deionized water in the Gilford 222 at 580 nm. using 1.0 cm. square cuvettes.

6. CALCULATIONS:

   Subtract the A of the Reagent Blank plus the A of the Serum Blank from the A of the Test. Determine the corrected LAP G-R units from the Calibration Curve.

NORMAL VALUES:		70 - 200 G-R units
	Borderline:	200 - 250 G-R units
	Elevated:	Over 250 G-R units

REFERENCES:

1. Goldbarg, Julius A. and Rutenburg, Alexander: Cancer, 11:283, March-April, 1958.

2. Sigma Technical Bulletins No. 250 and 251.

3. Bergmeyer: METHODS OF ENZYMATIC ANALYSIS, Academic Press, 1965.

CLINICAL INTERPRETATION:

Amino peptidases hydrolyze amino acids containing alpha-amino groups. The highest concentration of leucine amino peptidase is usually found in the pancreas and liver.

LAP, Goldbarg and Rutenburg Method (Continued):

NORMAL RANGE: Newborn 1 - 3 months: 40 G-R units.
Adults: 200 G-R units.

Increased levels are found in:

1. Carcinoma of the head of the pancreas
2. Diffuse carcinoma, lymphoma, and granulomata involving liver
3. Choledocholithiasis
4. Acute pancreatitis, causing obstruction of common bile duct
5. Pregnancy

The measurement of serum leucine amino peptidase activity was introduced in the late 1950's by Rutenburg, et. al., who utilized the substrate 2-leucyl-beta-naphylamide. The enzyme is found in renal tubules, mucosa of small intestine and colon, stomach, bile duct epithelium and pancreatic acini. Originally the determination for leucine amino peptidase was proposed as a laboratory test for the detection of pancreatic carcinoma. However, since introduction of the test, it was subsequently observed that only patients with carcinoma of the head of the pancreas caused an elevation of this enzyme in the serum. Furthermore, it was also observed that any condition causing biliary obstruction, either intra or extra-hepatic would cause an elevated serum leucine amino peptidase level. Thus, metastatic carcinoma to the liver would elevate the serum leucine amino peptidase. Since leucine amino peptidase is not affected by bone disease, this enzyme determination is a useful one to employ when there is an elevated serum alkaline phosphatase, and if one is not certain whether bone or liver disease is present.

The determination of this enzyme has a special place in Pediatric Clinical Chemistry for the differentiation of neonatal hepatitis and biliary atresia. In this clinical situation, this determination ranks with I^{131} labeled Rose Bengal Test as a diagnostic test to differentiate biliary atresia from neonatal hepatitis presenting with jaundice. In patients with neonatal hepatitis, the values for leucine amino peptidase are generally lower than 200 G-R units. However, in patients who have biliary atresia, the levels are greater than 500 units. Leucine amino peptidase will at times be elevated before an elevation of alkaline phosphatase is observed in patients suffering from obstructive jaundice. It may be elevated without a simultaneous elevation of alkaline phosphatase in non-jaundiced individuals. In this situation, it strongly suggests the presence of a carcinoma of the head of the pancreas. Patients with carcinoma of the body or the tail of the pancreas usually do not demonstrate any elevation in the serum of this enzyme.

LAP, Goldbarg and Rutenburg Method (Continued):

Two amino peptidases may be elevated in the serum in patients who are pregnant. Leucine amino peptidase and cystine amino peptidase, which is also known as oxytocinase, have been found to be elevated especially in the latter phases of pregnancy. An increase in serum leucine amino peptidase activity during the course of pregnancy apparently is a summation of the activity of both cystine amino and leucine amino peptidase and can only be differentiated by the employment of specific substrates for these two enzymes. Two forms of cystine amino peptidase exist and they have been differentiated by starch gel electrophoresis. They have been designated as cystine amino peptidase 1 and cystine amino peptidase 2. Cystine amino peptidase 1 has its greatest activity early in pregnancy, but decline as the pregnancy progresses; in contrast to the activity of cystine amino peptidase 2 which has its greatest activity during the latter part of pregnancy in the third trimester.

The increasing activity of cystine amino peptidase throughout pregnancy has caused speculation that this enzyme is related to the quiescence of the uterus during pregnancy. Hilton has speculated that a decrease of cystine amino peptidase prevents the inactivation of oxytocin which thereby initiates labor. This concept is in keeping with an investigation where he demonstrated a rapid decline of cystine amino peptidase at the onset of labor. Cystine amino peptida is absent in fetal serum.

The chief value of determining LAP is its ability to discriminate between hepatobiliary tract and other diseases. LAP elevations are limited to hepatobiliary disease and pregnancy. Thus, LAP has advantages over BSP, GOT, GPT; the flocculation tests and alkaline phosphatase.

The determination of the isoenzymes of leucine amino peptidase may be helpful in diagnosis of biliary atresia. In neonatal giant cell hepatitis, electrophoresis gives one zone of LAP activity corresponding to the post-albumin zone, while in contrast in biliary atresi the electrophoresis gives two zones; an intensified zone as in giant cell hepatitis, which extends to the alpha globulin zone, and a second band between the $alpha_2$ and globulin zones.

LEUCINE AMINO PEPTIDASE (LAP)
(Kinetic Method After Method of Nagel)

PRINCIPLE: The ability of LAP to split the substrate leucine-p-nitroanilide leucine and p-nitroaniline is utilized. With the LAP activity being proportional to the intensity of the yellow color of p-nitroanilien.

The absorbance of p-nitroaniline is very high at 405 nm. (9.9×10^3 moles/Liter x cm.), while the substrate has minimum absorbance at this wavelength, making possible the direct kinetic measurement of LAP.

SPECIMEN: 0.1 ml. of serum. May be stored at 4°C. up to 5 days without significant loss of activity.

REAGENTS: (Biochemica Test Combination Kit, TC-LAP, Cat. No. 15952)

1. <u>Stock Buffer</u> (Bottle No. 1)
   0.05 M. phosphate buffer, pH 7.2.

2. <u>L-leucyl-p-nitroanalide Substrate</u> (Bottle No. 2)
   0.25 M. leucine-p-nitroanalide

3. <u>Working Buffer</u>
   Dissolve contents of Bottle No. 1 in 100 ml. of redistilled water. Stable for 1 year at 4°C.

4. <u>Working Substrate</u>
   Dissolve contents of Bottle No. 2 in 3.0 ml. of methanol. Stable for 6 months at 4°C. in a brown bottle.

5. <u>Buffered Substrate</u>
   Add 30 parts of Solution in Step 3 to 1 part of Step 4, should be discarded after 1 day's use. Very sensitive to light.

   Solution in Step 4 can easily be made by dissolving 125.6 mg. of anhydrous leucine-p-nitroanalide (mol. wt. 251.3) in 20 ml. of methanol. Enough for approximately 200 tests.

PROCEDURE:

1. Dispense 3.0 ml. buffered substrate into 10 mm. square cuvettes. Allow to equilibrate to 30°C. in a water bath for 5 minutes.

LAP, Kinetic Method (Continued):

2. Add 100 microliters of serum. Mix by inversion. Place in a 30°C. thermostated cuvette well. Instrument should be set at 405 nm.

3. Record the reaction until at least two consecutive minutes are linear, using either the Beckman DBG with recorder or the Gilford 300 N with digital print-out.

   Method linear up to 100 mU/ml. if greater, dilute 1:10 with saline. Multiply results x 10.

4. CALCULATION:

   Calculations are based on mean ΔA/minute.

$$\frac{\Delta A}{\epsilon \times d} \times 10^6 \times \frac{TV}{SV} \times \frac{1}{Time} = \text{I.U./Liter or mU/ml.}$$

ϵ of p-nitroaniline at 405 nm. = 9.9×10^3 Liters/Mole x cm.

$$\Delta A = \frac{1}{9.9 \times 10^3 \times 1} \times 10^6 \times \frac{3.1}{0.1} \times \frac{1}{1}$$

$\Delta A \times F$ = mU/ml.
F = 3131

NORMAL VALUES: 15 - 33 mU/ml. at 30°C.

REFERENCES:

1. Nagel, W., Willig, F. and Schmidt, F. H.: Klin. Wachs., 42:447, 1964.

2. Szasz, G.: Am. Jour. Clin. Path., 47:607 - 613, May, 1967.

CLINICAL INTERPRETATION:

Refer to Interpretation section under LAP Procedure (Goldbarg and Rutenburg Method) on page 260.

SERUM LIPASE DETERMINATION
(Turbidimetric)

PRINCIPLE: The serum lipases are enzymes which hydrolyze long chain fatty acid esters of glycerin. The methodology measures enzyme concentration via its activity upon a nearly pure triglyceride emulsion. The amount of turbidity emulsion is made with purified olive oil and utilizing a natural emulsifying agent in the bile salt, sodium deoxycholate. Olive oil is reported to be the most specific substrate for "pancreatic" lipase activity. The pH 9.1 of Tris buffer, as well as the concentration of the bile salt present, is optimal for the activity of "pancreatic" lipase, but not for normal serum lipase. The original method indicates a reaction temperature of 38°C., however, 37°C. may be used with temperature correction.

SPECIMEN: 0.2 ml. of unhemolyzed serum. Hemolysis lowers the results. Plasma from calcium-binding anticoagulants should not be used; EDTA is inhibitory to lipase of pancreatic origin; the other anticoagulants are uncertain as to effect.

REAGENTS:

1. 0.05 M. Tris Buffer, pH 9.1

2. Buffer Diluent
   This contains 0.35% sodium deoxycholate in the above buffer; re-pH, after dissolving the bile salt, to 9.1 at 38°C. if being made from dry reagents. Buffer diluent also available as Harleco Item No. 64196-B. Good for one month or until pH varies by 0.3 units of stated value.

3. Purified Olive Oil, 1% Solution (Alcoholic)
   Available as Harleco Item No. 64196-A

4. 0.04% Triglyceride Emulsion
   Measure 25 ml. of buffer diluent into a wide-mouth flask. While mixing the buffer on a magnetic mixer, add 1.0 ml. of the 1% Olive oil dropwise. The emulsion is usable as long as the initial absorbance has not decreased by more than 10% of its original value. Use for one day only.

COMMENTS ON PROCEDURE:

1. The original methodology recommends making the turbidity determination at 400 nm. However, the specific absorbance peak must be determined for the instrument being used. (Beckman instruments use 450 nm.)

Serum Lipase Determination (Continued):

2. In a triglyceride emulsion undergoing the action of lipase, the rate of removal of triglyceride will decrease as the reaction proceeds. As fatty acids are individually removed from triglyceride, the enzyme may also act on di-and monoglyceride in the presence of the inhibitory effect of fatty acids. For this reason, the reaction is a non-linear one, and further dilutions must be made on sera of high activity.

3. The 1:5 serum dilution, reaction temperature, pH, and deoxycholate concentration should significantly diminish normal serum lipase activity, making the method more sensitive to that lipase present in pancreatitis.

4. The original methodology utilizes "lipase units" derived from the absorbance change (A) multiplied by 1000. The Harleco method utilizes a factor (34 or 35 depending upon reaction temperature) which attempts to correlate the turbidimetric procedure values with those of the Cherry-Crandall Method.

5. Stability of substrate (1% alcoholic olive oil) is a problem.

PROCEDURE:

1. Dilute all serum specimens (1:5) by delivering 0.2 ml. of serum into 0.8 ml. of the buffer diluent.

2. Pipette 4.0 ml. of triglyceride substrate (0.04%) into sufficient tubes for reagent blank, sera and controls. Bring all tubes to temperature for five minutes in a 37°C. water bath (or a 38°C. water bath if available).

3. Remove the reagent blank from the water bath, mix <u>gently</u> by inversion and determine the absorbance against <u>buffer</u> diluent at 450 nm. DO NOT SHAKE WHEN MIXING. Replace tube in water bath.

4. Remove a tube containing substrate from the water bath, deliver 0.2 ml. of buffered serum into it and mix gently by inversion. DO NOT SHAKE THE TUBE. Determine the absorbance <u>immediately</u> against buffer diluent and return tube to water bath. Note time.

5. At timed intervals, remove the other tubes from the water bath and repeat Step No. 4.

6. At exactly 20 minutes incubation time, mix and determine the absorbance value on the first serum specimen against buffer

Serum Lipase Determination (Continued):

diluent. Determine the absorbance values on all subsequent specimens in timed sequence.

7. Mix and determine the absorbance value for the reagent blank against buffer diluent.

8. If the absorbance of any of the tubes has dropped by more than one-third the initial absorbance, the determination must be rerun on a dilution. Prepare dilution by delivering 0.2 ml. of the original serum-buffer diluent into 0.8 ml. of buffer diluent. The final lipase units must be multiplied by 5.

9. CALCULATION:

   A. $$\frac{\text{(Absorbance of initial reading)} - \text{(Absorbance of second reading)}}{\text{Absorbance of initial reading}}$$

   B. The above factor is multiplied by 34 (at 38°C.) or 35 (at 37°C.) to get Units of Lipase activity.

NORMAL VALUES: 0.1 - 1.5 Units

Most normal sera will give no reaction with this test. A few normal sera will give up to 1 Unit of lipase activity.

NOTE: Values in the lower range correspond to Cherry-Crandall Units: Values greater than 10 Units appear elevated in relation to the Cherry-Crandall Units.

REFERENCE: Vogel, W. C., and Zieve, L.: "A Rapid and Sensitive Turbidimetric Method for Serum Lipase Based upon Differences Between the Lipases of Normal and Pancreatitis Serum", Clin. Chem., 9:168, 1963.

CLINICAL INTERPRETATION:

The purpose of determining serum lipase is in the diagnosis of acute pancreatitis. It is not as popular as amylase determination. It becomes elevated in 24 hours and may persist for 2 weeks after the onset of acute pancreatitis. Lipase is present in the lung and serum lipase may rise with fat embolism to the lung. Furthermore, duodenal ulcers which perforate into the pancreas will elevate the serum lipase.

Serum Lipase Determination (Continued):

Since the enzyme is excreted by the kidney, renal failure will result in elevated levels. Lipase should not be determined in the urine. There are anti-lipases in the urine and there is a frequent dissociation of serum and urine lipase activity.

Serum lipase is a stable enzyme and activity remains unchanged for 7 days at 25°C. or 3 weeks at refrigerator temperature. Lipase should not be determined on a hemolyzed specimen since hemoglobin is inhibitory.

TOTAL LIPIDS

PRINCIPLE: Complete extraction of lipids can be achieved with a minimal contamination by non-lipid materials in a two phase system of chloroform-methanol-water. The extract is washed with dilute sulfuric acid to remove the non-lipid reducing substances. An aliquot of the chloroform phase is evaporated to dryness and the lipid is determined gravimetrically.

SPECIMEN: 2.0 ml. of unhemolyzed plasma or serum. Fasting specimen is not required. Specimen should be removed from contact with red cells within one-half hour of collection and frozen. Stable up to three months if frozen. If a serum or plasma is very lipemic or is known to be very high in lipid content, 1.0 ml. of specimen may be used. When using 1.0 ml. of specimen, 5.0 ml. of dilute sulfuric acid must be substituted into the methodology.

REAGENTS:

1. <u>Chloroform-Methanol</u>
   2:1, volume for volume. Use distilled solvents.

2. <u>Sulfuric Acid, dilute</u>
   0.5 ml. concentrated H_2SO_4 to 1000 ml. distilled water.

PROCEDURE:

1. Measure, with a graduated cylinder, 24 ml. of chloroform-methanol solution and place it in a 40 or 50 ml. conical centrifuge tube.

2. With a volumetric pipette, blow in (with tip under the surface of the solvent mixture), 2.0 ml. of serum. Be sure to set up a control with each set of determinations. (Hyland Special Chemistry)

3. Stopper and let stand without agitation at room temperature for 5 minutes.

4. Shake vigorously for 30 seconds and let stand an additional 5 minutes.

5. Add 4.0 ml. of dilute sulfuric acid, stopper, and invert gently 10 times without shaking. (Gentle inversion is necessary to prevent the formation of a heavy emulsion.)

Total Lipids (Continued):

6. Let stand 10 minutes for separation of the phases. This is complete when methanol combines with the water and leaves 18 ml. of chloroform below. It may be necessary to invert a few more times to accomplish a good separation.

7. Centrifuge at 2,000 rpm for 20 minutes.

8. Weighing vials are carefully wiped with Kimwipes or gauze and placed in a vacuum dessicator over Tel-Talc Silica Gel dessicant for about 45 minutes, and then weighed. Note number and weight of each vial in workbook.

9. Remove the aqueous phase by suction and carefully pipette a 10 ml. portion of the chloroform phase from beneath the protein layer and transfer to a tared weighing vial.

10. Evaporate the extracts in the weighing vials to dryness under a stream of nitrogen in a water bath no higher than 60°C. The use of nitrogen prevents oxidation from occurring at this elevated temperature.

11. Wipe the vials carefully after evaporation and place back into the vacuum dessicator for 45 minutes. Then weigh and note in workbook.

12. CALCULATIONS:

Weight of vial + Extract — Weight of empty vial = Weight of Lipids.

$$\text{Total Lipids in Mg\%.} = \text{mg. (in vial)} \times \frac{100}{\text{ml. serum used}} \times \frac{\text{Total ml. } CHCl_3 \text{ Extract}}{\text{Amt. in ml. of Extract used}}$$

$$\text{Total Lipids in Mg\%.} = \text{mg. of lipid} \times \frac{100}{2} \times \frac{18}{10}$$

$$\text{Total Lipids in Mg\%.} = \text{mg. of lipid} \times 90$$

REFERENCES:

1. Sunderman & Sunderman: LIPIDS AND THE STEROID HORMONES IN CLINICAL MEDICINE, Lippincott, pg. 6 - 14, 1960.

Total Lipids (Continued):

2. Sperry, W. M. and Brand, F. C.: "The Determination of Total Lipids in Blood Serum", J. Biol. Chem., 213:69 - 76, 1955.

CLINICAL INTERPRETATION:

Triglycerides, free and esterified cholesterol, phospholipids, especially lecithin and small amounts of unesterified fatty acids, are the most important and abundant lipids in the plasma. In addition, glycolipids, certain hormones and vitamins with a lipid moiety, are also present. The substances which thus compose the blood lipids are heterogenous, have different metabolic origins and fates and should be considered individually. These lipid substances frequently are complexed to plasma proteins. The plasma lipids begin to rise within two hours after a mixed meal, reaching a maximum concentration in six to eight hours. Pregnancy is associated with an increase in total lipids. Other causes for an increase in total plasma lipids are uncontrolled diabetes mellitus, obstructive biliary tract disease, nephrosis, hypothyroidism, pancreatitis, gout, Gaucher's disease, and inherited hyperlipemia. Decreased total plasma lipids occur in acute infections, malabsorption states, hyperthyroidism, anemia, hepatic necrosis, inanition and terminal states relating to malignancy.

LITHIUM BY ATOMIC ABSORPTION

PRINCIPLE: A diluted serum sample may be analyzed by atomic absorption spectrophotometry using resonant energy at a wavelength of 6708 Angstroms. Lithium carbonate has been of value in treating manic-depressive psychoses during the acute mania episodes. Since lithium therapy has been widely used, its serum level must be monitored to safeguard against untoward side effects.

SPECIMEN: At least 0.5 ml. of serum or plasma is used for the determination. Lithium is stable in the frozen state for many months and for a week at 4°C. Blood should be drawn in the morning before patients receive their first dose of lithium.

EQUIPMENT AND REAGENTS:

1. Perkin-Elmer Model 403 Atomic Absorption Spectrophotometer
   Equipped with a three-slot burner head.

2. Stock Lithium Standard, 10 mEq./L.
   Lithium carbonate (Li_2CO_3) was chosen as the standard because it is non-hygroscopic. Dry the lithium carbonate for 4 hours at 200°C. and cool to room temperature in a dessicator. Dissolv 369.45 mg. Li_2CO_3 in 250 ml. of water and 100 ml. of 0.1 N. HCl in a one liter volumetric flask. Mix well and dilute to the mark with deionized water. This standard is stable indefinitely

3. Working Lithium Standards
   The working standards are made up by diluting definte amounts of Stock Standard to 500 ml. with deionized water.

Volume of Stock Standard	Total Volume	Concentration
5.0 ml.	500 ml.	0.5 mEq. Li./L.
10.0 ml.	500 ml.	1.0 mEq. Li./L.
15.0 ml.	500 ml.	1.5 mEq. Li./L.

   Note that these dilutions correspond to patient samples diluted 1:5.

PROCEDURE:

1. Sample and a control are prepared by diluting 0.5 ml. of specimen with 2.0 ml. of deionized water.

2. Lithium cathode lamp should be warmed up for at least 15 minutes.

Lithium by AA (Continued):

3. Instrument Settings:
   Range - VIS
   Wavelength - 335
   Slit - 4
   Fuel - Acetylene (oxidizing flame)
   Sensitivity - 0.04 micrograms/ml. lithium for 1% absorption

4. Set the machine for concentration mode. Set to zero with deionized water.

5. Run middle standard (1.0 mEq./L.) and set proper reading with Concentration Potentiometer.

6. Check linearity by running the other two standards.

7. Recheck Auto Zero with water.

8. Recheck middle standard.

9. Run control and unknowns. Report in mEq./L.

10. Instrument Shutdown

    A. Aspirate the following each at least one minute, deionized water then 0.1 N. HCl, and then deionized water again.

    B. Turn gas switches off. The flame will go out.

    C. Turn acetylene and air main value off.

    D. Turn gas switches back on, and hit the Sensor override button.

    E. When control unit gauges have dropped to zero, turn Sensor Override off, and turn gas switches off.

NORMAL VALUES: None normally present.
Therapeutic Range: 0.6 - 1.2 mEq./L.

REFERENCES:

1. STANDARD METHODS OF CLINICAL CHEMISTRY, 6:189 - 192, Academic Press, 1970.

Lithium by AA (Continued):

2. Little, B. R., Platman, S. R., and Fieve, R. R.: Clin. Chem., 14:1211 - 1217, December, 1968.

3. Perkin-Elmer Manual, September, 1968.

CLINICAL INTERPRETATION:

Lithium Carbonate is employed in the treatment of manic-depressive states. It should be carefully monitored since toxic levels may result in a state of semiconsciousness or coma. Therapeutic levels are 0.6 to 1.2 mEq./Liter. Levels above this are toxic. Lithium is transported across the placenta and may result in cerebral depression in the newborn infant. Extreme care in the use of this compound should be used when this drug is used during pregnancy.

Toxic levels of lithium may also result in edema or anasarca in patients with cardiovascular failure since the ion is handled in a similar manner as sodium.

MAGNESIUM BY ATOMIC ABSORPTION

PRINCIPLE: A simple water dilution of a blood serum sample or other body fluid may be analyzed by using the atomic absorption spectrophotometer. Magnesium atoms absorb resonant energy of a wavelength of 2852 Angstroms.

SPECIMEN: 40 microliters of non-hemolyzed serum or plasma run in duplicate is required. The serum or plasma should be separated from the cells as soon as possible and frozen if determination is to be run at a later time.

EQUIPMENT AND REAGENTS:

1. Microdilutor
   Equipped with sample syringe set to pick-up 40 microliters of specimen and flush syringe set to dispense 2.0 ml. of deionized water.

2. Perkin-Elmer Atomic Absorption Spectrophotometer Model 403
   Equipped with 3-slot burner head and magnesium cathode lamp (15 minute warm-up time). Acetylene-air fuel mixture.
   a. Range: UV
   b. Wavelength: 285
   c. Slit: 5 (3.0 mm., 20 A.)
   d. Flame: Reducing (yellow traces)

3. Stock Magnesium Standard
   100 mEq./L. Dissolve 1.216 gm. magnesium metal in 10 ml. of concentrated HCl, and dilute to 1.0 liter with deionized water.

4. Intermediate Magnesium Standard
   Make 1:100, 1:50, and 1:25 dilutions of the Stock Standard with deionized water. This will correspond to concentrations of 1.0 mEq./L., 2.0 mEq./L. and 4.0 mEq./L.

5. Working Magnesium Standards
   Dilute each of the intermediate standards 1:50 with deionized water.

COMMENTS ON PROCEDURE:

1. Unlike calcium, nagnesium analysis does not suffer significant anionic chemical interferences.

Magnesium by A.A. (Continued):

2. A reducing flame is used to reduce the amount of magnesium oxides and hydroxides which interfere with accurate analysis.

PROCEDURE:

1. Dilution
   Dilute controls and specimens in duplicate or triplicate, using autoAnalyzer cups.

2. Instrument Set-Up

   A. Turn power to "ON". Warm-up time is 15 minutes.

   B. Check beam position over burner with white card. (Alighment of lamp should be necessary only when lamp has been moved or changed.

   C. Check amperage to lamp and adjust if needed. (Amperage requirements and limits are marked on lamps. Proper amperage setting will approach upper limit as lamp ages.)

   D. Open acetylene cylinder and air valve on wall.

   Acetylene: Do not run tank below 100 lbs. pressure. Secondary gauge on tank set to 15 psi.

   Air: Wall gauge set at 60 psi.

   E. Turn GASES switch to "ON", Acetylene and Air switches up.

   F. Fuel Flow Check: 8 psi. 45 on flow gauge.

   G. Oxidant Check: 29 - 30 psi. 70 on flow gauge. (Perform fuel check first, so that accumulated acetylene will be flushed out when followed by oxidant check.)

   H. Ignite: Have aspirator tip in water!

   I. Check for "reducing flame" and adjust if needed, with acetylene. (Thin whitish band should be present above center burner cones. If white band is absent, the flame is too hot; if white bands are too high, the flame is too cool.)

   J. Check aspiration rate: It should be 5.3 ml./minute. Turn knurled knob counterclockwise to reduce, clockwise to increase. Knob is very sensitive. Turn very slowly.

Magnesium by A.A. (Continued):

K. Set proper wavelength for test. Peak energy level is as follows:

1). Set needle to peak area with GAIN.

2). Find point of maximum deflection using Fine Adjust Knob on wavelength selector.

3). Set needle to center of Red area with GAIN.

L. Check for proper SLIT, VIS or UV range.

3. <u>Run</u>

A. Aspirate water and set Automatic Zero.

B. Check reading for absence of contamination. Reset autozero.

C. Run middle standard (20 mEq./L. Magnesium). Set proper reading with "Concentration Potentiometer".

D. Check Linearity: Depress "Absorbance" button. Read standards in series. Plotted graph of absorbance vs. concentration should be linear on regular graph paper.

E. Recheck Autozero with blank. (Concentration button depressed.)

F. Recheck middle standard.

G. Run unknowns.

4. <u>Instrument Shut-down</u>

A. Aspirate water, then a few minutes of 0.1 N. HCl (for cleaning), then water again.

B. Turn GASES switch OFF. The flame will go out.

C. Turn main acetylene tank OFF and house air OFF.

D. Turn GASES switch back ON and hit the SENSOR OVERRIDE button. (This bleeds the lines to the Control unit.)

Magnesium by A.A. (Continued):

E. When Control unit gauges have dropped to zero, turn SENSOR OVERRIDE back OFF, and turn GASES switch to OFF.

NORMAL VALUES: 1.3 - 2.5 mEq./L.

REFERENCES:

1. Perkin-Elmer Manual 403 Spectrophotometer.

2. Elwell, W. T., and Gridley, J. A. F.: ATOMIC ABSORPTION SPECTROPHOTOMETRY, 2nd revised Edition, 1967.

3. Howe, Sister M. Martin: "Atomic Absorption Spectrophotometry, Theory, Instrumentation and Application", Am. J. of Medical Technology, Vol. 33, No. 2, March-April, 1967.

CLINICAL INTERPRETATION:

Refer to Interpretation section under Magnesium (Titan Yellow) Procedure on page 281.

MAGNESIUM IN SERUM
(Titan Yellow)

PRINCIPLE: The serum proteins are precipitated by means of trichloroacetic acid. An aqueous solution of Titan Yellow is added to the filtrate, and in the presence of strong base, (sodium hydroxide) forms with magnesium, a red lake called Magnesium Hydroxide-Titan Yellow Complex. Polyvinyl alcohol is used to stabilize the color lake.

SPECIMEN: 2.5 ml. of unhemolyzed serum. Should be frozen if it is not done right away.

EQUIPMENT AND REAGENTS:

1. Gilford 300 N Spectrophotometer
   Fitted with microaspiration cuvette.

2. Trichloroacetic acid, 5.0% (W/V)
   Dissolve 50 gm. of reagent grade trichloroacetic acid in deionized water and make up to 1 liter. The normality should be 0.306 N.

3. Sodium hydroxide, 2.5 N.
   Dilute 100 ml. of a 50% (W/V) solution of NaOH to 500 ml. with boiled and cooled distilled water.

4. Polyvinyl alcohol, 0.1% (W/V)
   Suspend 1.0 gm. of polyvinyl alcohol (PVA) in 40 - 50 ml. of 95% ethyl alcohol and pour the mixture into 500 ml. of swirling distilled water. Warm the solution until it is clear. Cool and bring to volume.

5. Titan Yellow Stock Solution
   Dissolve 75 mg. of Titan Yellow in 0.1% PVA and make up to 100 ml. Store in a brown bottle at room temperature.

6. Titan Yellow Working Solution
   Make a 1:10 dilution of the Stock with 0.1% PVA.

7. Magnesium Stock Standard, 20 mEq./L.
   Dissolve 243 mg. of magnesium metal turnings (reagent grade, suitable for Grignard reaction) in a liter volumetric flask with 100 ml. of deionized water and 4.0 ml. of concentrated HCl. After the vigorous reaction, add sufficient HCl to

Magnesium in Serum (Continued):

dissolve the magnesium completely. Dilute to volume and store in borosilicate bottle. Solution is stable indefinitely in the refrigerator.

8. Working Standards
The Working Standards are made up by diluting definite amounts of Stock Standard to 100 ml. with deionized water.

Volume of Stock Standard	Total Volume	Concentration mEq.
2.5 ml.	100 ml.	0.5 mEq./L.
5.0 ml.	100 ml.	1.0 mEq./L.
10.0 ml.	100 ml.	2.0 mEq./L.
15.0 ml.	100 ml.	3.0 mEq./L.

PROCEDURE:

1. Pipette 5.0 ml. of 5% Trichloroacetic acid into a 15 ml. test tube and slowly add 1.0 ml. of serum. Set up the test in duplicate if there is enough specimen.

2. Mix gently and centrifuge at high speed for 10 minutes.

3. Transfer 3.0 ml. of clear supernatant into clean test tubes.
BLANK - 0.5 ml. of deionized water plus 2.5 ml. of 5% TCA.
STANDARDS - in duplicate. 0.5 ml. of Standards plus 2.5 ml. of 5% TCA.

4. Add 2.0 ml. of Titan Yellow.

5. Add 1.0 ml. of 2.5 N. NaOH.

6. Mix thoroughly and read on the Gilford Spectrophotometer 300 N at 540 nm. against the Reagent Blank.

CALCULATIONS:
Magnesium values are calculated from the formula:

$$\text{mEq./L. of Mg.} = \frac{\text{A unk.} \times \text{C Std. (mEq./L.)}}{\text{A Std.}}$$

A unk. - Absorbance of unknown.

A Std. - Absorbance of Standard

C Std. - Concentration of Standard Magnesium solution to most nearly corresponding to the value of the unknown.

Magnesium in Serum (Continued):

Another way is to plot concentration of Standards versus absorbance on a graph paper and read values for the unknowns from the Standard Curve.

NORMAL VALUES: 1.3 - 2.5 mEq./L.

NOTES:

1. Stock Titan Yellow will keep for at least 2 months.

2. The polyvinyl alcohol is obtained from DuPont under the label "Elvanol, Grade 70-05."

3. The reaction of Magnesium with Titan Yellow is instantaneous and readings may be made right away. The lake is stable within a half-hour period.

REFERENCES:

1. Basinski, D.: STANDARD METHODS OF CLINICAL CHEMISTRY, 5:137 - 142, Academic Press, 1965.

2. Orange, M. and Rhein, H. C.: Journ. of Biol. Chem., 189:379 - 386, 1951.

CLINICAL INTERPRETATION:

The most common cause for hypermagnesemia is renal failure. Drugs or compounds containing magnesium should be utilized judiciously in patients who have defective renal function. Hypermagnesemia may result in central nervous system depression with muscle weakness.

Hypomagnesemia may result from hypoalbuminemia since magnesium is bound to albumin and alpha-globulin. The commonest etiologies for hypomagnesemia are post surgical conditions where magnesium is not added to intravenous fluids. Malnutrition, malabsorption, acute pancreatitis, alcoholism and diabetic acidosis may result in excess magnesium loss or inadequate replacement. Other rare causes are hyperthyroidism, hyperparathyroidism, and hyperaldosteronism.

Hypomagnesemia may present with clinical signs and symptoms of muscular irritability with tetany, weakness and ataxia.

URINARY MYOGLOBIN
(Qualitative)

PRINCIPLE: Myoglobin is a respiratory pigment found in muscle cells and is functionally and structurally related to hemoglobin. It consists of a single polypeptide chain and an iron-porphyrin-heme group, while hemoglobin has four such peptide chains and four heme groups. The primary function of myoglobin is oxygen storage.

The molecular weight of myoglobin (16,700) is one one-fourth that of hemoglobin and, unlike hemoglobin, myoglobin is not bound significantly by plasma proteins. These two factors account for a low renal threshhold for myoglobin and its rapid excretion in the urine.

SPECIMEN: 5.0 ml. of urine from a random collection. The specimen should be fresh. A fresh morning voiding or a voiding just after exercise is preferred.

REAGENTS:

1. <u>Ammonium Sulfate</u> $(NH_4)_2SO_4$

2. <u>Sulfosalicylic Acid</u>, 3%

COMMENTS ON PROCEDURE:

1. Urine containing myoglobin will have a reddish-tinge when fresh, which gradually turns black on standing.

2. First test the urine for protein with sulfosalicylic acid. If the pigment is precipitated, it is protein-bound and therefore not myoglobin. If the pigment persists in the supernatant it may be myoglobin or hemoglobin.

3. The heat and acetic acid test for protein does not precipitate either myoglobin or hemoglobin and should not be used in this instance.

PROCEDURE:

1. Dissolve 2.8 gm. of ammonium sulfate in 5.0 ml. of urine, mix well, and centrifuge. Compare the color of the supernatant fluid to that of the untreated urine sample. If the pigmentation persists in the supernatant, then myoglobin (or met-

Urinary Myoglobin (Continued):

myoglobin) is present. If the pigmentation has been precipitated or removed from the supernatant, then hemoglobin is present.

2. This presumptive differential test can be confirmed by spectrophotometric scan or the myoglobin identified by electrophoresis (myoglobin migrates only half the distance of hemoglobin). Immunologic identification is also reliable.

NORMAL VALUES: Negative Urine

REFERENCES:

1. Henry, R. B.: CLINICAL CHEMISTRY: PRINCIPLES AND TECHNIQUES, Hoeber, pg. 733, 1964.

2. Blondheim, S. H., Lathrop, D., and Zabrinskie: J. Lab. Clin. Med., 60:31, 1962.

3. Lehninger, A. L.: BIOCHEMISTRY, pg. 116 - 117, 1970.

4. Davidsohn, I. and Henry, J.: TODD-SANFORD: CLINICAL DIAGNOSIS BY LABORATORY METHODS, 14th Edition, pg. 68, 1969.

CLINICAL INTERPRETATION:

Myoglobinuria results from necrosis of skeletal or cardiac muscle. Patients complain of painful skeletal muscle weakness. The urine is dark and the Benzidine test is positive with the absence of RBC's in the urine. Myoglobin is toxic to renal tubules and acute tubular necrosis occurs with renal failure.

The commonest causes for myoglobin to be present in the urine are skeletal muscle necrosis. Some common etiologies are crush injury, alcoholic myopathy, extensive burns, electric shock, viral influenza myositis, and dermatomyositis.

Acute myocardial infarction may be diagnosed by the presence of cardiac muscle myoglobin in the urine.

GUAIAC FILTER PAPER TEST FOR FECAL BLOOD
(Occult Blood Test)

PRINCIPLE: Hemoglobin catalytically decomposes hydrogen peroxide with the liberation of oxygen which in an acid solution oxidizes gum guaiac, a colorless phenol, to a blue-colored derivative. In feces, hematin, the iron-containing complex of protoporphyrin catalyzes the guaiac reaction but free iron and iron-free porphyrins do not.

REAGENTS:

1. Gum Guaiac, 20%
   20 gm. guaiac dissolved in 100 ml. 95% ethanol.

2. Glacial Acetic Acid

3. Hydrogen Peroxide, 3%
   Usable as long as it effervesces when placed in contact with blood.

4. Positive Control
   0.05 ml. venous blood to 50 ml. water. Approximately 1+ reaction will occur with guaiac.

COMMENTS ON PROCEDURE:

1. Test is merely qualitative.

2. Lots of gum guaiac vary in sensitivity.

3. Alcoholic solutions of gum guaiac and 3% hydrogen peroxide are unstable and need to be checked with known positive specimen or control.

PROCEDURE:

To a thin fecal smear and a positive control on filter paper, add 2 drops each of glacial acetic acid, gum guaiac and 3% hydrogen peroxide.

NORMAL VALUES: Report as Follows:

Trace	-	Faint blue or greenish-blue appearing in one minute.
1+	-	Light blue appearing slowly.
2+	-	Clear blue appearing fairly rapidly.

Guaiac Filter Paper Test (Continued):

3^+ - Deep blue appearing almost immediately.
4^+ - Deep blue appearing immediately.

REFERENCES:

1. Post Grad. Medicine, May, 1962.

2. Ham, T.: A SYLLABUS OF LABORATORY EXAMINATIONS IN CLINICAL DIAGNOSIS, 1950.

CLINICAL INTERPRETATION:

Loss of 50 ml. of blood from the upper gastrointestinal tract causes the stool to be dark black and tarry. Bleeding from lower gastrointestinal tract will cause a red stool. Smaller amounts of blood will not cause a color change in the stool, and bleeding in the upper or lower gastrointestinal tract will remain undetected unless an Occult Blood Test is performed. The test may be applied to stool, urine or gastric juice. It is positive for hemoglobin or myoglobin. For the stool test, the patient should not eat meat for 3 days since meat contains myoglobin and hemoglobin. The reagents differ chiefly in sensitivity. Orthotolidine is 1 to 10 times more sensitive than benzidine; benzidine 10 to 1000 times more sensitive than Guiac. False negative reactions may occur if excess reducing substances such as Vitamin C are ingested.

SERUM AND URINE OSMOLALITY
(Fiske Freezing Point Osmometer)

PRINCIPLE: When a solute is dissolved in a solvent, four properties are changed: 1). The freezing point is lowered, 2). The boiling point is raised, 3). The osmotic pressure is increased, and 4). The vapor pressure is lowered. The Fiske freezing point Osmometer measures the total solute concentration in body fluids by comparing their freezing point depression against that of NaCl standards of known osmolality. A molal solution represents one mole of solute plus one kilogram of solvent (weight-to-weight); osmolality varies in a linear function with respect to the freezing point. When a liquid is cooled below its freezing point, it is super-cooled; the amount of super-cooling is the difference in temperature between freezing point of the substance and the temperature at which crystal formation is initiated. Mechanical vibration of a super-cooled solution with a uniform temperature initiates crystal formation. Heat of fusion is given off as crystal formation occurs, and the temperature rises to the freezing point of the solution. Once a stable plateau is reached, the freezing point may be measured at any empirically determined time as long as it is kept constant for both standards and unknowns.

The chart on the following page indicates a typical cooling and freezing cycle for the Fiske Osmometer:

SPECIMEN:

1. 2.0 ml. well mixed and centrifuged urine; specific gravity of original specimen should be taken.

2. 0.5 ml. unhemolyzed serum. Even a small amount of hemolysis significantly affects the accuracy of the determination. Serum should be removed from clot within 45 minutes of collection and centrifuged twice to avoid any cellular contamination.

3. Serum or urine specimens should be stored, tightly capped, at refrigerator temperatures until specimen is processed. If determination is not to be made on the day of collection, serum or urine specimens should be immediately frozen until just prior to use.

REAGENTS AND EQUIPMENT:

1. <u>100 milliosmol/L. Standard</u>
   Weigh out 0.771 gm. dried, dessicated NaCl on an analytical balance. Add this to a 250 ml. volumetric flask already

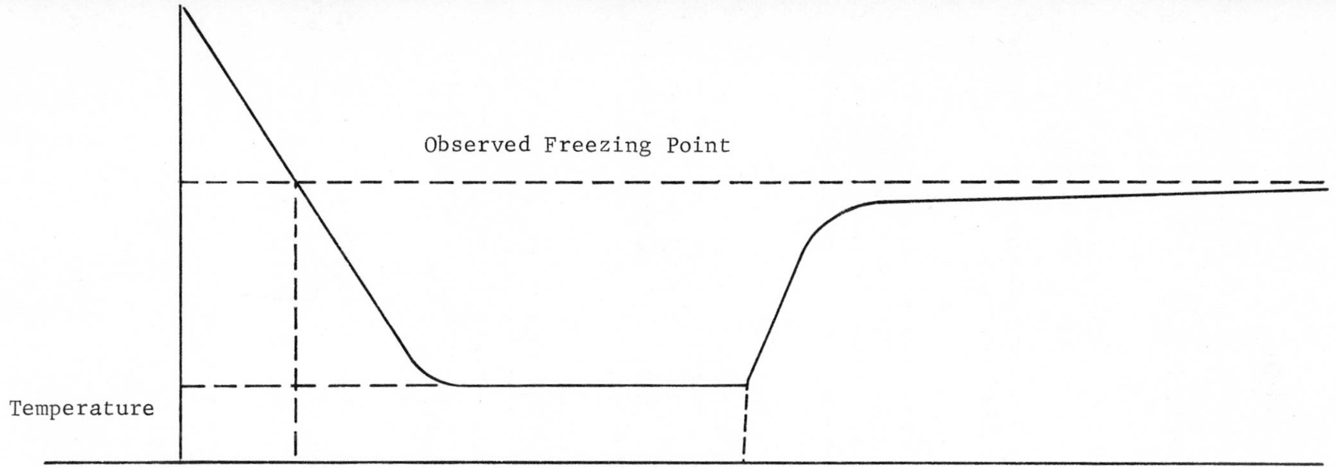

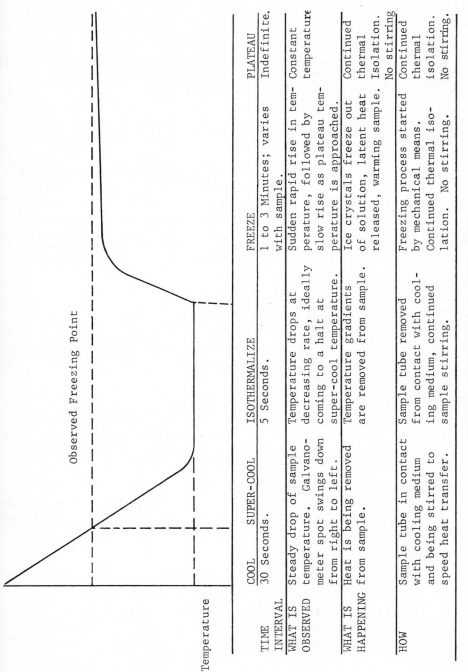

	COOL / SUPER-COOL	ISOTHERMALIZE	FREEZE	PLATEAU
TIME INTERVAL	30 Seconds.	5 Seconds.	1 to 3 Minutes; varies with sample.	Indefinite.
WHAT IS OBSERVED	Steady drop of sample temperature. Galvanometer spot swings down from right to left.	Temperature drops at decreasing rate, ideally coming to a halt at super-cool temperature.	Sudden rapid rise in temperature, followed by slow rise as plateau temperature is approached.	Constant temperature
WHAT IS HAPPENING	Heat is being removed from sample.	Temperature gradients are removed from sample.	Ice crystals freeze out of solution, latent heat released, warming sample.	Continued thermal Isolation. No stirring
HOW	Sample tube in contact with cooling medium and being stirred to speed heat transfer.	Sample tube removed from contact with cooling medium, continued sample stirring.	Freezing process started by mechanical means. Continued thermal isolation. No stirring.	Continued thermal isolation. No stirring.

TYPICAL COOLING AND FREEZING CYCLE

containing deionized water to volume. Invert flask and mix thoroughly. Store in a tightly capped reagent bottle at refrigerator temperatures. Stable indefinitely.

2. <u>300 milliosmol/L. Standard</u>
Weigh out 2.364 gm. dried, dessicated NaCl on an analytical balance. Add this to a 250 ml. volumetric flask already containing deionized water to volume. Invert flask and mix thoroughly. Store in a tightly capped polyethylene reagent bottle at refrigerator temperature. Stable indefinitely.

3. <u>500 milliosmol/L. Standard</u>
Weigh out 3.98 gm. dried, dessicated NaCl on an analytical balance. Add this to a 250 ml. volumetric flask already containing deionized water to volume. Invert flask and mix thoroughly. Store in a tightly capped polyethylene reagent bottle at refrigerator temperature. Stable indefinitely.

4. <u>750 milliosmol/L. Standard</u>
Weigh out 6.025 gm. dried, dessicated NaCl on an analytical balance. Add this to a 250 ml. volumetric flask already containing deionized water to volume. Invert flask and mix thoroughly. Store in a tightly capped polyethylene reagent bottle at refrigerator temperature. Stable indefinitely.

5. <u>1000 milliosmol/L. Standard</u>
Weigh out 8.057 gm. dried, dessicated NaCl on an analytical balance. Add this to a 250 ml. volumetric flask already containing deionized water to volume. Invert flask and mix thoroughly. Store in a tightly capped polyethylene reagent bottle at refrigerator temperature. Stable indefinitely.

6. <u>Fiske Osmometer</u>, Model G-62
The instrument may be left on continually or, if used infrequently, turned on when needed. Allow 45 - 60 minutes for temperature equilibration and instrument stabilization.

    A. <u>Thermostat Setting</u>:
    The thermostat is located on the upper left hand corner of the back of the osmometer housing. The direction which the knob is to be turned for adjustment is indicated on its face. The temperature reading is indicated in Farenheit degrees. A setting of approximately 18°F. should give the desired cooling bath temperature of -7°C.

Serum and Urine Osmolality (Continued):

B. Cooling Solution:
Prepare a 1:3 dilution of propylene glycol or standard anti-freeze for use in both the pre-cooling bath and the cooling bath. The cooling bath should be filled until the fluid level reaches the top of the textolite post in the center of the coolant pump tube. When in position, the bottom of the sample tube should be 1/8 of an inch above the surface of the bath.

C. Vibrator Setting:
The vibrator knob should be turned approximately ¼ of a complete turn from the "ON" position, when using 2.0 ml. of specimen. The knob should be set at the OFF position when 0.2 ml. of specimen is being used.

D. Instrument Calibration:
Knobs "A" and "B" adjust the span and slope of the standard curve and should not be touched unless one desires to re-calibrate the instrument. For the procedure of instrument calibration, refer to the instrument manual. Knobs 1, 2, and 3 are calibration knobs for ranges above 1000 mOsm. Refer to supplementary sheet in manual if these ranges are desired.

E. Range Extender:
This switch extends the range of sensitivity of the milli-osmol dial to the desired range. For most routine specimens a range of 0 - 1000 is adequate. The instrument must be re-calibrated with higher standards if other ranges are to be used. Consult instrument manual for standard solution preparation.

F. Temperature Probe:
The tip of the probe is extremely sensitive and must be handled gently to avoid chipping or cracking of the bead. Both the probe and the vibrator should be rinsed with de-ionized water after each specimen, and should be stored in a tube full of deionized water.

G. Sample Tubes:
2.0 ml. size may be purchased directly from Kimax, and are culture tubes, without lip, 16 x 100 mm.; matching is un-necessary. The 0.2 ml. small sample adaptor tubes must be purchased from Fiske, and are also 16 x 100 mm. When order-ing, indicate volume that the tube is to contain.

COMMENTS ON PROCEDURE:

1. The observed freezing point is directly related to the amount of super-cooling. The greater the concentration of the specimen, the more the effect of super-cooling on the freezing point. It is essential to reproducibility that the degree of super-cooling be rigidly controlled, and that standards and unknowns be super-cooled to the same degree.

2. Too rapid a rate of super-cooling may result in spontaneous freezing.

3. Use of the precooling bath shortens the time required for freezing in the main bath, but is not mandatory. This may be used for 2.0 ml. specimens but should not be used for 0.2 ml. specimens (produces too fast a rate of super-cooling).

4. The vibrator acts to eliminate temperature gradients from the center of the tube to the outer rim by mixing the contents during the freezing process. In general, the more active the vibration, the more rapid the super-cooling. Too active vibration may cause spontaneous freezing of the sample. When using 0.2 ml. of sample the vibration should be turned OFF.

5. Only deionized water should be used with this instrument, both in reagent preparation and final rinsing of glassware. Glassware must be scrupulously clean and stored in a dust-free environment. Contamination of glassware or specimens with particulate matter may result in alteration of the freezing point depression. Contamination of glassware with oil or grease will affect the super-cooling phase.

6. Standard milliosmol solutions should be poured into clean test tubes and capped until aliquoted with a pipette. Under no circumstances should a pipette be introduced into the stock standard solutions; contamination and evaporation of standards and specimens is one of the greatest sources of error in this methodology.

7. The volume of specimen or standard is critical to reproducibility; measurement should be made quantitatively and volumetrically.

8. Even a low degree of hemolysis significantly increases a serum osmolality measurement. The specimen should be removed from the clot within 45 minutes of collection, and serum should always be centrifuged twice to avoid any cellular contamination.

Serum and Urine Osmolality (Continued):

9. The microhead apparatus for 0.2 ml. specimens does not display the same degree of reproducibility as the 2.0 ml. sample tube. However, duplicate specimens should agree within 4.0 milli-osmoles/Liter.

10. When working routinely in osmolality ranges above 1000, it is helpful to adjust the cooling bath temperature to -12°C.

PROCEDURE:

1. Before initiating specimen measurements, the temperature of the cooling bath should be checked for -7°C. temperature. Range selector switch should be at 0 - 1000 milliosmols.

2. Urine Osmolality:

   A. Volumetrically measure 2.0 ml. of deionized water into the appropriate size sample tube and label as "Blank".

   B. Volumetrically measure 2.0 ml. of 100, 300, 500, 750, and 1000 milliosmol standards into appropriately labelled sample tubes.

   C. Volumetrically measure 2.0 ml. of frozen urine control into a sample tube labelled "Control".

   D. Volumetrically measure 2.0 ml. of patient's urine specimen into an appropriately labelled sample tube in duplicate. A specific gravity determination should be noted in both the Work Book and on the Requisition.

   E. Place all prepared tubes in pre-cooling bath.

   F. Fit "Blank" sample tube onto the operating head and lower into cooling bath.

   G. With vibrator positioned at ¼ of a full turn in the "ON" position, turn "operation" switch from "0" or OFF to position 1 or "COOL". Cooling pump should be circulating bath solution up and around the end of the sample tube. Galvanometer light should come on.

   H. Place the galvanometer sensitivity switch on "LOW". The galvanometer light will swing to the far right of the scale.

Serum and Urine Osmolality (Continued):

This allows one to follow the speed with which the sample is super-cooling; as the sample cools the galvanometer light swings to the left of the scale.

I. When galvanometer light reaches 20 degrees to the left of "0", turn "operation" switch to position 2 or "Super-cool". Super-cool until galvanometer light comes to rest on 30 degrees to the left of "0". It is essential that this endpoint of super-cooling be consistent. One may momentarily return operation switch to position 1 to bring temperature down to 30 degrees.

J. Switch to position 3 or "Freeze"; this activates buzzer and initiates crystal formation. Set timer for 1 minute.

K. Position galvanometer sensitivity on central "Zero Set" and adjust galvanometer hair-line on "0" if necessary using fine adjustment knob on face of galvanometer.

L. Depress galvanometer sensitivity switch and rotate milliosmol knob to zero galvanometer. Take reading at exactly 1 minute. Duplicate determinations should not vary by more than 2 - 3.0 milliosmoles.

M. Turn operation switch to "OFF" position, remove operating head from bath and disengage sample tube. Rinse vibrator and thermister probe with deionized water and blot off excess moisture with clean tissue.

N. Continue procedure from "F." above with all of the prepared sample tubes. Very elevated samples may be diluted 1:1 with deionized water with fair reproducibility; no greater dilution should be made.

O. Plot blank and standard on Linear Graph Paper with observed milliosmoles on the Y axis and actual milliosmoles on the X axis. Determine value for urine control and specimens from the graph. (If necessary, multiply this value times 2.)

3. <u>Serum Osmolality</u>:

A. Quantitatively deliver 0.2 ml. of deionized water into the bottom of a "Microhead" sample tube, and label as "Blank".

Serum and Urine Osmolality (Continued):

B. Quantitatively deliver 0.2 ml. of 100, 300, and 500 milli-osmol standards into the appropriately labelled microhead sample tubes.

C. Quantitatively deliver 0.2 ml. of frozen serum control into a microhead sample tube labelled "Control".

D. Quantitatively deliver 0.2 ml. of serum specimen into duplicate microhead sample tubes.

E. Do <u>not</u> place these tubes in a pre-cooling bath.

F. Fit "Blank" tube onto the operating head and lower into cooling bath.

G. With vibrator turned in the "OFF" position, turn operation switch from "O" or "OFF" to position 1 or "COOL". Cooling pump should be circulating and galvanometer light should come on.

H. Place the galvanometer sensitivity switch on "LOW". The galvanometer will swing to the far right of the scale, and then drift to the left as the specimen cools.

I. When the galvanometer light approaches approximately 10 degrees to the left of zero, turn operation switch to position 2 or "Super-cool". Super-cool until galvanometer hair-line comes to rest upon 30 degrees to the left of zero. This endpoint of super-cooling must be consistent. The operation switch may momentarily be returned to the "Cool" position to bring the specimen within the 30 degree range.

J. Switch to position 3 or "freeze"; this activates buzzer and initiates crystal formation.

K. Immediately place galvanometer sensitivity switch in the "Zero-Set" position and adjust the hair-line on "0" if necessary, using the fine adjustment knob on the face of the galvanometer.

L. Immediately depress galvanometer sensitivity switch to "Fine" and rotate milliosmol knob until galvanometer hair-line rests on zero. These last two steps should be performed as quickly as possible to prevent thawing of the sample. Take blank reading from milliosmol potentiometer knob.

Serum and Urine Osmolality (Continued):

M. Turn operation switch to "OFF" position, remove operating head from bath and disengage sample tube. Rinse vibrator and thermister probe with deionized water and blot off excess moisture with clean tissue.

N. Continue procedure from "F." above with all of the prepared tubes.

O. Plot blank and standards on Linear Graph Paper with observed milliosmoles on the Y axis and actual milliosmoles on the X axis. Determine value for serum control and specimens from the graph.

NORMAL VALUES:

Urine Osmolality: Average normal urines range from 500 - 800 mOsm./Kg. water. This may vary greatly depending upon the dietary intake of the patient.

Serum Osmolality: Average normal sera are around 285 - 295 mOsm./Kg. water. Again, this may vary depending upon the dietary intake.

REFERENCES:

1. Holmes, Joseph: "Measurement of Osmolality in Serum, Urine, and Other Biological Fluids by the Freezing Point Determination", ASCP, WORKSHOP MANUAL OF URINALYSIS AND RENAL FUNCTION STUDIES, (Pre-Workshop Manual), 1962.

2. Holmes, Joseph: "Measurement of Solute Concentration in Body Fluids by the Freezing Point Technique", ASCP, WORKSHOP MANUAL ON URINALYSIS AND RENAL FUNCTION STUDIES, (Technique Manual), 1962.

3. Henry, R. B.: CLINICAL CHEMISTRY: PRINCIPLES AND TECHNIQUES Hoeber, pg. 884, 1965.

CLINICAL INTERPRETATION:

Osmometry is a valuable procedure in the study of patients with fluid, electrolyte and renal problems.

Serum and Urine Osmolality (Continued):

Normal serum osmolality has a normal range between 275 and 295 mOsm. per Kg. water. The serum osmolality is proportional to electrolytes especially sodium. The osmolarity of normal urine varies widely with a range of 50 mOsm./Kg. to 1400 mOsm./Kg.

In hyponatremia, the serum sodium is low and the osmolality of serum is low indicating dilutional hyponatremia or inappropriate ADH associated with such diseases as bronchogenic carcinoma of undifferentiated type. Hypernatremia may result in an elevated serum osmolarity. This may occur in dehydration or excessive intravenous infusion of saline.

Hyperosmolar hyperglycemic nonketotic diabetic acidosis has a marked elevation in serum osmolarity and must be treated rapidly with hypotonic intravenous solutions.

The serum osmolarity may increase in acute alcoholism due to the rise in serum osmolarity induced by ethyl alcohol.

Patients with polyuria due to diabetes insipidus or psychogenic polydipsia have low urine osmolarity in the range of 50 to 200 mOsm. per Kg. The urine of a dehydrated individual has a osmolarity of approximately 800 mOsm./Kg. Acute tubular necrosis results in a urine osmolarity of approximately 200 - 300 mOsm./Kg.

When one screens for renal disease and institutes a fluid deprivation test, the urine osmolarity must reach at least 850 mOsm./Kg. The failure to achieve this degree of concentration is regarded as a reliable indication of renal damage.

As tubular function of the kidney deteriorates in chronic kidney disease, there is loss of concentrating function and low urine osmolarity. As tubular function recovers, the urine osmolarity improves and the urine osmolarity rises above 500 mOsm./Kg.

In the inappropriate ADH syndrome, the serum osmolarity is reduced while the urine osmolarity is elevated.

PANDY'S QUALITATIVE GLOBULIN TEST FOR CSF

PRINCIPLE: This is a rapid, qualitative test for increased globulin concentration in spinal fluid. This may detect an increase of actual globulin protein, or an apparent increase of globulin protein due to a decrease in albumin concentration. Hence, the test is sensitive to an alteration of the normal albumin-to-globulin ratio, resulting in an increased globulin portion. The total protein concentration does not affect the results of the test; it is possible to have a positive Pandy with a normal total protein concentration.

SPECIMEN: One drop of a cell-free or centrifuged spinal fluid specimen.

REAGENT:

Phenol Reagent, saturated, aqueous solution
Transfer 10 ml. of melted phenol into a large, clean stoppered bottle; add 90 ml. distilled water. Incubate this bottle in a 37°C. water bath for several days, shaking contents periodically during this time. Only the clear supernatant solution is used in the determination.

PROCEDURE:

Pipette 1.0 ml. of the clear supernatant solution into a clean, unscratched test tube. With a Pasteur pipette, add one drop of spinal fluid.

NORMAL VALUES:

1. Normal fluids may show a faint turbidity or no turbidity. This is reported as negative.

2. The formation of a bluish-white cloud immediately around the drop is indicative of a positive test.

REFERENCES:

1. Hepler, O.: MANUAL OF CLINICAL LABORATORY METHODS, p. 152.

2. Annino, Joseph: CLINICAL CHEMISTRY, p. 361, 1964.

CLINICAL INTERPRETATION:

Pandy's Test (Continued):

The Pandy Phenol Test is a qualitative protein test for elevation of spinal fluid protein. The increased globulins in the spinal fluid are precipitated by the phenol solution and is a rapid qualitative test to ascertain a pathologic increase in spinal fluid protein. It is a qualitative test and is not considered too accurate.

PHENYLALANINE
(Automated, Fluorometric Method)

PRINCIPLE: Phenylalanine produces a fluorescent substance when heated with ninhydrin in the presence of L-leucine-L-alanine. Whole blood obtained from newborns by heelstick is stored dried on filter paper discs. Phenylalanine is dialyzed from the diluted sample into a recipient stream of ninhydrin buffered to pH 5.8 $\pm$ 0.1 by succinate buffer. L-leucine-L-alanine is added and the reaction mixture is heated to 95°C. for 6 minutes. The reaction mixture is cooled, alkaline copper tartrate is added and the fluorescence produced by excitation at 360 nm. is measured at 485 nm. Fluorescence is directly proportional to the phenylalanine concentrat

SPECIMEN COLLECTION AND PREPARATION:

1. Equipment Needed:

   A. 20 microliter pipettes (Sahli hemoglobin pipettes may be used.)

   B. Whatman #3MM filter paper, cut to about 1 mm^2 in size and placed in auto-analyzer cups with caps.

   C. Labels.

2. Place 20 microliters whole blood obtained by heelstick on filter paper disc and dry immediately for storage.

3. Whole blood, serum, or heparinized plasma may be used. Phenylalanine in whole blood is stable if dried immediately after collection. Phenylalanine in serum or heparinized plasma is stable when frozen or dried. Specimens left moist at room temperature tend to give increased values.

EQUIPMENT AND REAGENTS:

1. Turner fluorometer with flow-cuvette for auto-analyzer. The instrument should contain the following:
   a. Primary filter 7 - 60 (360 nm.)
   b. Secondary filter: 8, 65-A (510 nm.)
   c. Range selector: 30 X
   d. Slit: None
   e. Sampler cam: 50/hour (45" sample, 35" wash)

Phenylalanine (Continued):

2. Succinate Buffer, 0.04 M., pH 5.8

Succinic acid disodium salt (M.W. 270) (Eastman Kodak #1219)		10.8 gm.
Hydrochloric acid, concentrated		0.5 ml.
Deionized water	q.s.	1000.0 ml.

Place 500 ml. deionized water in a 1 liter beaker. Set on magnetic stirrer and begin agitation. Slowly add succinic acid disodium salt. Add 0.5 ml. HCl. Check pH. Use 0.1 N. HCl or 0.1 N. NaOH to bring pH to 5.8. Transfer solution to a 1 liter volumetric flask, rinsing beaker and adding rinsings to flask also. Bring to volume and store in polyethylene in refrigerator. Add 1.0 ml. Brij-35 prior to use. Stable for about 1 week. Deterioration causes loss of sensitivity.

(72 ml. required per hour)

3. Saline Diluent, 0.9%

Sodium chloride		36 gm.
Chloroform		8 ml.
Deionized water	q.s.	4000 ml.

Add sodium chloride to a 4 liter flask. Add chloroform. Bring to volume with deionized water.

Prior to use, pour off amount needed, avoiding chloroform at bottom and add Brij-35 (0.5 ml./liter). Chloroform will damage lucite dialyzer plates if it is pumped into dialyzer.

(25 - 50 ml. required per hour)

4. Ninhydrin

Ninhydrin (#1831 Mann Research Labs)		1.3 gm.
Deionized water	q.s.	250.0 ml.

Place ninhydrin in a 250 ml. volumetric flask containing about 200 ml. deionized water. Stir with magnetic mixer until material goes into solution. This may take some time. Bring solution to mark with deionized water, mix, and transfer to amber bottle. Keep refrigerated when not in use. This solution deteriorates on standing, causing loss of sensitivity.

(36 ml. required per hour)

Phenylalanine (Continued):

5. L-leucyl-L-alanine, 80 mg%.

L-leucyl-L-alanine (#244 Mann Research Labs)	100 mg.
Deionized water	125 ml.

Store in refrigerator when not in use.

(19 ml. required per hour)

6. Copper Reagent

Sodium carbonate	16.00 gm.
Potassium sodium tartrate	0.65 gm.
Cupric sulfate ($CuSO_4 \cdot 5H_2O$)	0.60 gm.
Deionized water	q.s. 1000.00 ml.

Dissolve each of the dry reagents in separate aliquots of about 300 ml. water. Combine the sodium carbonate and potassium sodium tartrate solution in a liter volumetric flask and mix. Add the cupric sulfate solution and mix. Bring to volume with deionized water and mix. Before use, check solution for turbidity. Salts may precipitate out on standing.

(19 ml. required per hour)

7. Stock Phenylalanine Standard, 100 mg%.
Quantitatively prepare the following:

Phenylalanine (DL-phenylalanine, Eastman Kodak #894)	100 mg.
Chloroform	2 ml.
Deionized water	q.s. 100 ml.

Store in refrigerator.

8. Working Phenylalanine Standards
Prepare a 1:100 dilution of the Stock Standard. This will contain 0.01 mg./ml. From this dilution prepare the following in volumetric flasks for use as working standards:

Phenylalanine Concentration	Ml. of 1:100 Dilution of Stock	Plot as
0.02 mg%.	2.0 ml.	1 mg%.
0.04 mg%.	4.0 ml.	2 mg%.
0.06 mg%.	6.0 ml.	3 mg%.
0.08 mg%.	8.0 ml.	4 mg%.
0.10 mg%.	10.0 ml.	5 mg%.

Phenylalanine (Continued):

Bring each flask to 100 ml. with deionized water saturated with 2.0 ml. chloroform per liter.

Standard concentrations are based on the use of a 1:50 dilution of specimen for the determination. Stable if refrigerated.

COMMENTS ON PROCEDURE:

1. The reaction is carried out at pH 5.8 to minimize interference from tyrosine glutamic acid, leucine, lysine, arginine and methionine. Copper reagent decreases the interference from histidine.

2. Each new batch of filter paper must be checked for contaminating fluorescence.

3. When eluting specimens from filter paper discs, avoid shredding off fibers from the filter paper as they will clog up the sample line and channel I of the dialyzer.

4. Specimens must be run immediately after elution. If allowed to stand over 1 hour, the values will be significantly elevated.

5. Deterioration of the succinate buffer, L-leucine-L-alanine or ninhydrin will decrease the sensitivity of the reaction.

6. Avoid getting chloroform from the saline diluent or standards into the lucite dialyzer plates, as it will etch them.

PROCEDURE:

1. Just prior to run, add 1.0 ml. deionized water to each specimen in sample cup and cover with cap. Allow specimen to elute for 10 minutes. This results in a 1:50 dilution of the blood sample. Rotate the cup gently and remove the filter paper with an applicator stick. Recap the specimen and run within 1 hour of elution.

2. Warm up Turner Fluorometer for 15 minutes.

3. Place zero aperture in front of cuvette. Set recorder pen on zero with Blank Knob on Turner. Remove zero aperture.

4. Start all reagents pumping.

Phenylalanine (Continued):

5. Aspirate 5.0 mg% standard continuously for 5 minutes. Reading should be about 20 divisions above baseline with a noise level no greater than ± 1.5 divisions.

6. Run water wash through sample-pick up line for about 2 minutes.

7. Begin sampling, running all standards in duplicate. Follow with Control Sera and a previously determined high specimen. Intersperse a water and two 2.0 mg%. standards after each fifth unknown.

8. At the end of the run, wash the system thoroughly with deionized water.

NORMAL VALUES: Less than 4.0 mg%.

REFERENCES:

1. Hill, Summer, Pender, and Roszel: Clin. Chem., 11:541 - 546, 1965.

2. McCaman and Robins: J. Lab. & Clin. Med., 59:885 - 890, 1962.

CLINICAL INTERPRETATION:

An inborn error of metabolism results in an elevated blood phenylalanine above 4.0 mg%. The blood phenylalanine elevation occurs several days after birth and with persistent elevation damage of the brain occurs. The underlying defect is congenital deficiency of phenylalanine hydroxylase, the enzyme involved in conversion of phenylalanine to tyrosine. As a result, phenylalanine accumulates and is metabolized by a minor pathway to phenylpyruvic acid. Mental deficiency may result. Thus, it is necessary to institute a rigorous diet with restriction of phenylalanine to prevent the damage to the brain. The special diet must be maintained for many years.

AutoAnalyzer ®

Methodology

Phenylalanine

Autoanalyzer Flow Diagram

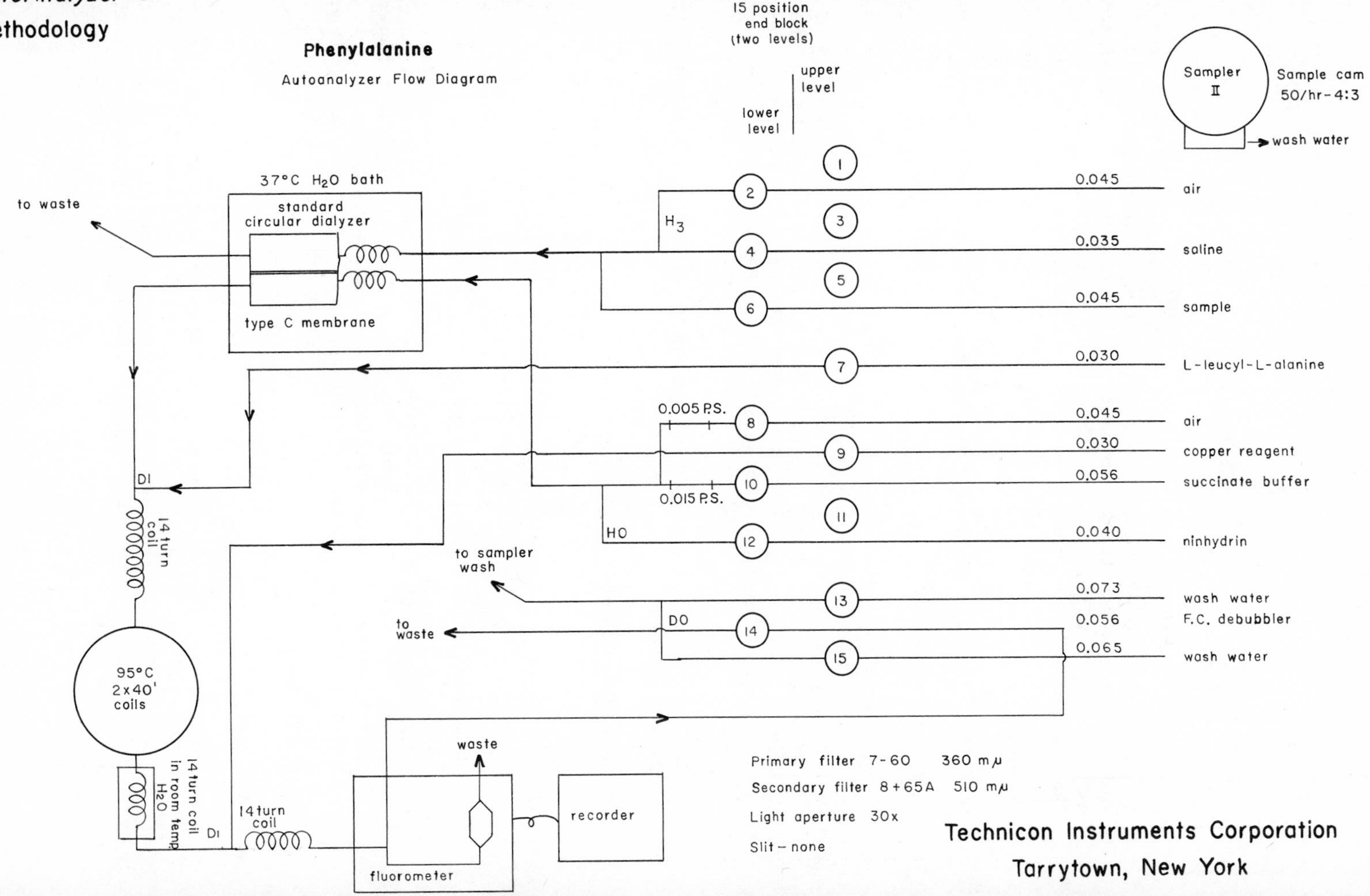

Primary filter 7-60 360 mμ

Secondary filter 8+65A 510 mμ

Light aperture 30x

Slit – none

Technicon Instruments Corporation

Tarrytown, New York

PHOSPHOLIPIDS

PRINCIPLE: An aliquot of the chloroform layer of the lipid extract is evaporated to dryness. The lipids present are digested with sulfuric acid and hydrogen peroxide. The phosphorus in the digested extract reacts with acid molybdate to form phosphomolybdic acid. The phosphomolybdic acid is reduced by Elon to form a blue color, the intensity of which, is proportional to the amount of phosphorus present.

SPECIMEN: 3.0 ml. of chloroform-methanol extract from "Total Lipids"

REAGENTS:

1. Sulfuric Acid, 5.0 N.
   Place approximately 600 ml. of deionized water in a 1 liter volumetric flask, which is in an ice bath. Slowly add 145 ml. of concentrated H_2SO_4 (s. g. 1.84) to the flask and swirl the flask. Dilute to the 1 liter mark when mixture is cool. Titrate and adjust to a normality of 4.8 - 5.2.

2. Hydrogen Peroxide, 30%
   Highest purity and essentially free of phosphates. Store in the refrigerator.

3. Ammonium Molybdate Solution
   2.5% in 4.0 N H_2SO_4.

4. Elon Solution (p-methylamino-phenol sulfate)
   1.0 gm. Elon and 3.0 gm. of sodium bisulfite are added to a 100 ml. volumetric flask. This is then diluted to volume with deionized water. Prepare fresh every 2 weeks.

5. Phosphorus Standard, 0.08 mg. P/ml.
   Dissolve 0.3514 gm. of pure dry monopotassium dihydrogen phosphate (Burea of Standards Reagent) in deionized water in a 1 liter volumetric flask. Add 20 ml. of 5.0 N. H_2SO_4 and dilute to the 1 liter mark. 1.0 ml. = 0.08 mg. P. (P = 30.9, K_2PO_4 = 136.091).

PROCEDURE:

1. Pipette 3.0 ml. of chloroform-methanol extract into a digestion (NPN) tube calibrated at 25 ml. (If 1.0 ml. of serum was used for the Total Lipid, use 6.0 ml. for the Phospholipids or use 3.0 ml. and double the results.) Evaporate to dryness with air in warm water.

Phospholipids (Continued):

2. Set up a tube for each of the 3 working standards. Pipette 1.0 ml. of each of the working standards into the three tubes which represent 0.02 mg.; 0.04 mg.; and 0.08 mg. standard respectively.

3. Pipette 2.5 ml. of 5.0 N. H_2SO_4 into a Blank tube, 3 Standards, each of the Test tubes.

4. Digest all tubes at approximately 210° - $225^{o}C$. in the aluminum blank for 45 minutes (or until the organic material is completely charred).

5. Add 5 drops of hydrogen peroxide to each tube allowing the drops to fall directly into the digestion mixture.

6. Rotate the tubes at a horizontal angle so that the viscous acid will wash down the sides of the tubes.

7. Continue to digest for at least 10 minutes after adding the hydrogen peroxide.

8. The contents of the tubes should be colorless, if not, continue to add peroxide and heat for 10 minutes after each addition of peroxide.

9. Dilute to the 25 ml. mark with deionized water and mix.

10. Into Coleman cuvettes, pipette 8.0 ml. of each digestate, 1.0 ml. of ammonium molybdate solution, and 1.0 ml. Elon and mix. Let stand for 25 minutes.

11. Read at 660 nm. in a spectrophotometer in 19 x 105 mm. cuvettes.

12. CALCULATIONS:

$$\frac{\text{O.D. Unk.}}{\text{O.D. Std.}} \times \text{Conc. Std.} \times \frac{100}{\text{ml. serum used}} \times \frac{\text{Total vol. Extract}}{\text{Vol. Extract used}}$$

$$= \text{Mg\%. P}$$

$$\frac{\text{O.D. Unk.}}{\text{O.D. Std.}} \times 0.04 \times \frac{100}{2} \times \frac{18}{3} = \text{Mg\%. P}$$

$$\frac{\text{O.D. Unk.}}{\text{O.D. Std.}} \times \frac{12}{1} = \text{Mg\%. P}$$

Phospholipids (Continued):

Mg%. P x 25 = mg% phospholipid

$0.02 \times \frac{100}{2} \times \frac{18}{3} = 6$ mg%.

$0.04 \times 300 = 12$ mg%.

$0.08 \times 300 = 24$ mg%.

NORMAL VALUES: 150 - 300 mg%.

REFERENCE: Sunderman and Sunderman: LIPIDS AND STEROID HORMONES IN CLINICAL MEDICINE, Lippincott, pg. 28 - 30, 1960.

CLINICAL INTERPRETATION:

Phospholipids are complex lipids which contain phosphate and a nitrogenous base. The major phospholipids are lecithin and sphingomyelin. The phosphate and nitrogenous bases are water soluble, which is important in lipid transport. Phospholipids are widely distributed in all tissues and are associated with mitochondrial metabolism. Dietary phospholipids may be absorbed because of their solubility, and the phospholipids of plasma may be derived primarily from liver synthesis. The function of plasma phospholipids is not definite. They are concerned with transport of lipids and blood coagulation.

INORGANIC PHOSPHORUS
(Gomori, Modified)

PRINCIPLE: The phosphorus present in a protein-free trichloroacetic acid filtrate reacts with acid molybdate to form phosphomolybdic acid. The phosphomolybdic acid is in turn reduced by Elon to form a "molybdenum blue" color, the intensity of which is proportional to the amount of phosphorus present.

SPECIMEN: 1.0 ml. of an unhemolyzed serum specimen. The serum should be removed from contact with the red cells as soon as possible. Serum is stable at refrigerator temperatures for several days. Use 1.0 ml. of a 1:10 dilution of well-mixed uncentrifuged urine.

REAGENTS:

1. 4.0 N. H_2SO_4
   Into a 1000 ml. volumetric flask measure approximately 700 ml. of deionized water. Add 112 ml. of concentrated Sulfuric Acid (S. G. 1.84; 96%) and allow the contents to cool. Bring to volume with deionized water.

2. Acid Molybdate
   Into a 1000 ml. volumetric flask, dissolve 25 gm. of ammonium molybdate in about 600 ml. of 4.0 N. Sulfuric Acid. Bring to volume with the acid.

3. Elon
   In a 100 ml. volumetric flask, dissolve 1.0 gm. of Elon (p-methylaminophenol sulfate) and 3.0 gm. of sodium metabisulfite into 80 ml. of water. Bring to volume with deionized water. Make fresh every two weeks.

4. Trichloroacetic acid, 20%
   In a 1000 ml. volumetric flask, dissolve 200 gm. of trichloroacetic acid in about 700 ml. of deionized water, and bring to volume. Stable indefinitely.

5. Stock Phosphorus Standard, 1.0 mg./ml.
   In a 250 ml. volumetric flask, dissolve 1.098 gm. of potassium phosphate monobasic (KH_2PO_4) in deionized water and bring to volume. Store over long periods of time in the refrigerator.

6. Working Phosphorus Standard, 0.02 mg./ml.
   In a 100 ml. volumetric flask, pipette 2.0 ml. of stock phosphorus standard and bring to volume with deionized water.

Inorganic Phosphorus (Continued):

7. Working Phosphorus Standard, 0.04 mg./ml.
In a 100 ml. volumetric flask, pipette 4.0 ml. of stock phosphorus standard and bring to volume with deionized water.

8. Calibration Curve
Prepare ten test tubes in duplicate according to the following chart: Standard measurements should be made volumetrically.

Tube Number	1	2	3	4	5	6	7	8	9	10
Ml. Working Standard, 0.02	1.0	2.0	3.0	4.0	5.0					
Ml. Working Standard, 0.04						3.0	3.5	4.0	5.0	
Ml. Deionized Water	5.0	4.0	3.0	2.0	1.0	3.0	2.5	2.0	1.0	6.0
Ml. 20% TCA	4.0	4.0	4.0	4.0	4.0	4.0	4.0	4.0	4.0	4.0
Plot as: mg%.	1.0	2.0	3.0	4.0	5.0	6.0	7.0	8.0	10	Blk.

Mix each tube well after addition of the TCA. Pipette 2.0 ml. from each tube into a clean test tube and follow procedure for color development from Step "9" beneath. Record absorbance and plot against concentration on linear graph paper.

COMMENTS ON PROCEDURE:

A blue color in the blank suggests either a contaminated test tube or contamination in the water or reagents. The blank should read approximately 0.000 Absorbance against water. If the blank gives a reading of greater than 0.005 Absorbance, and if the assay includes Control sera and standards, determine the Absorbance of the blank and all other tubes against water. Subtract the Absorbance of the blank from that of all other tubes and calculate the results from the standards included in the analysis.

PROCEDURE:

1. Into a series of appropriately marked test tubes, pipette 1.0 ml. of serum specimen, control serum or diluted urine.

Inorganic Phosphorus (Continued):

2. Pipette 2.0 ml. of 0.02 mg./ml. Working Standard into a tube.

3. Pipette 2.0 ml. of 0.04 mg./ml. Working Standard into a tube.

4. Pipette 3.0 ml. deionized water into a tube as a "Blank".

5. Add 2.0 ml. of deionized water to the tubes containing specimens or controls.

6. Add 1.0 ml. of deionized water to the tubes containing Standards.

7. Add 2.0 ml. of 20% TCA to all tubes. Mix thoroughly.

8. Wait 20 minutes for complete precipitation. Mix again and filter.

9. Volumetrically transfer 2.0 ml. of filtrate from each of the preceding tubes into clean test tubes.

10. Add 6.0 ml. of deionized water to all test tubes.

11. To all test tubes add 1.0 ml. of Acid Molybdate and mix. To all tubes add 1.0 ml. of Elon and mix thoroughly.

12. Allow 25 minutes for color development and determine the absorbance for each tube within 40 minutes of the time that color development was initiated. 780 nm. is the optimal wavelength, but the upper 600 nm. range may be used.

Ml. of:	Specimen or Control	0.02 mg./ml. Standard	0.04 mg./ml. Standard	Blank
	1.0	2.0	2.0	
Water	2.0	1.0	1.0	3.0
20% TCA	2.0	2.0	2.0	2.0
Filtrate	2.0	2.0	2.0	2.0
Water	6.0	6.0	6.0	6.0
Molybdate	1.0	1.0	1.0	1.0
Elon	1.0	1.0	1.0	1.0

13. CALCULATION:

$$\frac{\text{Mg\%. Standard}}{\text{O.D. Standard}} \times \text{O.D. Unknown} = \text{Mg\%. Unknown}$$

Inorganic Phosphorus (Continued):

14. Specimens with an absorbance (O. D.) higher than the greatest Standard should be diluted at the filtrate level and rerun with appropriate standards and controls.

NORMAL VALUES: 2.0 - 5.0 mg%. Serum Phosphorus

REFERENCE: Gomori: J. Lab. Clin. Med., 27:955, 1941 - 42.

CLINICAL INTERPRETATION:

The inverse relationship to calcium is well known and the etiology for hyperphosphatemia is frequently related to hypocalcemia. A physiologic cause for hyperphosphatemia is bone growth in children present until age fourteen. In a hospital population, the most common cause is renal failure with inability of the diseased kidney to excrete phosphate. Other causes are healing fractures and bone growth in acromegaly. Hypoparathyroidism due to accidental surgical removal of the parathyroids during thyroidectomy or pseudohypoparathyroidism are associated with an increase in serum phosphate.

The commonest cause for hypophosphatemia is hypercalcemia due to various etiologies. Hyperparathyroidism, osteomalacia relating to metastatic bone cancer or multiple myeloma may cause hypophosphatemia A common cause for hypophosphatemia is continuous use of intravenous glucose in a non-diabetic patient. This results from a need to phosphorylate glucose. Low serum potassium results in renal tubular wasting of phosphate with low serum phosphate. Other rare causes for low serum phosphate are rickets, vitamin D resistant rickets, Fanconi Syndrome, and excessive use of antacids which bind phosphates in the gastrointestinal tract and decrease phosphate absorption.

ULTRAMICRO SERUM PHOSPHORUS
(American Monitor)

PRINCIPLE: This method of phosphorus analysis is rapid, due to the elimination of the classical TCA filtrate, and is performed under conditions which may make the analysis more specific for inorganic phosphorus. The procedure is based on the formation of a phosphomolybdate complex and subsequent reduction with hydroxylamine in the presence of an alkaline buffer. Polyvinylpyrrolidone is incorporated with the reducing agent in this procedure and allegedly catalyzes the formation of a phosphomolybdate polymer. This polymer is then reduced to a molybdenum blue complex for measurement. The mechanism of PVP is unclear, however, and it may simply serve as a suspending agent for the phosphomolybdate complex and protein. The phosphomolybdate complex formation requires the presence of sulfuric acid, however, strongly acid conditions at this stage may result in hydrolysis of organic phosphate esters. The method presented attempts to avoid this source of error by performing the reaction under relatively mild conditions and within a short exposure time. Protein is precipitated by the molybdate ion, but this effect is pH dependent, being absent under strongly acid or alkaline conditions. Hence, after the phosphomolybdate complex is allowed to form for two minutes, sodium hydroxide in carbonate is added to the reaction mixture; the resulting alkaline pH reverses the protein precipitating effect of the molybdate ion.

SPECIMEN: 40 microliters of unhemolyzed serum. A fasting blood specimen is preferred; ingestion of glucose lowers the inorganic phosphorus level. Serum should be removed from the red cells as soon as possible. Urine must be diluted 1:10 before being analyzed.

REAGENTS AND EQUIPMENT:

Reagent available in Kit form or as separate items from American Monitor Corporation, Indianapolis, Indiana.

1. <u>Reductant - Catalyst Reagent</u>
   0.143 M. Hydroxylamine HCl in 0.001 M. Polyvinylpyrrolidone (PVP). Stable indefinitely at room temperature.

2. <u>Molybdate Reagent</u>
   0.163 M. ammonium molybdate in 0.175 M. sulfuric acid. Transfer the vial of dry reagent supplied into the bottle of sulfuric acid; mix contents well until dissolved. Rinse the remaining crystals in the vial with molybdate reagent, and return the contents to

the molybdate reagent bottle. Stable indefinitely at room temperature.

3. Buffer
0.05 M. sodium carbonate in 10 M. sodium hydroxide. Stable indefinitely at room temperature.

4. Brij.-Water, approximately 0.1%
Add 1.0 ml. of Brij.-35 to 1000 ml. deionized water.

5. Stock Phosphorus Standard, 1.0 mg. P/ml.
Place 1.098 gm. KH_2PO_4 into a 250 ml. volumetric flask. Dissolv and bring to volume with deionized water. Add a few drops of chloroform as preservative and store in the refrigerator.

6. Working Phosphorus Standards
a). 2.0 mg%. = 2.0 ml. Stock Standard to 100 ml. with deionized water.
b). 5.0 mg%. = 5.0 ml. Stock Standard to 100 ml. with deionized water.
c). 10.0 mg%. = 10 ml. Stock Standard to 100 ml. with deionized water.

Stable for an indefinite period of time at refrigerator temperature. Avoid chemical contamination.

7. Aliquotor
The 250 microliter Hamilton syringe is set with multiple stops for sequential delivery of 40 microliters of "Buffer".

8. Hamilton Gas-Tight Syringe
A 1.0 ml. syringe with Chaney adapter is set to repeatedly dispense 0.6 ml. of fluid.

9. Gilford 300 N Spectrophotometer
Equipped with a microaspiration cuvette.

COMMENTS ON PROCEDURE:

1. In order to correct for absorbance contributed by lipemia in specimens or control sera, a saline specimen blank must be included with each sample.

Ultramicro Serum Phosphorus (Continued):

2. One should proceed through the test once it is begun. Long delays after sample addition may result in hydrolysis of organic phosphate esters (i.e. this is not a point at which the procedure may be stopped for an indefinite time period).

3. Once all test reagents have been added and the cups capped and mixed, those cups containing protein should not be remixed. Mixing shortly before reading does not give time for the bubbles formed to leave the somewhat viscous solution.

PROCEDURE:

1. Immediately before use, prepare working molybdate reagent by mixing two volumes of "reductant-catalyst reagent" with one volume of "molybdate reagent" in quantities sufficient for all of the tests to be performed.

2. Deliver 0.6 ml. of the working molybdate into an A.A. cup for each test. Deliver 0.6 ml. of saline into A.A. cups for each specimen blank.

3. Rinse 20 microliters of sample, standard, or water (reagent blank) into an aliquot of working molybdate. Rinse 20 microliters of each specimen into the prepared aliquots. Allow the tests to stand for 2 minutes. Specimen cups will have a precipitate at this point.

4. Deliver 40 microliters of "buffer" into each cup containing working molybdate. Cap all cups, and mix well by inversion. Allow 5 minutes for color development. Do not remix before reading.

5. Determine the absorbance of all tests at 670 nm. in the Gilford 300 N with microaspiration cuvette, using reagent blank as reference. Readings should be made within 10 minutes of buffer addition; color fades after this time period. Determine the absorbance of all saline blanks at 670 nm. against a Brij.-water blank.

6. <u>CALCULATION</u>
   Subtract the absorbance of any saline specimen blanks from the test absorbance for that specimen. Check standards for linearity, and calculate test concentration according to the following formula:

$$\text{Specimen Conc. (mg\%.)} = \frac{\text{A. of specimen}}{\text{A. of standard}} \times \text{Conc. of standard}$$

Ultramicro Serum Phosphorus (Continued):

NORMAL VALUES: Fasting Levels in mg%.

1. Healthy Premature Infants

At Birth:	6.2 - 7.4
6 - 10 days:	7.5 - 10.3
20 - 25 days:	7.3 - 8.7

2. Full-Term Newborns

At Birth:	5.7 - 7.0
3 days:	6.6 - 8.2
6 - 12 days:	5.8 - 7.9

3. Normal Males

1 Year:	4.5 - 5.6
10 Years:	4.2 - 5.2
20 Years:	3.7 - 4.6

4. Normal Females
   Tend to have values about 0.1 mg%. lower than Males.

REFERENCES:

1. Baginski, E., Foa, P., and Zak, B.: "Determination of Phosphate: Study of Labile Organic Phosphate Interference". Clin. Chim. Acta, 15:155, 1967.

2. Crock, S. and Malmstadt, H.: "A Mechanistic Investigation of Molybdenum Blue Method for Determination of Phosphate". Analytic Chem., 39:1084, 1967.

3. Goodwin, J.: "Quantification of Serum Inorganic Phosphorus, Phosphatase, and Urinary Phosphate without Preliminary Treatment". Clin. Chem., 16:776, 1970.

4. O'Brien, D., Ibbott, F., and Rodgerson, D.: LABORATORY MANUAL OF PEDIATRIC MICROBIOCHEMICAL TECHNIQUES, Hoeber, 4th Ed., pg. 253 - 254, 1968. (Normal Values)

5. Modification by American Monitor Corporation. (Method)

CLINICAL INTERPRETATION:

Refer to Interpretation section under the Inorganic Phosphorus Procedure on page 310.

QUALITATIVE URINARY PORPHOBILINOGEN

PRINCIPLE: Porphobilinogen reacts with Ehrlich's aldehyde reagent to form a red compound, porphobilinogen aldehyde, which is soluble in aqueous solutions but not in n-butyl alcohol. Urobilinogen reacts with Ehrlich's reagent also but is readily extracted by butyl alcohol.

SPECIMEN: Urine sample must be a freshly voided specimen as porphobilinogen is unstable. May be spun down and frozen to preserve.

REAGENTS:

1. Modified Ehrlich's Reagent
   Ingredients include: 0.7 gm. Paradimethylaminobenzaldehyde, 150 ml. concentrated HCl, and 100 ml. deionized water. The solution is mixed and stored in a brown bottle.

2. Sodium Acetate, saturated aqueous solution
   In a 100 ml. volumetric flask, weigh out 175 gm. of sodium acetate. Dilute to volume with deionized water, using heat to dissolve contents.

3. N-Butyl Alcohol

PROCEDURE:

1. To 2.5 ml. of well-mixed urine add 2.5 ml. Ehrlich's reagent and mix.

2. Add 5.0 ml. of saturated sodium acetate. Shake vigorously.

3. Add 5 - 10 ml. n-butyl alcohol and mix.

4. Examine the lower aqueous layer for red color of porphobilinogen aldehyde.

5. The aldehyde compound of urobilinogen will be extracted by the n-butyl alcohol while the insoluble porphobilinogen will remain in the lower aqueous layer.

6. Report as Negative, trace or positive.

NORMAL VALUES: Normal porphobilinogen is a negative reading.

REFERENCE: Schwartz, S., et. al., METHODS OF BIOCHEMICAL ANALYSIS, Vol. 8.

Qualitative Urinary Porphobilinogen (Continued):

CLINICAL INTERPRETATION:

Porphobilinogen is a monopyrrole formed by condensation of two molecules of delta-amino-levulinic acid. Porphobilinogen is colorless and converts to a red porphobilin and uroporphyrin after excretion. Urinary excretion of porphobilinogen is greatly increased in acute intermittent porphyria. These individuals exhibit acute abdominal pain and neurologic symptoms or peripheral neuropathy. When the patient is asymptomatic, the test for porphobilinogen may be negative.

Patients with Porphyria Cutanea Tardea may excrete Porphobilinogen occasionally.

PORPHYRINS

PRINCIPLE: Most porphyrins give a bright red fluorescence in ultra-violet light. This characteristic and the different solubilities of coproporphyrins and uroporphyrins allow extraction and qualitative separation in this screening test.

SPECIMEN: 25 ml. of urine that has been protected from light. Specimen may be stored in the refrigerator.

REAGENTS:

1. Glacial Acetic Acid
2. Ether
3. HCl, 5%

COMMENTS ON PROCEDURE:

Small amounts of coproporphyrins are present in normal urine. Uroporphyrins are not normally present in urine.

PROCEDURE:

1. In a separatory funnel, place 25 ml. of urine.
2. Add 10 ml. of glacial acetic acid and shake.
3. Extract this mixture twice with 25 ml. portions of ether and combine the extractions.
4. Wash the combined ether extracts with 10 ml. of 5% HCl.
5. Examine the urine residue after extractions, the HCl washings, and the washed ether extract under ultra-violet light.

NORMAL VALUES: Uroporphyrins 10 - 30 micrograms/day
Coproporphrins up to 160 micrograms /day

REFERENCE: Miale, John B.: LABORATORY MEDICINE: HEMATOLOGY, 3rd. Edition, Mosby Co., 1967.

Porphyrins (Continued):

CLINICAL INTERPRETATION:

A red fluorescence in the washings or extract is evidence of coproporphyrins. A red fluorescence in the urinary residue is evidence of uroporphyrins.

Porphyrins are tetrapyrrole pigments which are precursors of the hemoglobins and cytochromes. Porphyrins exist in two isomeric forms: Types I and III, mainly coproporphyrin and uroporphyrin.

Uroporphyrins are tetrapyroles formed as a by-product of heme synthesis. Its excretion is greatly increased in acute porphyrias and may be increased in lead poisoning, cirrhosis of liver and hemochromatosis. Normal urinary uroporphyrin levels are 10 to 30 micrograms per day.

Coproporphyrin excretion in the urine up to 160 micrograms per day is normal. Diseases associated with elevated excretion are listed below.

Coproporphyrins I and III are extracted from urine with ethyl ether. Type I is increased in vitamin deficiency (Pellagra) in acute hepatitis, obstructive jaundice, and in acute porphyria. Type III is increased in toxic states such as lead poisoning, after salvarsan, quinine, and sulfonamide therapy, in acute poliomyelitis, and in acute intermittent porphyria, and in alcoholic cirrhosis.

Uroporphyrins I and III are left in urine after ether extraction, but are extracted with ethyl acetate. Type I is increased in acute porphyria and type III is increased in acute intermittent porphyria, both in greater quantities than coproporphyrins.

All the above porphyrins are distinguished by their solubilities in organic solvents, their absorption spectra, and by the melting points of the crystalline materials.

1. Porphyria Erythropoietica
   Increase in urine of Uroporphyrin and Type I Coproporphyrin

2. Acute Intermittent Porphyria
   Increased excretion of Coproporphyrin and Uroporphyrin similar to Uroporphyrin I

3. Porphyria Cutanea Tarda
   Uroporphyrins and Coproporphyrins are present in increased levels in the urine

PROTEIN, LIPOPROTEIN AND IMMUNOELECTROPHORESIS

PRINCIPLE: Electrophoresis is the physical separation of a mixture of charged particles under constraint of a superimposed current. The net charge of proteins depends on the summed amide and carboxyl charges and varies with the pH of the medium. The wider the separation of the medium pH and the pI or isoelectric point of the protein, the greater the impetus on the protein to migrate to the oppositely charged pole. The migration rate is directly proportional to the voltage on the system. Should the amide and carboxyl charges be equal, the molecule is neutral.

SPECIMEN: 0.5 ml. of unhemolyzed serum, 50 ml. aliquot of a 24 hour urine collection preserved with a few crystals of thymol, 5.0 ml. cerebrospinal fluid specimen, preserved with a small crystal of thymol, and the quantity of other body fluids varies in proportion to their protein content.

EQUIPMENT:

1. Power Supply, Beckman Duostat
2. Microzone Cell
3. Microzone Sample Applicator
4. Stainless steel forceps
5. Bard-Parker Knife
6. Washbottle
7. Sample Covers (8)
8. Glass Drying Plates (2)
9. Covered Staining Trays (5)
10. Squeegee
11. Ventilated Oven capable of maintaining 100 - 110°C.
12. Microzone R-110 Densitometer with R-111 Digital Integrator

Protein, Lipoprotein and Immunoelectrophoresis (Continued):

SUPPLIES:

1. Acetate membranes (Gelman Sepraphore)
   2.25 x 5.687 inch with perforations to fit Beckman Support.

2. Blotters

3. Sample Board
   A piece of ¼ inch plywood or hardboard, 4 x 16 inches is adequate and can be easily marked off to hold two or three series of 8 samples.

4. Parafilm
   To cover above sample board, a 4 inch width roll is ideal.

5. Barbituric-Barbiturate Buffer, aqueous (Beckman B-2), 0.075 M.
   2.76 gm./L. Diethyl barbituric acid
   15.40 gm./L. Sodium Diethyl barbiturate
   Together should provide a pH of 8.6.

6. Ponceau S Dye, aqueous
   0.2% Ponceau S
   3.0% TCA
   3.0% Sulfosalicylic Acid

7. Normal Saline, 0.9%

8. Absolute Methanol

9. Glacial Acetic Acid

10. Plastic Protective Envelopes

11. Labels:
    a). Narrow strips for the protective envelopes
    b). Wide strips for densitometer tracings.

COMMENTS ON PROCEDURE:

1. This methodology makes use of the Beckman Microzone System and all equipment mentioned is available through their Spinco Division in Stanford Industrial Park, Palo Alto, California, unless otherwise noted.

   The cellulose acetate is available from Gelman Instrument Co., Ann Arbor, Michigan.

Protein, Lipoprotein and Immunoelectrophoresis (Continued):

Special supplies other than chemicals and common laboratory materials are available from Beckman.

2. Much detail has been incorporated into the procedure description which follows, particularly with respect to handling of the acetate film and the samples. Adherence to these techniques have served our laboratory:
   a). To provide films which are optically clean and free from distortion.
   b). Minimize variation in results due to ambient atmospheric conditions such as humidity and heat.
   c). Reduce danger of confusion in sample identification.
   d). Increase reproducibility.

PROCEDURE:

1. <u>Prepare the five staining trays as follows</u>:
   a). B-2 Buffer (#1)
   b). Working Ponceau S Dye (#2)
   c). 5% Acetic Acid (#3) (50 - 75 ml. of solution is adequate)
   d). Absolute Methanol (#4)
   e). 10% Acetic Acid made up in absolute methanol (#5)

   Place an immaculately clean and unscratched glass plate in the bottom of the last (#5) tray. Keep all trays covered when not in use.

   The dye is adequate for staining about three membranes and then should be replaced. The acetic acid and methanol solution should be fresh.

2. Fill the Microzone Cell to the level line with refrigerated B-2 Buffer (approximately 200 ml.)

3. <u>Preparation of samples for loading</u>
   Up to eight samples can be run at one time, there being eight positions on the membrane. Conventionally the outer positions, No. 1 and No. 8 are reserved for Control samples.

   A. Cover the sample board with a clean strip of parafilm.

   B. Assemble the samples to be analyzed in the order in which they are to be loaded onto the membrane.

Protein, Lipoprotein and Immunoelectrophoresis (Continued):

C. Using a clean Pasteur pipette for each sample, place a drop or two of specimen onto the appropriate position on the sample board and cover immediately with a sample cover to prevent evaporation.

D. Prepare the label for the envelope cover, including all pertinent data (i. e. run number, date, etc.). Most important, indicate the position number for each sample as well as its identification data in such a way that the label cannot be inadvertently reversed or read upside-down.

E. This is a good time to also prepare the labels for the densitometer tracings. Strip labels, torn off in groups of eight, are most convenient. Again designate all pertinent information, i. e., run number, position, date, sample identification number, name, etc.).

F. Check the sample applicator, to be sure that it is clean and in good working condition.

G. Have on hand a few pieces of clean, lintless tissue (Kimwipes) folded into a small pad to be used during application of the samples.

4. <u>Preparation of the Cellulose Acetate Membrane</u>

A. Always use forceps and not fingers to handle the membrane.

B. Take up one sheet of acetate membrane from between the protective sheets and carefully float in on the B-2 Buffer (Tray #1).

C. Allow the buffer to penetrate the membrane from below. After it has become uniformly saturated, submerge it briefly by rocking the tray.

D. Grasp the membrane with forceps near one corner and drain it briefly against the side of the tray. Then carefully place the membrane between clean blotters. Blot firmly and then transfer to the bridge support of the Microzone Cell. Note that the acetate has a double perforation on one corner. Line this perforation up with the lid position #1.

E. Carefully place the bridge into the cell, making sure that the acetate membrane is hanging straight into the buffer and is not caught up on the cell sides.

F. Put on the numbered cover and lid. Attach to the Duostat before application of the samples.

5. Application of Samples to Membrane

Apply specimens without delay, matching specimen number on sample board with position number on application slot.

The applicator loop is "seasoned" (i.e. rinsed with a droplet of sample), prior to each application. The following procedure is recommended:

A. Release the applicator loop and skim it across the top of the droplet on the sample board. Allow the loop to fill by capillary attraction. Penetrating too far into the droplet will cause excess sample to adhere to the loop ends, yielding an undesirable dumb-bell shaped application.

B. Retract the loop and set the applicator on a piece of lint-less tissue or blotting paper.

C. Release the loop. A visible pattern of sample should be evident.

D. Repeat the loading procedure and gently retract the loop.

E. Set the applicator on the cover guide with its central pin in the slot lined up with the appropriate position number. Rock slightly to check positioning and release the loop. The loop will drop to the cellulose acetate, which will flex visibly.

F. On the count of four seconds, retract the loop and remove the applicator. Check visually to see if the sample pattern appears.

G. Release the loop and wash gently with 0.9% saline from a washbottle. Touch gently to tissue paper or blotting paper to dry. Use great care as the platinum loop is very frail and must not be distorted.

Protein, Lipoprotein and Immunoelectrophoresis (Continued):

H. Repeat from (A). with the next sample until the cell is filled.

6. <u>Applying Current to Cell</u>

Turn Duostat "ON". Wait for needle deflection before advancing the Output Adjust. The Voltmeter should read 180 Volts. Use constant voltage on the 0 - 300 setting and read the voltmeter on the 0 - 500 setting. When equilibrated, shift to ammeter 0 - 15 ma scale. Amperage starts at 4.5 - 5.5 ma and may reach 7.5 ma by conclusion of the twenty minute run. If the initial amperage is higher, reduce the time of the run.

7. <u>Staining the Membrane</u>

Recheck the location of the double perforation to insure positiv identification.

A. Turn off the Duostat and unplug the cell. Remove the bridge and carefully lift the membrane from its supports.

B. Float the membrane on Ponceau S Dye Solution. Avoid trappin bubbles underneath the membrane. Submerge by rocking staini tray and stain for ten (10) minutes.

C. Destain with many scanty rinses of 5% acetic acid. Use about 20 ml. solution at a time and decant between each rinse. Repeat rinses until background of membrane is clear, and the rinse solution no longer is tinged pink. Drain membrane against side of tray.

D. Float membrane onto absolute methanol, again taking care not to entrap bubbles underneath. Submerge for one minute. Drain.

E. Float onto surface of 10% acetic acid in methanol solution. Immerse for no longer than 20 seconds. (Timing may vary slightly with different batches of cellulose acetate).

F. Lift membrane out, using the glass plate that had been place in the tray. At this stage the acetate is very soft and easily distorted. Line the membrane up parallel to the plate edges and drain excess clearing solution. Moisten the squeegee by dipping into the clearing solution and then gently squeegee the membrane, taking care to avoid uneven pressure in the area of the patterns. Wipe off bottom and edges of plate and flatten any air bubbles or puddles since these will interfere with uniform drying and cause peeling.

G. Suspend the glass plate between two inverted beakers in a 110°C. oven for 10 - 15 minutes. Be sure that nothing damp (i.e. washed glassware), is in the oven since extraneous moisture will cause premature peeling of the membrane. Check occasionally and if peeling starts, remove from oven before end of 15 minutes.

H. At the end of 15 minute drying time, remove the plate from the oven and slip between two sheets of clean paper (a 3 x 5 scratch pad works well), and allow it to cool until it can be handled. The film is now cellophane-like and clear. Label it with a felt tip pen specifying date and run number. Keep the film covered to protect it from lint and accidental damage until it can be mounted into the plastic envelope.

8. <u>Mounting Membrane and Preparation for Scanning</u>

A. Ventilate and pre-warm a plastic protective envelope. Prop the "jacket" open, handling it by its edges or with a clean tissue to avoid soiling the optical surface. Place it in a warm, dry place (i.e. on top of the drying oven) for a few minutes. This procedure seems to reduce the tendency to form a "Newton-Ring effect" which results in noisy tracings.

B. If the cooled film has not begun to peel from the glass plate, release it by making a cut parallel to one end with a Bard-Parker blade. The natural curling tendency of the film may be assisted by gentle pulling and the blade used to "tease" stuck spots.

C. Slip the released film immediately into the prepared plastic envelope, keeping the double perforation towards the open side of the jacket. Use the excess end-laps to manipulate the film into position so that the patterns are parallel to the long sides of the jacket. Attach the prepared label, making sure that the double perforation coincides with the number one position of the label.

9. <u>Tracing and Calculating</u>

Read and graph on the Beckman R-110 Densitometer. Calculations are done by the R-111 Digital Integrator Attachment. The Total

Protein, Lipoprotein and Immunoelectrophoresis (Continued):

Protein values must be available at the time of scanning if manual calculation is to be avoided.

Graph positions one through eight, attaching the prepared labels to each tracing.

PROCEDURE FOR CONCENTRATION OF SPECIMENS:

For concentration of dilute samples, we use Union Carbide Dialysis Tubing, supplied in 100 foot rolls. For volumes up to 10 ml. the 8 mm. diameter is satisfactory. For larger volumes, the 20 mm. diameter tubing is used. The average specimen requires about 1 foot of tubing.

A. Presoak a section of dialysis tubing in water until it is pliable. Tie one end with two overhand knots; avoid undue pulling so as not to distort pore size. Leave enough space between the knots to attach a small gummed label bearing the sample identification and indicating the volume to be used. Protect the label from smearing during subsequent handling by covering with a layer of transparent cellophane tape.

B. Insert the tapered end of a small plastic funnel into the open end of the tubing to facilitate transfer of the sample.

C. Cerebrospinal fluid or similar fluids should be centrifuged. For urines, filtration through Whatman #1 filter paper has been found to be more satisfactory, although some protein may be lost on the filter paper. Be sure that a few crystals of thymol are present.

D. The filled tubing is then tied off with two knots. Urines are first dialyzed against warm running tap water for half to three-quarters of an hour. The tubing is then taped horizontally in a convenient location, such as the bottom edge of a hanging cabinet or shelf and evaporated by the air current from an electric fan.

E. As the volume decreases, occasionally soften the tubing by immersing in tap water. Massage gently to extract any protein from the walls of the tubing and tie off the empty end. This avoids denaturation due to overdrying.

Protein, Lipoprotein and Immunoelectrophoresis (Continued):

F. When the tubing is flat and the material very viscous or almost dry, place the tubing in a test tube of B-2 Buffer. Agitate until a very small amount of buffer has entered the tubing. Massage tubing with the buffer to extract any protein adhering to the walls of the tubing and then squeeze the material to one end. Cut the tubing on a long diagonal, well above the liquid and carefully deliver into a small sealable container, such as an AutoAnalyzer sample cup. Transfer the label as well and indicate the estimated concentration factor. 10 ml. reduced to 0.1 ml. is quite feasible. CSF specimens of 5.0 ml. or less can be taken to less than 0.1 ml.

G. For specimens whose total protein concentration is within the normal range, a triple application of the concentrated material will usually provide an adequate traceable pattern. A urine with 4+ protein will often give a traceable pattern using a triple application of unconcentrated sample.

H. Whenever possible, the total protein is determined on all samples prior to electrophoresing. A negative qualitative result on a urine sample should not discourage one. Most urines on concentration from 20 ml. to 0.1 ml., will yield a distinct pattern. 50 ml. is the maximum useful sample volume.

LIPOPROTEIN ELECTROPHORESIS

SUPPLIES:

1. Buffer
IL/vial; 4.0 liters/run. Replace completely once a week.

2. Agarose-Albumin

	Single Row of Troughs	Double Row of Troughs	
		1 Sheet	2 Sheets
Agarose	0.21 gm.	0.35 gm.	0.7 gm.
Buffer	30 ml.	50 ml.	100 ml.
Albumin	0.60 ml.	1.0 ml.	2.0 ml.

3. Sheet
7½ x 10 - 12 inches (Cronar Clear Base)

4. Templates
These are hand-made by Dr. Peter Wood and his staff.

Protein, Lipoprotein and Immunoelectrophoresis (Continued):

4. Single Row of Troughs
   190 x 130 mm. agarose area
   25 ml. of agarose - albumin
   Troughs 25 mm. from cathodal edge of gel.

5. Double Row of Troughs
   190 x 207.5 mm. agarose area
   40 ml. of agarose - albumin
   Troughs 35 mm. and 115 mm. from cathodal edge of gel.

6. Gel 15 minutes at room temperature with templates in place.

PROCEDURE:

1. 13 microliters of plasma or serum are applied to troughs with sheet in place on electrophoresis platform.

2. Run at 50 ma (about 300 Volts) for 65 minutes.

3. Fix for 30 minutes in 500 ml. of 70% ethanol + 37.5 ml. glacial acetic acid.

4. Dry in oven at 90°C. for 30 minutes.

STAINING PROCEDURE:

Using 0.1% Sudan Black "B" (1.0 gm./L. in 60% Ethanol), stain 60 - 90 minutes. Store stain in tightly closed bottle.

REFERENCES:

1. Instruction Manual RM-IM-3 for Beckman Model R-101 Microzone Electrophoresis Cell, 1965.

2. Instruction Manual RMTB-005 for Beckman Model R-110 Microzone Densitometer, 1967.

3. Preliminary Instruction for Model R-111 Microzone Digital Integrator, Beckman, 1971.

4. Beckman Microzone Electrophoresis System, Price List for Bulletins #7704 & #6301, July, 1971.

Protein, Lipoprotein and Immunoelectrophoresis (Continued):

All of the above available through:
Beckman Instruments, Inc.
Fullerton, California, 92634

5. Fredrickson, D. S., Lees, R. S. & Levy, R. I.: "Genetically Determined Abnormalities in Lipid Transport", PROGRESS IN BIOCHEMICAL PHARMACOLOGY, Vol. 3, Basel, Karger, 1967.

IMMUNOELECTROPHORESIS

PRINCIPLE: Two methods are combined: Gelectrophoresis followed by immunodiffusion, using "Ion Agar".

REAGENTS:

1. Veronal Buffer, pH 8.2, 0.05 M.
   Usually 8 liters is prepared at one time for convenience. This required 63.51 gm. sodium barbital and 92 ml. of 1.0 N. HCl.

2. 1.5% Buffered Agar Gel

3. Glass Slides, 1 x 3 inches

PROCEDURE:

1. Melted agar is poured on 1 x 3 inch glass slides, 2.5 ml. gives approximately 2mm. thickness.

2. Allow plates to stand at least one hour before using.

3. Using a 13 gauge needle, cut wells for test material.

4. Longitudinal trenches are cut parallel to electrical flow, about 4 mm. from circular wells.

5. Slides are placed as a bridge between the anode and cathode reservoirs.

6. Filter paper is used for the connection to the electrode vessels.

7. Current flows through a continuous slab of agar.

8. Perform electrophoresis in a closed box to maintain constant humid atmosphere.

Protein, Lipoprotein and Immunoelectrophoresis (Continued):

9. Run at 50 ma amps and less than 150 volts to get a separation of protein in one and one-half hours.

10. After electrophoresis, the trenches of agar are removed and 100 microliters of antibody are added.

11. Slides are allowed to stand overnight for double diffusion to take place. In addition to the 100 microliters of antibody that were added, the amount of antigen added is approximately 3.0 to 4.0 microliters.

NORMAL VALUES:	IgA:	59 - 489 mg%.
	IgG:	711 - 1536 mg%.
	IgM:	37 - 212 mg%.

CLINICAL INTERPRETATION:

Lipoprotein Electrophoresis

Lipoprotein electrophoretic patterns are divided into five types. Definitions of hyperlipoproteinemias are listed below:

Type I: Lipoprotein-lipase deficient familial hyperchylomicronemia. Hyperglyceridemia or hyperlipemia fat induced. Familial hyperchylomicronemia.

Type II: Elevation of normal density beta-lipoprotein with or without an elevation of pre-beta lipoproteins. Hypercholesterolemia. Essential familial cholesterolemia. Hereditary Xanthomatosis. Xanthelasma.

Type III: Elevation of abnormally low-density beta-lipoprotein and and a pre-beta lipoprotein elevation. Hyperglyceridemia or hyperlipemia that is carbohydrate induced (sensitive) with hypercholesterolemia.

Type IV: Hyperpre-beta lipoproteinemia. Hyperglyceridemia (hyperlipemia) that is carbohydrate induced (sensitive) without hypercholesterolemia.

Type V: Hyperchylomicronemia and hyperpre-beta lipoproteinemia. Hyperglyceridemia (hyperlipemia), combined fat and carbohydrate induced.

Protein Electrophoresis and Immunoelectrophoresis

Refer to Interpretation section under Total Serum Protein and Albumin Determination (Biuret Reaction) on page

PRINCIPLE: The transfer of hydrogen catalyzed by lactic dehydrogen-ase (LDH) proceeds as follows:

$$
\begin{array}{ccc}
\begin{matrix} \text{COO}^- \\ | \\ \text{C}=\text{O} \\ | \\ \text{CH}_3 \end{matrix} + \text{DPN-H}^+ & \underset{\xrightarrow{\hspace{1cm}}}{\overset{\text{LDH}}{\rightleftharpoons}} & \begin{matrix} \text{COO}^- \\ | \\ \text{CHOH} \\ | \\ \text{CH}_3 \end{matrix} + \text{DPN}^+
\end{array}
$$

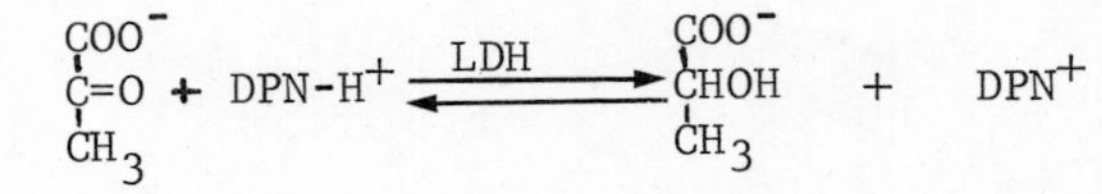

Pyruvate   Lactate

The equilibrium of this reaction is almost completely on the side of lactate and DPN^+. This reaction proceeds very rapidly in the presence of increased H^+ concentration and, within certain limits, excessive DPNH. Also, under these conditions, pyruvate is converted completely into lactate. DPNH which is the hydrogen donor for the above reaction is consumed in a stoichiometrical ratio (1 mole DPNH per 1 mole pyruvate). It can be measured spectrophotometrically on the Beckman DU at a wavelength of 340 nm. using a slit of 0.17 mm. and 10 mm. pyrex cuvettes.

SPECIMEN: Pipette into the proper number of tubes 7% cold perchloric acid, equal volumes of acid and blood. Mix well and centrifuge the precipitated protein for 10 - 15 minutes at 3500 rpm. Transfer the supernatant to another tube and recentrifuge in order to obtain an absolutely clear filtrate.

It cannot be overemphasized that 1). The blood is drawn without the use of a tourniquet, 2). The sample should be kept cold at all times necessitating a refrigerated centrifuge at 4°C., 3). If it is necessary to keep the sample longer than the day on which it is drawn, the filtrate should be frozen. When frozen, the filtrate can be kept for at least 6 weeks without change.

REAGENTS AND EQUIPMENT:

1. 7% $HClO_4$
    Dilute 70% $HClO_4$ 1:10 with distilled water.

2. KOH 5.0 N.
    Dissolve 280.54 gm. in a liter volumetric flask and dilute to volume to deionized water.

Blood Pyruvic Acid (Continued):

3. Potassium Phosphate Buffer, 0.1 M.
Dissolve 8.7 gm. of K_2HPO_4 in 500 ml. of deionized water. Dissolve 3.4 gm. of KH_2PO_4 in 250 ml. of deionized water. Mix these two solutions together and adjust the pH to 7.4 - 7.5. A large amount of buffer may be prepared at once if it is kept in the refrigerator and the required amount removed from the bottle each day throwing all unused solution away.

4. DPNH (Diphosphopyridine nucleotide, reduced form = 709 gm. - anhydrous weight).
Make a solution of 5.0 mg./ml. in water. Dilute this with phosphate buffer when analysis is made. For 6.0 ml. of buffer use 0.2 ml. stock DPNH. To a Beckman cuvette add 0.1 ml. of the buffered DPNH plus 2.9 ml. water and read on DU against a cuvette with water. This generally reads 0.040 - 0.050 O.D. Units. This times 10 will give approximate unknown readings since in the actual analysis 1.0 ml. of the buffered DPNH is used. To each vial of Sigma DPNH (5.0 mg./vial), add 30 ml. of potassium phosphate buffer.

5. LDH (Lactic Dehydrogenase)
Order concentrated stock from Sigma in an ammonium sulfate suspension. Keeps months or even a year under refrigeration although it gradually loses its strength. Concentration of stock should be about 25000 Vester Units/ml. or more when fresh. Dilute the stock 1:20 in water for the analysis and continue to keep in refrigerator. (This is a large excess which is desirable.)

6. Standard, Lithium Pyruvate (Store in desiccator in refrigerator)

   A. Stock Standard (1.0 mg./ml.
   Stable for months if stored in refrigerator.
   Pyruvic acid: 88.06 gm.
   Lithium pyruvate: 112.01 gm.
   1.2719 x 25 ml. = 31.7975 mg./25 ml. = 1.0 mg./ml. pyruvic acid. The theoretical O.D. reading for 0.1 micromole pyruvic acid (8.806 ugm.) is 0.207 (by definition). Therefore 5.0 ugm. gives an O.D. reading of 0.118.

   B. Working Standard
   Dilute the stock 1:20 in water. 0.1 ml. contains 5 ugm. Make fresh daily.

Blood Pyruvic Acid (Continued):

7. Neutralized Mixture
Dilute 7% $HClO_4$ with an equal volume of water (simulates 7% $HClO_4$ with an equal volume of blood). Prepare 10 ml. at a time, neutralize to pH 3.5 - 4, centrifuge to remove precipitate and run along with samples. (Used in preparing standard cuvette and in making neutralized filtrates to an even volume.)

PROCEDURE:

1. Preparation of the filtrate for analysis
Take a convenient aliquot (we generally use 4.0 ml.) for analysi and neutralize to a pH of 3 - 4 using pH paper. Use short disposable pipettes to remove negligible amounts of the solution while trying to obtain the correct pH. Calculate the amount of 5 N. KOH needed to neutralize the amount of acid in the filtrate This is equal to approximately 0.1 ml. the volume of filtrate taken. Since this amount does not give the desired pH, it is necessary to have available 0.1 N. KOH and 0.1 N. HCl. Keep a careful record of the volumes added.

   After neutralization it is generally convenient to bring the volume to an even figure (such as 4.5 or 5.5 ml.) before proceeding. This is done by adding a 1:1 mixture of 7% $HClO_4$ and water, the mixture having been neutralized in the same manner as the filtrates. It is important to have pH close to 4 using 5.0 N. KOH.

   Mix well, then allow to stand in refrigerator (stoppered) for 10 - 15 minutes to complete the precipitation. Spin for 10 minutes at 3500 rpm and pour off supernate. Do not allow the filtrate to stay in contact with precipitate for longer than necess If necessary, this precipitation can be done one day, the supernatant frozen, and the analysis performed the following day.

2. Do in duplicate, samples less than 1.0 ml. ust be made to this volume with the neutralized mixture, and do standards with samples.

   Pipette the following directly into the cuvettes:

Blood Pyruvic Acid (Continued):

Reagent	Blank	Std.	Sample	Recovery
1:1 mixture of (neutralized mixture)				
7% $HClO_4$	—	1.0 ml.	—	—
Blood Filtrate	—	—	1.0 ml.	1.0 ml.
Std. (1:20 Dil.)	—	0.1 ml.	—	0.1 ml.
DPNH (Buffered)	—	1.0 ml.	1.0 ml.	1.0 ml.
Water	3.0 ml.	0.9 ml.	1.0 ml.	0.9 ml.

Stir with plastic stirrer and get initial readings on Beckman DU.

LDH	0 ml.	0.02 ml.	0.02 ml.	0.02 ml.

Stir and take final 30 minutes reading.

Timing is important. Three cuvettes can be run at once. After adding the DPNH, do not delay carrying the analysis to completion. Read at 340 nm. Using a stop watch, take 0 - 3 minute readings. Add the LDH, stir and read at 30 minutes again. The readings used are those taken immediately before addition of LDH and the final reading, the difference representing the amount of pyruvate in the sample.

CALCULATIONS:

In calculating, a correction must be made for the change in volume after neutralization and for the aliquot taken for analysis. Calculate the amount in 1.0 ml. of blood or serum and correct to mg%.

Since blood is 80% liquid, this must be taken into account in the calculation. One ml. of blood weighs 1.06 gm. Because of this, 1.0 ml. of blood plus 1.0 ml. of 7% $HClO_4$ yields 1.85 rather than 2.0 ml. of liquid.

$$1.06 \times 80/100 + 1.0 = 1{,}848 \text{ ml.}$$

For Whole Blood, the calculation is as follows:

$$\text{ugm./cuvette} \times \frac{\text{Vol. after neut.}}{\text{Vol. analyzed in cuvette}} \times \frac{1.85}{\text{Vol. Blood Filtrate}} \times 0.1 = \text{mg\% Pyruvate}$$

Blood Pyruvic Acid (Continued):

For serum, the calculation will be similar except that 1.85 will be replaced by 2. If blood is weighed instead of measured, use the following calculations:

$$\text{ugm./cuvette} \times \frac{\text{vol. after neut.}}{\text{vol. analyzed in cuvette}} \times \frac{(\text{wt.} \times 0.8) + \text{vol. 7\% } HClO_4}{\text{vol. blood filtrate}}$$

$$\times \frac{1.06}{\text{Wt.}} \times 0.1 = \text{mg\%. pyruvate}$$

Range of method = 0 - 25 ugm.

NORMAL VALUES: 0.4 - 2.0 mg%.

REFERENCES:

1. Cantarow and Trumper: CLINICAL BIOCHEMISTRY, 6th Ed., Saunders, pg. 60, 1962.

2. Hoffman, W. S.: THE BIOCHEMISTRY OF CLINICAL MEDICINE, 3rd Ed., Year Book Publishers, pg. 101 - 103, 1966.

3. Lundholm, L., Mohme-Lundholm, E. and Svedmyr, N.: "Comparative Investigation of Methods for Determination of Lactic Acid in Blood and in Tissue extracts", J. Clin. & Lab. Investigation, 15:311 - 316, 1963.

4. Olson, G. F.: "Optimal Conditions for the Enzymatic Determination of L-Lactic Acid", Clin. Chem., 8:1, 1962.

5. Barker, S. B. and Summerson, W. H.: "The Colorimetric Determination of Lactic Acid in Biological Material", J. Biol. Chem., 138:535, 1941.

CLINICAL INTERPRETATION:

Refer to Interpretation section under Blood Lactic Acid Procedure on page 256.

SERUM QUINIDINE

PRINCIPLE: Quinidine, a stereoisomer of quinine, is a drug used for management of cardiac arrhythmias. Due to the wide variation in individual metabolism, dosage is set by blood levels checks. The proteins are precipitated from the serum with trichloroacetic acid. The fluorescence of the supernatant is then compared with serum (free of quinidine) and given concentrations of quinidine.

SPECIMEN: 100 microliters of serum or plasma is required. Fasting specimens are preferred to avoid lipemia. Once the serum is separated from the cells, it should be frozen. The specimen is quite stable for a long period of time in the freezer.

REAGENTS AND EQUIPMENT:

1. Turner Fluorometer
   The machine should be warmed for at least 30 minutes before use.
   Filter Selection:
   Primary Wratten filter #7-60 (360 nm.)
   Seconary Wratten filter #2A (415 nm.)
   Range Selector: 1 Slit: None

2. Stock Trichloroacetic Acid, 20%
   Weigh out 200 gm. of reagent grade trichloroacetic acid into a liter volumetric flask. Dissolve in deionized water and bring to volume.

3. Trichloroacetic Acid, 5%
   Dilute 25 ml. of the 20% acid to 100 ml. with deionized water. This solution should be made up fresh every week.

4. Sulfuric Acid, 0.10 N. (for Quinidine Stock Standard)
   Carefully add 2.8 ml. of concentrated (36 N.) H_2SO_4 to around 600 ml. of deionized water in a liter volumetric flask. Bring to volume with deionized water.

5. Quinidine Stock Standard, 100 mg./L.
   Dissolve 120.7 mg. of Quinidine sulfate ($C_{20}H_{24}O_2N_2 \cdot H_2SO_4 \cdot 2H_2O$ in 0.1 N. sulfuric acid and make up to 1 liter. It should be stored in the refrigerator in an amber-colored bottle. This solution is stable indefinitely.

6. Quinidine Working Standards, 2.0 mg./L., 5.0 mg./L. and 8.0 mg./L.
   2.0 ml., 5.0 ml., and 8.0 ml. of the Stock Standard are diluted to 100 ml. with deionized water. These solutions are stable for

Serum Quinidine (Continued):

one week. These Working Standard Solutions are used to reconstitute a lyophilized quinidine-free control serum. Aliquots of these working serum standards are frozen and used as needed. They are stable indefinitely.
NOTE: A blank serum must be made by reconstituting the same lot number of lyophilized control serum with water and aliquots are also frozen.

PROCEDURE:

1. Using 16 x 100 mm. test tubes, set up tests, standards and blank in duplicate.

2. Using an Eppendorf pipette, add 0.05 ml. (50 microliters) of test or serum standards to 5.0 ml. of 5% TCA. The blank is 0.05 ml. of serum. With the aid of a Vortex mixer, mix immediately for at least 15 seconds. This step should produce a very fine precipitate.

3. Centrifuge at fast speed for 15 minutes.

4. Reading of fluorescence:

   A. Set the fluorometer to zero with the black "Dummy" cuvette.

   B. The clear supernatants are transferred into 12 x 75 mm. cuvettes or culture tubes. Readings are taken of the blank, standards and tests. Readings higher than the highest standard should be diluted with 5% TCA. If disposable 12 x 75 mm test tubes are used, they should be checked for uniformity of readings (empty) before use.

5. CALCULATIONS:

   Subtract the average blank reading from the standards and tests. Using linear graph paper, plot the concentration of Standards versus Fluorescence Units. This curve is used to determine patient quinidine levels by their relative fluorescence to the standards.

REFERENCES:

1. Gelfman, M. and Seligson, D.: "Quinidine", The Amer. J. Clin. Path., Vol. 36, No. 6, pg. 390-392, 1961.

Serum Quinidine (Continued):

2. Brodie, B. and Udenfriend, S.: "The Estimation of Quinine in Human Plasma with a Note on the Estimation of Quinidine", Jour. Pharm. and Exp. Therap., 78:154 - 158, 1943.

CLINICAL INTERPRETATION:

Quinidine is an excellent drug for the management of certain types of cardiac arrhythmias. Dosage may be based on the clinical response or on the serum levels. Quinidine is completely absorbed after oral administration. Maximal effects occur within 1 to 3 hours and persist for 6 to 8 or more hours. When cumulative effects are sought, repeated doses are given at intervals of 2 to 4 hours. Quinidine is rapidly bound by plasma albumin. When the total plasma concentration of the drug is in the therapeutic range of 3 to 6 mg. per liter, approximately 60% of the alkaloid is in the bound form. Therapeutic responses can rarely be expected below 3.0 mg. per liter of plasma and toxic reactions are almost certain to occur at levels above 10 mg. per liter.

ULTRAMICRO SERUM SODIUM AND POTASSIUM
(I. L. Flame Photometer)

PRINCIPLE: Under the action of energy, such as heat, the electrons of an atom can be caused to move to a higher energy level (energy is absorbed). When these electrons return to their normal state, the absorbed energy is released in the form of light. The light emitted is characteristic of the atom. The measurement of this emitted light is called emmision spectroscopy. When a flame is used to excite the atoms, the term "flame photometry" is used.

The suction feed atomizer is enclosed in a spray chamber which allows the large droplets to settle out and carries the finely dispersed aerosol into the flame. It is essential that the sample be introduced into the flame at a constant rate. Natural gas (propane) with compressed air will produce a flame with a temperature of 1700 - 1800°C. This is adequate for sodium and potassium. Too hot a flame will give interference from the excitation of other elements or the fuel breakdown products. Too cool a flame will decrease sensitivity.

The spectral lines emitted are isolated by interference filters which are placed before each of three photocells: one photomultiplier tube and a filter for sodium, one for potassium, and one for lithium, which is utilized as the internal standard. A ratio detector analyzes the difference in light intensity between the sodium and/or potassium and the lithium. This instrument also contains a simultaneous digital readout. The internal standard significantly increases precision and accuracy by minimizing variation in flame intensity, viscosity, electronic components, and gas pressure between specimens. (See I. L. Instruction Manual for the mathematical interpretation).

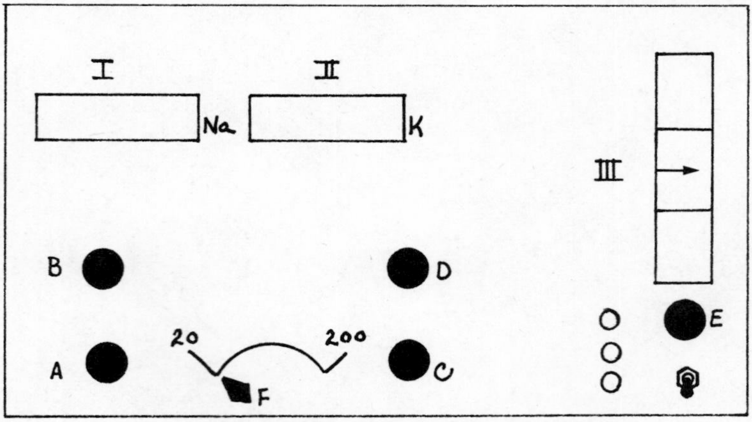

INSTRUMENT FACE

Ultramicro Serum Sodium and Potassium (Continued):

SPECIMEN: 10 microliters of unhemolyzed serum. Plasma may be used in place of serum of the anticoagulant used does not contain sodium, potassium, or lithium.

It is essential to the validity of the results that the serum or plasma be separated from the cells as soon as possible after specimen collection. A gross increase in the potassium concentration and decrease in sodium concentration will be noted if cells are not removed. Hemolysis of red cells will yield false elevation of potassium.

EQUIPMENT AND REAGENTS:

1. I.L. Flame Photometer, Model 143
   Aspiration rate should be 1.3 ml./minute

2. Ultramicrodilutor
   Sample syringe set to take up 10 microliters, and flush syringe set to deliver 2.0 ml. of working lithium diluent (1:200). The dilutor should be flushed out and cleaned periodically.

3. Standards for serum
   a. 120/2 mEq./liter of Na/K (I.L. Cat. #35120, 16 fluid ounces)
   b. 140/5 mEq./liter of Na/K (I.L. Cat. #35140, 16 fluid ounces)
   c. 160/8 mEq./liter of Na/K (I.L. Cat. #35160, 16 fluid ounces)

4. Standards for urine
   a. 50/50 mEq./liter of Na/K (I.L. Cat. #35050)
   b. 100/100 mEq./liter of Na/K (I.L. Cat. #35100)

5. Stock Lithium Diluent, 1500 mEq./liter
   I.L. Cat. #35000

6. Working Lithium Diluent, 150 mEq./liter
   Dilute 100 ml. of stock lithium diluent to one liter with deionized water. Good for one week.

COMMENTS ON PROCEDURE:

1. The same solution of working lithium diluent must be used for blank, standards, controls, and unknowns within a given run.

2. If either gas or air is not reaching the ignition chamber properly, the "no gas" and "no air" light will turn on. If this occurs, turn the air off and determine the cause of the insufficiency. When no air is reaching the ignition chamber, the

air safety valve automatically shuts off the flow of propane to the chamber.

3. Careful precise measurement of all solutions and dilutions is mandatory for accurate results, as is scrupulously clean glassware. Contamination of specimen or standards is one of the most frequent causes of inaccuracy.

4. The I.L. Flame Photometer may be calibrated with any available standard, but several standards must be used to cover the range of sample readings and to establish linearity.

PROCEDURE:

A. Ignition of Flame Photometer

1. Turn the propane tank valve completely counter-clockwise; the gas pressure must be at least 25 PSI.

2. Turn on the compressed air line from the wall completely; the air pressure must be at least 25 PSI and should not exceed 50 PSI.

3. Turn the "ON-OFF" toggle switch to the "ON" position. Allow 10-15 seconds for ignition to take place.

4. Check the drain from the atomizer section; it must be completely filled with fluid and must be thoroughly purged of air.

5. Fill a plastic vial with working lithium diluent, and place it on the sample stand directly beneath the atomizer. Raise the stand so that the tube is just clear of the bottom of the vial. Aspirate for at least five minutes before attempting zero calibration.

6. Take up 10 microliters of standard or sample, and flush with 2.0 ml. of lithium diluent into large AutoAnalyzer cups. Cap and mix. Prepare a duplicate dilution for each standard and sample.

B. Instrument Calibration and Operation

7. Using the working lithium diluent "blank", set the lithium indicator (III) in the center of the arrow head. The needle should not deflect beyond the range indicated by the parallel lines. Adjustment is made with knob "E".

Ultramicro Serum Sodium and Potassium (Continued):

8. Adjust the sodium digital readout (I) to 000.0 with knob "A", and lock the control. Adjust the potassium digital readout (II) to 000.0 with knob "C", and lock the control.

9. Sample analysis is essentially the same for serum or urine. However, when analyzing serum, the potassium range selector "F" is placed on '20', and when analyzing urine the range is extended to '200'.

10. Using the appropriate standards for the type of sample being analyzed, aspirate one of the standards, and set the digital readout for sodium and potassium with the respective balance knobs "B" and "D" to the known values for that standard.

11. Check the other standard or standards for linearity. All standards should agree within ± 1.5 mEq./L. for Na^+ and ± 0.5 mEq./L. for K^+.

12. Determine the control value and if within the range of standard deviation, continue with the analysis of the unknown specimens. Instrument calibration should be rechecked with standards after every fifth unknown.

13. After completion of the analysis, aspirate deionized water through the atomizer system for a minimum of five minutes.

C. Analysis of Urine Samples

14. Urine samples are diluted and analyzed in the same manner as sera except that "50/50" and "100/100" standards are used, and the potassium range selector "F" is extended to the "200" position.

D. Turning Off the Instrument:

15. After rinsing the unit with deionized water, turn off the Propane gas. It is not necessary to remove the beaker containing the deionized water.

16. When the "flame on" indicator light goes out, the solenoid valve will shut off the compressed air. Allow the automatic valve to purge moisture and oil from the air filter.

17. Turn-off the instrument and the air compressor by placing the "ON-OFF" switch to the "OFF" position.

18. Turn off the compressed air line from the wall.

Ultramicro Serum Sodium and Potassium (Continued):

NORMAL VALUES: Serum Sodium 138 - 146 mEq./Liter
Serum Potassium 3.7 - 5.3 mEq./Liter

REFERENCES:

1. Instrumentation Laboratory, Inc., Instruction Manual for Flame Photometer, Model 143.

2. E. A. Boling (Medical Service and Research Laboratory, Boston Veterans Administration Hospital, and the Department of Medicine, Tufts University School of Medicine): "A Flame Photometer with Simultaneous Digital Readout for Sodium and Potassium", J. Lab. and Clin. Med., pg. 501 - 510, March, 1964.

CLINICAL INTERPRETATION:

The common causes for hypernatremia are severe dehydration, excess intravenous utilization of saline and brain damage due to trauma or hypothalamic lesions. A high serum sodium may occur in Cushing's and Conn's Syndrome.

Hyponatremia is found in patients who receive excess intravenous glucose and thus suffer from water intoxication. Severe diarrhea may cause hyponatremia. Total body sodium is increased in heart failure but serum sodium is paradoxically lowered. Addison's disease is characterized by hyponatremia and hyperkalemia. The syndrome of inappropriate ADH which may be caused by an undifferentiated bronchogenic carcinoma is associated with hyponatremia.

Hyperkalemia is usually found in renal failure or inappropriate excessive administration of potassium. Excessive destruction of tissue and acidosis results in high serum potassium. Hyperkalemia is found in Addison's disease. In addition, excessive destruction of erythrocytes and platelets releases potassium into the plasma.

Hypokalemia results from severe diarrhea or utilization of diuretics or cortisone. Conn's and Cushing's Syndrome are associated with a hypokalemic metabolic alkalosis. A fall in serum potassium occurs with the administration of intravenous glucose and insulin since potassium is shifted from the serum into the cell. Finally, periodic paralysis is related to a decrease in serum potassium.

The usual ratio between urine sodium and urine potassium is approximately two parts sodium to one part potassium. With increased

Ultramicro Serum Sodium and Potassium (Continued):

aldosterone excretion, the urinary sodium-potassium ratio reverses. In Addison's Disease, the ratio of urinary sodium and potassium increases.

SORBITOL DEHYDROGENASE
(UV Methodology)

PRINCIPLE: The enzyme sorbitol dehydrogenase is found in its highest concentration in liver. The serum and hemolyzed blood of health individuals contains very little sorbitol dehydrogenase. However, with liver cell damage (toxic infectious or hypoxic in origin), the enzyme becomes demonstrable in serum specimens.

The reaction utilized by the methodology is stated below, with the loss of NADH being a measure of enzyme activity. NADH will absorb and can be measured at 340 nm. or 366 nm.

$$\text{d-Fructose} + \text{NADH} + H^+ \underset{}{\overset{\text{SDH}}{\rightleftharpoons}} \text{d-Sorbitol} + \text{NAD}^+$$

SPECIMEN: 1.0 ml. fresh serum, free from hemolysis. Heparinized plasma does not affect the reaction, but the presence of EDTA is detrimental to enzyme activity.

REAGENTS:

Reagent Kit available as Biochemica Test Combination (TCN #15960 TSAB).

1. Triethanolamine Buffer, 0.2 M., pH 7.4
   Dissolve the contents of Bottle No. 1 in 60 ml. deionized water. Stable for one year at room temperature; prevent evaporation.

2. NADH, 0.015 M.
   Dissolve the contents of Bottle No. 2 in 1.5 ml. of deionized water. Stable for 4 weeks at approximately 4°C.

3. Fructose, 1.65 M.
   Dissolve the contents of Bottle No. 3 in 6.0 ml. of deionized water. Stable for four weeks at approximately 4°C.

COMMENTS ON PROCEDURE:

1. Pre-incubation of specimen with NADH allows exhaustion of metabolites in serum which oxidize NADH (with the aid of dehydrogenases also in serum).

2. Explanation of factor for 340 nm. wavelength:

   A. $\frac{\text{Absorbance}}{\text{Absorptivity}} \times \frac{1}{t} \times \frac{\text{T.V.}}{\text{S.V.}}$

Sorbitol Dehydrogenase (Continued):

B. $\frac{0.001}{6.22} \times \frac{1}{10} \times \frac{3.25}{1.00} = 52.1$

PROCEDURE:

1. Into a test tube, pipette 2.0 ml. of triethanolamine buffer (Bottle No. 1), 0.05 ml. NADH (Bottle No. 2), and 1.0 ml. serum specimen.

2. Mix contents well by inversion and preincubate in 25°C. water bath for approximately 45 minutes.

3. At the end of preincubation, add 0.2 ml. of fructose substrate (Bottle No. 3). Mix contents and pour into a glass cuvette with a 1.0 cm. light path.

4. Take an absorbance reading (E_1) at 340 or 366 nm. Replace the cuvette in the 25°C. water bath.

5. Take a second absorbance reading (E_2) exactly 10 minutes after the first reading.

6. With ten minute absorbance differences greater than 0.300 (366 nm.) or 0.600 (340 nm.), prepare a dilution of the specimen. Mix 0.2 ml. of serum with 1.8 ml. physiological saline and repeat the assay on 1.0 ml. of this dilution. Multiply the result by 10.

7. CALCULATION:

   $E_1 - E_2 = \Delta E$

   ΔE (at 340 nm.) x 52.1 = milliunits/ml. serum

   ΔE (at 366 nm.) x 98.5 = milliunits/ml. serum

NORMAL VALUES: Less than 0.4 milliunits/ml. serum

REFERENCES:

1. Bergmeyer: METHODS OF ENZYMATIC ANALYSIS, Academic Press, 1965.

2. Asada, M. and Galambos, J.: Gastroenterology, 44:578, 1963.

Sorbitol Dehydrogenase (Continued):

CLINICAL INTERPRETATION:

Sorbitol is a polyhydric alcohol derived from glucose. Conversion of sorbitol to glucose takes place in the liver catalyzed by the action of the reductase sorbitol dehydrogenase. Sorbitol dehydrogenase is mainly a liver enzyme. Minimal serum sorbitol dehydrogenase activity is present in the healthy adult. Myocardial infarction, skeletal muscle disease or kidney necrosis does not lead to elevated serum levels.

Sorbitol dehydrogenase is usually elevated to the greatest degree in acute hepatitis. Biliary tract obstruction does not cause elevated levels unless it is longstanding and then only when it is associated with parenchymal cell damage. The liver specificity of this enzyme is clinically emphasized by the presence of normal levels of sorbitol dehydrogenase in myocardial infarction. Even in disease of the prostate and kidney where there is an appreciable amount of tissue sorbitol dehydrogenase, serum levels are not increased.

Sorbitol dehydrogenase is easy to assay and shows promise as a valuable enzyme determination in clinical diagnosis of liver disease.

SPINAL FLUID PROTEIN
(Turbidimetric)

PRINCIPLE: Turbidimetric determination of proteins is a relatively old technique for the estimation of the protein content of body fluids. Optimally, one wishes to form a fine protein precipitate of uniform particle size in an acid solution. The acid solution should have the lowest concentration which will still give sensitivity. It is essential that the particle size be small and uniform. For this reason, turbidimetric techniques are best used on solutions of low protein concentration such as CSF. Trichloroacetic acid precipitates albumin and globulins with nearly equal efficiency and has a relatively good range of linearity when used in a final solution concentration of 5%. The methodology is preferred over other more accurate ones, due to the speed and simplicity of the measurement. Turbidity may be measured at any wavelength, with sensitivity increasing toward the ultraviolet end of the spectrum.

SPECIMEN: 95 microliters of centrifuged spinal fluid. Hemolysis from a bloody tap will increase the protein 1.0 mg%. for every 750 RBC/cu. mm.

REAGENTS AND EQUIPMENT:

1. Trichloroacetic Acid, Stock Solution, 20% (W/V)

2. Trichloroacetic Acid, 5% (W/V)
   Dilute one volume of Stock TCA (20%) with three volumes of water. Make fresh weekly and store at room temperature.

3. NaCl, 0.9%
   In a liter volumetric flask dissolve 9 gm. NaCl in approximately 300 ml. deionized water and dilute to volume.

4. Stock Protein Standard
   Prepare a serum pool collected from several clear, nonicteric, fresh sear. Mix well, and assay in triplicate according to the biuret method.

5. Working Protein Standards
   The following saline dilutions of Stock Standard (about 6.0 gm%.) are stable for several days at refrigerator temperature.

Spinal Fluid Protein (Continued):

A. ± 180 mg%. Protein - Place 3.0 ml. Stock Protein Standard into a 100 ml. volumetric flask, and dilute to volume with saline.

Exact mg%. = Gm%. Stock Standard x 30.

B. ± 120 mg%. Protein - Place 2.0 ml. Stock Standard into a 100 ml. volumetric flask, and dilute to volume with saline.

Exact mg%. = Gm%. Stock Standard x 20.

C. ± 60 mg%. Protein - Place 1.0 ml. Stock Standard into a 100 ml. volumetric flask, and bring to volume with saline.

Exact mg%. = Gm%. Stock Standard x 10.

D. ± 30 mg%. Protein - Place 0.5 ml. Stock Standard into a 100 ml. volumetric flask, and bring to volume with saline.

Exact mg%. = Gm%. Stock Standard x 5.

Using ultramicro dilutor, analyze Standards in same manner as unknowns in following "Procedure" section. Prepare a calibration curve for daily use.

6. <u>Ultramicro Dilutor</u>
Flush syringe should be set to deliver 0.8 ml. of 5%. TCA. Sample syringe should be set to take up 95 microliters.

7. <u>Gilford 300 N</u>
Fitted with microaspiration cuvette.

COMMENTS ON PROCEDURE:

1. It is essential to prepare a calibration curve with serum rather than an albumin standard such as Armour. Dilute solutions of the latter are not stable over a one day period.

2. Since the accuracy of a turbidimetric method is directly affected by the size of the precipitate particle, spinal fluids with high protein content tend to flocculate and should be repeated on dilution.

Spinal Fluid Protein (Continued):

3. The temperature of the TCA affects the amount and type of protein precipitated. Keep at room temperature.

PROCEDURE:

1. Using the ultramicro dilutor, take up 95 microliters of CSF specimen and flush with 800 microliters of 5% TCA into A.A. cups. Immediately cap and mix each aliquot. Using the same technique, a control specimen and reagent blank, substituting water for sample, are prepared.

2. After 10 minutes, determine absorbance compared to a reagent blank at 350 nm. Remix each sample shortly before taking a reading in the Gilford microaspiration cuvette.

3. Take the protein value from the precalibrated curve, or calculate concentration according to the following formula:

$$\text{Conc. Sample mg\%.} = \frac{\text{A. of sample}}{\text{A. of standard}} \times \text{Concentration of Std. mg\%.}$$

Samples greater than 180 mg%. should be repeated on double dilution.

NORMAL VALUES: 20 - 40 mg%.

REFERENCES:

1. Henry, R., Sobel, C., Segalove, M.: "Turbidimetric Determination of Proteins with Sulfosalicylic and Trichloroacetic Acids", Proceedings: Soc. Exp. Biol. & Med., 92:748, 1956.

2. Meulemans, O.: "Determination of Total Protein in Spinal Fluid with Sulfosalicylic and Trichloroacetic Acid", Clin. Chim. Acta, 5:757, 1960.

3. Schriever, H., Gambino, R.: "Protein Turbidity Produced by Trichloroacetic Acid and Sulfosalicylic Acid at Varying Temperatures and Varying Ratios of Albumin and Globulin", Am. J. Clin. Path., 44:667, 1965.

4. Henry, R.: CLINICAL CHEMISTRY: PRINCIPLES AND TECHNIQUES, Hoeber, pg. 186 - 189, 1964.

Spinal Fluid Protein (Continued):

CLINICAL INTERPRETATION:

The protein of the CSF is 15 to 45 mg. per 100 ml. obtained by lumbar puncture and 10 to 25 mg. per 100 ml. from the cisternal fluid. The blood-brain barrier is immature at birth, and the CSF protein is approximately 100 mg. per 100 ml. up to the age of one month. The usual cause for an increase in the CSF protein is bacterial, viral, or fungal inflammation. Capillary and cell permeability are increased in these conditions. Part of the increased CSF protein in inflammation is secondary to the protein derived from the inflammatory exudate.

Other etiologies for an increase in CSF protein are neoplastic disease, cerebrovascular conditions, such as cerebral thrombosis and hemorrhage, toxic states as uremia, Grand Mal convulsions, demyelinating diseases as multiple sclerosis, myxedema, and following a myelogram.

TOTAL SERUM PROTEIN AND ALBUMIN DETERMINATION
(Biuret Reaction)

PRINCIPLE: Protein reacts with cupric ions in an alkaline sodium potassium tartrate solution to form a complex colored compound. Any compound that has in its molecular structure pairs of carbamyl groups linked through nitrogen or carbon (or peptide linkages) will show a positive biuret reaction. The name of the test is derived from the simplest of such compounds, biuret:

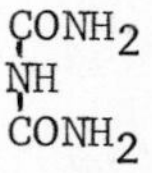

When biuret is treated with an alkaline potassium copper tartrate solution two biuret molecules are joined to form the following complex violet-colored compound:

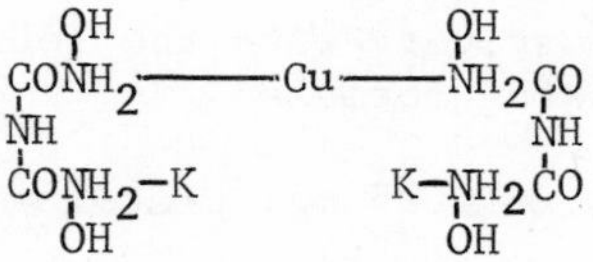

Since proteins contain these linkages, they react to give the characteristic color which is directly proportional to the amount of protein present. Total serum protein is determined by comparing the intensity of a colored solution of dilute protein with that of a known protein concentration. Albumin may be determined by fractionation of the serum before treatment with biuret reagent. The globulin portion is removed by precipitation with 26.88% sodium sulfite, and extraction with ether. The globulin is calculated as the difference between the albumin and the total protein concentrations.

SPECIMEN: 0.5 ml. of unhemolyzed serum is required for the Total Protein determination; however, a 0.1 ml. specimen may be used. 0.5 ml. of unhemolyzed serum is required for the albumin determination. Fasting specimens are preferred in order to avoid lipemia of the serum. Icterus does not influence the specificity of the color reaction up to 29 mg%. of bilirubin. Never use serum which contains BSP dye. Specimens may be stored up to a week at refrigerator temperatures or for several months frozen.

Total Serum Protein and Albumin Determination (Continued):

REAGENTS:

1. Saline, 0.9%

2. Biuret Reagent
Into a 4 liter volumetric flask, place 6.0 gm. cupric sulfate ($Cu_2SO_4 \cdot 5H_2O$) and 24 gm. sodium potassium tartrate ($NaKC_4H_4O_6 \cdot 4H_2O$), dissolving in 2 liters deionized water. Add 1.2 liters 10% NaOH and bring to volume with deionized water. Dissolution may be aided by a magnetic stirrer.

3. Sodium Sulfite, 26.88%
Bring 268.8 gm. Na_2SO_3 (anhydrous) to volume with deionized water in a 1000 ml. volumetric flask.

4. Calibration Curve - Total Protein
A sterile standard protein solution of bovine albumin (Armour) which has been standardized by assay of its protein nitrogen content is used in setting up the calibration curve. The assay of nitrogen content may vary with the lot, and is converted to protein by the following formula:

1.0 mg. protein N/ml. = 6.25 mg. protein/ ml.

5. Working Standard Solutions
Prepare the Working Standard Solutions in duplicate according to the table beneath:

Standard Number	1	2	3	4
Ml. Stock Standard	1.0	1.0		0.5
Ml. Standard 1			1.0	
Ml. Saline Diluent	4.0	5.0	2.0	3.0
Final Dilution	1:5	1:6	1:10	1:7

Vortex all tubes to insure thorough mixing.

6. Dilutions of Working Standards for Calibration Curve
Label a duplicate set of 7 test tubes for the calibration curve; include an extra tube which contains 1.0 ml. saline for a Reagent Blank. Prepare the final dilutions for the calibration curve from each set of Working Standards according to the following chart:

Total Serum Protein and Albumin Determination (Continued):

Tube Number	1	2	3	4	5	6	7	Blank
Ml. Working Std. 1.				0.5			1.0	
Ml. Working Std. 2.			0.5			1.0		
Ml. Working Std. 3.	0.5							
Ml. Working Std. 4.		0.5			1.0			
Ml. Saline	0.5	0.5	0.5	0.5				1.0
Ml. Biuret Reagent	4.0	4.0	4.0	4.0	4.0	4.0	4.0	4.0

Mix each tube thoroughly, and time the color development for 30 minutes. After 30 minutes, read the O.D. of each tube at 550 nm. in a 12 mm. cuvette. Average the readings for the two sets of standard solutions. Calculate the gm%. represented by each tube according to the following:

Gm%. Protein = mg./ml. Protein in Stock Std. x Dil. of Stock x Dil. of Wkg. Std. x 100

Plot the concentration against the optical density on linear graph paper.

7. Albumin Calibration Curve
   The same optical densities noted in the determination of the calibration curve for total protein (above) may be used for the calculation of the albumin calibration curve. One simply alters the formula to account for the difference in specimen dilution. Calculate as follows:

$$\text{Concentration of Total Protein} \times \frac{0.100}{0.132} \text{ (Ratio of Amts. of Sera used)} = \text{Gm\%. Albumin}$$

COMMENTS ON PROCEDURE:

1. Always check each tube for turbidity before reading in the spectrophotometer. Turbid solutions must be extracted with ether along with a Reagent Blank.

2. Each new reagent preparation must be checked against the old solution before being put into use.

3. The stability of the biuret reaction is questionable; read the determination as soon as possible after the 30 minute color development.

Total Serum Protein and Albumin Determination (Continued):

PROCEDURE:

Total Protein

1. Into a test tube measure 0.5 ml. of the serum specimen. If 0.1 ml. of specimen is to be used, rinse the specimen carefully into 0.9 ml. of saline and use the entire volume for the color reaction.

2. Into a second test tube measure 0.5 ml. of the Control serum.

3. Into a third test tube measure 1.0 ml. of saline for a "Reagent Blank".

4. Pipette 4.5 ml. saline into the tubes containing 0.5 ml. of the specimen or the control serum. Mix thoroughly; preferably by vortexing.

5. Transfer 1.0 ml. of the serum dilutions from Step No. 4 to clean tubes.

6. Add 4.0 ml. biuret reagent to the serum dilutions in Step No. 5 of both the specimen, the control serum, and to the saline "Blank". Mix all tubes thoroughly. Allow 30 minutes for color development.

7. After 30 minutes, read the optical density of each tube against the reagent blank at 550 nm. in a 12 mm. cuvette. CHECK FOR TURBIDITY!

8. Determine gm%. Total Protein from the calibration curve. The determination follows Beer's Law up to 15 gm%.

Albumin

1. Into a glass stoppered centrifuge tube pipette 0.5 ml. serum control.

2. Into a second glass stoppered centrifuge tube pipette 0.5 ml. specimen.

3. Pipette 2.0 ml. sodium sulfite into a test tube and set aside for the "Blank".

Total Serum Protein and Albumin Determination (Continued):

4. To the centrifuge tubes containing the control or the specimen, add 7.0 ml. sodium sulfite solution. Replace stoppers and mix gently 10 times by inversion. Place tubes in a rack for 5 minutes to allow for complete precipitation of the globulin.

5. Add 3.0 ml. ether to each of the tubes in Step No. 4 and shake gently 10 times.

6. Centrifuge these tubes for 10 minutes at 2000 rpm.

7. Remove the tubes from the centrifuge tubes and gently disengage the glass stoppers. Tilt the tube to the side to loosen the globulin plug and volumetrically pipette 2.0 ml. of the aqueous layer; transfer this solution into a clean test tube.

8. To the test tubes containing 2.0 ml. of the aqueous layer and the "Blank", add 3.0 ml. biuret reagent. Mix each tube thoroughly. Allow 30 minutes for color development.

9. After 30 minutes, read the optical density of each tube against the reagent blank at 550 nm. in a 12 mm. cuvette. CHECK FOR TURBIDITY!

10. Determine the gm%. albumin from the calibration curve for albumin.

11. To determine the gm%. globulin, subtract the gm%. albumin from the gm%. Total Protein.

NORMAL VALUES:

Total Protein:	6.0 - 8.0 Gm%.
Albumin:	3.5 - 5.0 Gm%.
Globulin:	1.5 - 3.0 Gm%.

REFERENCES:

1. Garnal, et. al.: "Total Serum Protein (modified Kingsley)", J. Biol. Chem., 177:751, 1949.

2. Campbell and Hanm: "Albumin Fractionation by the Sulfite Method of Campbell & Hanm", J. Biol. Chem., 119:9, 1937.

Total Serum Protein and Albumin Determination (Continued):

CLINICAL INTERPRETATION:

Total serum protein is made up of two fractions, the globulin fraction and albumin fraction. Each will be discussed individually:

Albumin

Elevated serum albumin is seldom encountered. Elevated serum albumin usually signifies that the patient is dehydrated.

The most common cause for a low serum albumin in a hospitalized patient is excessive intravenous infusion of glucose in water.

Liver cirrhosis results in hypoalbuminemia because of a decreased synthesis by the pathologic liver. Malnutrition may also serve as a major cause for low serum albumin.

Low serum albumin may result from loss or lack of absorption related to gastrointestinal disease such as sprue, ulcerative colitis, villous adenoma, or protein-losing enteropathy. Prominent proteinuria associated with the various causes for nephrosis results in marked hypoalbuminemia.

Finally a large amount of albumin occurs in diffuse bullous dermatitis or an extensive burn with loss of albumin from the skin lesion.

Globulins

An increase in $alpha_1$ globulin may occur in hepatoma or choriocarcinoma. The increase is due to the presence of a fetal protein produced by the neoplasm. Malignancy may result in an increase in glycoprotein which migrates as an alpha globulin. The hereditary lipid disorders may also cause an increase in the $alpha_1$ globulins. A decrease in $alpha_1$ globulin occurs in the nephrotic syndrome due to loss into the urine and in hereditary anti-trypsin deficiency associated with prominent emphysema of the lung.

$Alpha_2$ globulin acts as an acute phase protein. The main components are ceruloplasmin and haptoglobin. $Alpha_2$ globulin may increase with inflammation, tissue necrosis, immunologic disease and neoplasm including the various lymphomas. Pregnancy and use of oral contraceptives increases ceruloplasmin due to stimulation of hepatic synthesis. Low ceruloplasmin occurs in Wilson's hepatolenticular disease and with excessive urinary loss in nephrosis. A lower haptoglobin is found in hemolytic anemia. Haptoglobin combines with free plasma hemoglobin and protects the renal tubule. Severe disease and nephrosis may also cause a low haptoglobin.

The beta globulins consist of transferrin and apoprotein. An increase in beta lipoprotein occurs in the hereditary lipid disorders. Lipoprotein globulins also increase in obstructive jaundice, diabetes mellitus and nephrotic syndrome. Transferrin increases in iron deficiency anemia. Transferrin decreases with urinary loss in nephrosis or with decreased synthesis in chronic infections, neoplastic disease or chronic liver disease.

Hypergammaglobulinemia may present as a polyclonal or monoclonal gammapathy.

Polyclonal States

1. Chronic liver disease
2. Collagen diseases
3. Chronic infections
4. Malignancy including lymphomas

Monoclonal States

1. Plasma cell dyscrasias
2. Multiple myeloma
3. Solitary myeloma
4. Waldenstrom's Macroglobulinemia
5. Amyloidosis
6. Rarely chronic infections
7. Rarely epithelial neoplasms

Hypo or agammaglobulinemia may become symptomatic late in the first year of life with frequent bacterial infections. The reticuloendothelial system has decreased lymphocytes and plasma cells. The condition is hereditary. Acquired hypo or agammaglobulinemia may be acquired and is caused by lymphomas, metastatic malignancy to reticuloendothelial system, use of cytotoxic drugs, nephrosis, protein-losing enteropathy, exfoliative dermatitis, or malnutrition.

SERUM TOTAL PROTEIN
(Ultramicro)

PRINCIPLE: Protein reacts with cupric ions in an alkaline sodium potassium tartrate solution to form a complex colored compound. Any compound that has in its molecular structure pairs of carbamyl groups linked through nitrogen or carbon (peptide linkages) will show a positive biuret reaction, resulting in a violet-colored solution. Proteins contain these linkages and give the characteristic color which is directly proportional to the amount of protein present. Specimen lipemia of even moderate amount will give turbidity in the final solution and will significantly elevate the result. For this reason, all tubes are arbitrarily exposed to ether extraction.

SPECIMEN: 12.5 microliters of unhemolyzed serum.

REAGENTS AND EQUIPMENT:

1. <u>10% NaOH</u> (W/V)
   To 300 ml. deionized water in a 500 ml. volumetric flask, add 100 ml. 50% NaOH (W/V). Mix and bring to volume with deionized water.

2. <u>Biuret Reagent</u>
   Into a 1 liter flask, place 1.5 gm. cupric sulfate ($CuSO_4 \cdot 5H_2O$) and 6.0 gm. sodium potassium tartrate ($NaKC_4H_4O_6 \cdot 4H_2O$), and dissolve in approximately 500 ml. deionized water. Add 300 ml. of 10% NaOH and bring to volume with deionized water. Dissolution may be aided by a magnetic stirrer.

3. <u>Ether</u>

4. <u>Brij.-water</u>, approximately 0.1%.
   Add 1.0 ml. Brij.-35 to 1000 ml. deionized water.

5. <u>Ultramicro Dilutor</u>
   Sample syringe set to take up 12.5 microliters and dispensing syringe set to deliver 1.0 ml. of biuret reagent.

6. <u>Gilford 300 N Spectrophotometer</u>
   Equipped with micro evacuation cuvette.

Serum Total Protein - Ultramicro (Continued):

7. Protein Standards

   A. Human Albumin Fraction V (or Mercaptalbumin), approximately 10 - 11 gm%. protein. The exact concentration of protein must be determined from the Kjeldahl protein nitrogen assay for the lot. Convert by the following formula:

   1 mg./ml. protein nitrogen = 6.25 mg./ml. protein

   B. Bovine Albumin, Armour Pharmaceuticals; approximately 6.0 gm%. Convert nitrogen value to Gm%. according to above formula.

   Standards may be diluted quantitatively with 0.9%. saline for intermediate points on the curve.

   C. American Monitor Albumin and Total Protein Standards #1010.

PROCEDURE:

1. Using the microdilutor, dispense 1.0 ml. biuret reagent into a 10 x 75 mm. tube for a reagent blank.

2. Take up 12.5 microliters of standard, control or specimen and dilute with 1.0 ml. biuret reagent, dispensing into 10 x 75 mm. tubes.

3. Vortex all tubes thoroughly and allow 30 minutes for color development.

4. Extract lipids from all tubes by vortexing each solution with a small volume of ether. Centrifuge all tubes for a short period of time and suction off ether layer.

5. Determine the absorbance of all tubes at 540 nm. with the reagent blank as reference. Use the Gilford 300 N with micro evacuation cuvette; flush cuvette after each reading with dilute Brij. solution or other wetting agent.

6. Linearity of standards may be checked by plot on linear graph paper. Calculate the concentration of the controls and specimens from the following formula:

$$\text{Gm\%. Protein} = \frac{\text{Conc. Standard}}{\text{A. of Standard}} \times \text{A. of Specimen}$$

Serum Total Protein - Ultramicro (Continued):

NORMAL VALUES:

1. First Week (Full-term)
   Protein = 4.65 - 7.4 Gm%.

2. 12 Months (Full-term)
   Protein = 6.1 - 6.7 Gm%.

3. Four Years +
   Protein = 6.15 - 8.1 Gm%.

REFERENCES:

1. Henry, R. B.: CLINICAL CHEMISTRY: PRINCIPLES AND TECHNIQUE Hoeber, pg. 182 - 186, 1964.

2. Ibbott and O'Brien: LABORATORY MANUAL OF PEDIATRIC MICRO-BIOCHEMICAL TECHNIQUES, Hoeber, 1968. (Normal Values)

CLINICAL INTERPRETATION:

Refer to Interpretation section under Determination of Total Serum Protein and Albumin (Biuret Reaction) on page 358.

THYMOL TURBIDITY
(Shank-Hoagland)

PRINCIPLE: Serum added to a barbital buffer containing thymol will result in a turbidity which can be measured photometrically. The precipitate produced is a protein-thymol-phospholipid complex. The test is empirical, the turbidity reaction not being completely understood. It has been shown that the protein portion is probably a Gamma-globulin. The phospholipid contains both cholesterol and lecithin. The Shank-Hoagland modification uses a barium chloride (turbidity) standard. The test is read at 660 nm.; this wavelength is not the spectral absorbance peak but is less affected by icterus or hemolysis of the serum specimen. This is a relatively sensitive test of hepatocellular damage. Its primary value is in the differentiation of extrahepatic from parenchymal jaundice, and in the following of the course of active parenchymal damage.

SPECIMEN: 0.1 ml. serum. A fasting specimen is preferred. Lipemia may result in false positives.

REAGENTS:

1. Bacto-Thymol Turbidity Reagent
   This is prepared Reagent No. 328 from Difco Laboratories, pH 7.55. Filter before using if at all cloudy.

2. Stock $BaCl_2$ Turbidity Standard
   Place 5.0 ml. of 0.0962 N. $BaCl_2$ (1% solution using anhydrous $BaCl_2$) into a 100 ml. volumetric flask and dilute to volume with cold ($10^oC.$) 0.2 N. sufluric acid (1%). This temperature gives a stable suspension.

3. Calibration Curve
   Prepare the turbidity Standard Curve according to the following chart. Stock Standard should be mixed well between each pipetting, and the 0.2 N. sulfuric acid (1%) should be at 10^oC.

Units of Turbidity	Ml. Stock Standard	Ml. 0.2 N. H_2SO_4
5	1.35	8.65
10	2.70	7.30
15	4.05	5.95
20	5.40	4.60
25	6.75	3.25
30	8.10	1.90
35	9.45	0.55

Thymol Turbidity (Continued):

After adding the sulfuric acid to the Stock Standard, shake the mixture and read immediately in a 19 x 105 cuvette against a water blank at 660 nm.

COMMENTS ON PROCEDURE:

If the buffer is at all turbid, filter before using.

PROCEDURE:

1. Prepare three test tubes each containing 6.0 ml. of Bacto-Thymol Turbidity Reagent.

2. Pipette 0.1 ml. deionized water into one labelled "Blank".

3. Pipette 0.1 ml. serum specimen into one labelled "Patient".

4. Pipette 0.1 ml. serum control into one labelled "Control".

5. Mix each tube thoroughly and allow 30 minutes for development of turbidity.

6. Remix all tubes and read against the "Blank" at 660 nm. using 19 x 105 cuvettes. Determine units of turbidity from calibration curve.

Thymol Flocculation

7. If a thymol flocculation is desired, place the tubes in the dark for 18 hours and read as a Cephalin Flocculation.

NORMAL VALUES: Less than 5 Units.

REFERENCES:

1. Hepler, Opal, MANUAL OF CLINICAL LABORATORY METHODS, Ed. 4, Thomas, 1957, pg. 123.

2. DIFCO MANUAL, Ed. 9, Difco Laboratories, Detroit, Michigan.

3. Henry, J. B.: CLINICAL CHEMISTRY: PRINCIPLES AND TECHNIQUES Hoeber, pg. 562 - 569, 1965.

4. Sunderman, F. W., and Sunderman, F. W., Jr.: LABORATORY DIAGNOSIS OF LIVER DISEASES, pg. 252 - 257, 1968.

Thymol Turbidity (Continued):

5. ASCP Manual: LIVER FUNCTION TESTS, J. B. Fuller Editor, pg. 22 - 27, 1966.

CLINICAL INTERPRETATION:

This important test is based upon the observation that addition of thymol solution to serum containing an elevated globulin results in marked turbidity.

Little or no such reaction occurs in normal serum. The degree of turbidity expressed in units may be measured by comparison against a turbidity standard, e.g. Barium Sulfate. The normal range is 1 to 4 Units usually read at 30 minutes. The Thymol Turbidity Test is employed most extensively in the investigation of patients with hepatobiliary disease, but may be abnormal in non-hepatic diseases associated with hyperglobulinemia.

THYROXIN (T_4) BY COLUMN
(Using the Automated Bio-Bromine Method)

PRINCIPLE: Determination of serum levels of thyroxin (T_4) involves the use of ion-exchange column chromatography to separate T_4 from other serum iodine constituents. Thyroxin is quantitated in acetic acid column eluates. Eluates are added to 2 M. sulfuric acid and bromine working reagent and segmented with air. Bromine generated by the acid mixture reacts with the eluates and replaces the iodine on the thyroxin, releasing free iodide. Arsenious acid reagent is added and then ceric ammonium sulfate is added. The colored ceric ion is reduced to the colorless cerous ion by the catalytic amounts of iodide derived from the thyroxin:

$$\text{Bromination } T - I_4 + 2Br_2 \longrightarrow T - Br_4 + 2I_2$$

$$2Ce^{+4} + 2I^- \longrightarrow 2Ce^{+3} + I_2$$

$$AsO_2^- + I_2 + H_2O \longrightarrow AsO_3^- + 2I^- + 2H^+$$

After the reaction mixture passes through two 40 feet glass coils at 56°C., the mixture enters the colorimeter and measurements are made at 420 nm. with a 15 mm. flow cuvette.

The Bio-Rad T_4 Column Test contains four phases:

A. Adsorption:
Thyroid hormones are freed into solution when the serum is mixed with base. Optimum time for this reaction is 20 minutes. The solution is then applied to the column which has been primed with the base. The resin adsorbs the proteins and hormones, organic contaminents and inorganic iodide.

B. Non-Hormonal Elutions:
Proteins, iodotyrosines, and organic contaminants are washed off the column and discarded.

C. Hormonal Elutions:
The hormones to be measured are moved down the resin column, eluted and collected in two fractions; first elutions and second elution.

D. Analysis:
Quantitation of isolated thyroid hormone is accomplished by the method described above.

Thyroxin (T_4) by Column (Continued):

SPECIMEN: 1.5 ml. of unhemolyzed serum in a clean tube. Sera is refrigerated. Slight hemolysis does not interfere with test but marked hemolysis will result in low levels as the red cells contain no iodine. Specimen need not be collected in the post-absorptive state.

REAGENTS:

A. Column Chromatographic Procedure:

1. Sodium hydroxide, 0.15 N.

2. Sodium Acetate - Isopropyl alcohol

3. Acetic Acid, 0.2 N.

4. Concentrated Acetic Acid

5. Acetic Acid, 50%

B. Bio-Bromination:

1. 4.1 N. H_2SO_4

2. Bromide - Bromate Solution
Weigh 1.430 gm. Potassium Bromide, Mallinkrodt, Analytical Reagent; 0.449 gm. Potassium Bromate, Mallinkrodt, Analytical Reagent; 0.0528 gm. Reagent grade NaOH and add to 500 ml. deionized water, stirring continuously. Cool, and dilute to volume in a liter volumetric flask with deionized water.

3. Arsenious Reagent
Weigh 5.0 gm. of Arsenic trioxide, Matheson Coleman and Bell, Primary Standard, Reagent Powder and 3.0 gm. of Reagent Grade NaOH and add to 500 ml. of deionized water in a liter volumetric flask. Mix and dissolve. Add 274 ml. Reagent Grade Sulfuric acid, cool, and dilute to volume with deionized water.

4. Ceric Ammonium Sulfate
Weigh 5.34 gm. of Ceric Ammonium Sulfate (Frederick Smith) in a liter volumetric flask. Add 113 ml. of Reagent Grade sulfuric acid. Stir, break up clumps, and dilute to volume with deionized water. Let sit overnight, filter, and store in a brown bottle.

Thyroxin (T_4) by Column (Continued):

C. Preparation of Standard:

Use Hyland Special Chemistry Control Sera. Open six vials. Add 10 ml. deionized water into one vial. This will be S_o and will equal to one-half the value of the assayed mean. Add 5.0 ml. deionized water in each of 3 vials. Use one vial and label S_I, and it will equal the assayed mean. Pool 2 vials together and use 5.0 ml. of each to reconstitute 2 separate vials of Hyland Special Control. These will be labeled S_{II} and will equal two times the assayed mean. Label properly and refrigerate when not in use.

Use of a chemical standard is being investigated with the problem of recovery from aqueous or protein standard presenting a problem of stability and reproducibility.

COMMENTS ON PROCEDURE:

1. Sample at 50 per hour with 2 to 1 cam.

2. Transmission tubing can be either Tygon or Glass. Tygon will eventually stain after use and should be replaced.

3. If less sensitivity or a broader range Calibration Curve (up to 12 ug%.) is desired, decrease sampling tubing from yellow to white (0.04), increase sulfuric acid tubing from orange to white (0.04), and do not change other tubing sizes.

4. Adjusting the AutoAnalyzer
Operation of the colorimeter and recorder for the T-4 Automated Bio-Bromine Method is similar to any "Inverse Colorimetric Technique" as described in Technicon Methodology file, such as the Glucose System (N-2b I/II). The T-4 procedure measures the loss of color of the ceric reagent baseline when, catalyzed by thyroxin, it reduces to the colorless cerous ion.

   A. Insert 420 nm. filters in both reference and sample slots in the colorimeter. Turn on colorimeter lamp (30 minute warm-up) and recorder power (5 minute warm-up). Turn on recorder chart drive after colorimeter and recorder are warmed up.

   B. Immediately pump distilled water through reagent and sample lines. After five minutes make sure that the bubble pattern is even, all connections are tight with no leaks and the flow system is working properly.

C. Set the 100% T knob (Potentiometer) to midscale - 500. Find the aperture for the reference side of the colorimeter that gives a reading closest to but lower than 98% T. Fine-adjust the water baseline to 98% T by adjusting the 100% T knob. Do NOT remove the aperture; once determined, it should always be used for T-4 analysis.

D. Adjust the recorder gain according to Technicon Service Manual. Insert the 0 aperture in the sample side of the colorimeter and adjust the "Zero" knob until the recorder reads 0% T, then remove the 0 aperture.

E. Readjust the water baseline to 98% T by turning the 100% T knob. Repeat the 0% T and 98% T adjustments if necessary, but do not readjust the gain; once adjusted, the gain should not vary.

F. After a steady water baseline of 98% T ($\pm$ 2% T) is established place all reagent lines in the proper reagent bottles. The Automated Bio-Bromine reagents will establish the proper baseline of 15% T ($\pm$ 5% T) if the above procedure has been followed. If the baseline falls outside the specified range, check the flowcell for position and cleanliness.

5. Change deionized water every day.

6. Do not use dried-out columns.

7. Change manifold every 8 weeks.

8. When the alkali is added to the serum for the T-4 to be released, mix the contents of the tube gently to prevent any foaming.

9. Tobacco smoke contains iodine, so smoking in the laboratory should be minimized, particularly when solutions are being pipetted into the sample eluates.

Thyroxin (T_4) by Column (Continued):

PROCEDURE:

1. Screening Specimens
Normal T-4 values are considered by many to range from 3 to 6.5 micrograms per 100 ml. serum. Grossly elevated levels of iodine (17 micrograms/100 ml. or more) are often encountered due to contamination of specimen with iodine-containing compounds. To avoid introducing these samples into the analytical system, a rapid screening of samples is carried out. Into a heavy-walled ignition tube, place 2 drops of serum and 0.3 ml. of acid digestion mixture (1:1). Place in a digestion block for 10 minutes at 195°C. After specimen is completely digested and is clear, remove from the heating block and cool with a blowing fan. Work under hood and avoid breathing fumes. Then add 3.0 ml. of 1:1 dilution of 0.2 N. Arsenious acid and 0.04 M. Ceric ammonium sulfate solution. Place in a 45°C. water bath for 10 minutes and check for decolorization. Compare against 10, 20, 30 and 50 microgram standards that are set up the same way at the same time. Note samples that are high and make proper dilutions before serum incubation with base.

2. Column Chromatography
Set up a column for each Reagent Blank, Standards, Controls, and patient's specimens.

   A. Into clean 16 x 100 mm. tubes, place samples in the following order:

      Tube 1, Reagent Blank, 1.5 ml. deionized water
      Tube 2 0.75 ml. of S_o
      Tube 3 1.50 ml. of S_o
      Tube 4 1.50 ml. of S_I
      Tube 5 1.50 ml. of S_{II}
      Tube 6 2.00 ml. of S_{II}

      And 1.5 ml. of patient's serum and controls in appropriately labelled tubes. (Add 0.5 ml. serum if the screening gave high results.) To each tube add 5.0 ml. of 0.15 N. NaOH and mix thoroughly. Allow to stand 20 minutes.

   B. To prepare the column, shake each vigorously to completely resuspend the resin. Remove the cap, snap off the tip, place column in rack, and allow to drain. Discard the run-off.

Thyroxin (T_4) by Column (Continued):

C. Add 3.0 ml. of 0.15 N. NaOH to each column being careful to wash down into the resin bed any resin beads clinging to the inside of the top of the column. Allow to drain and discard the run-off.

D. Transfer contents of each tube into appropriately labelled columns and allow to drain. Discard the run-off. Be careful not to disturb the resin bed when samples are added to the columns.

E. Rinse each tube with 10 ml. of alcohol-acetate buffer and transfer contents to appropriate columns. Repeat procedure using 15% Acetic acid. Allow to drain. Discard run-off. Discard tubes already rinsed twice.

F. To each column pipette exactly 0.70 ml. of glacial acetic acid. Allow to drain. Discard run-off. It is extremely important that the exact amount be pipetted.

G. Pipette exactly 3.0 ml. of 50% Acetic acid into each column and collect the eluates in appropriately labelled, clean, 16 x 100 mm. test tubes. Mix the eluates well to distribute the thyroxin evenly and label as "first eluate". Here again, it is extremely important to pipette as accurately as possible.

H. Pipette exactly a second 3.0 ml. of 50% Acetic acid into each column and collect each eluate. Mix to distribute the thyroxin and label as "second eluate". The remaining thyroxin comes off the resin at this point, normally representing about 10% of the total of the two elutions. This second elution is made to check for possible sample contamination. Eluates are now ready for analysis. They may be capped and stored at room temperature, overnight if necessary, prior to sample analysis.

I. Load on sample wheel "first elution" followed by the second elution. Place Reagent Blank with its first and second elutions, then Standards in ascending order, each followed by its own second elutions. Add 2 cups of 50% acetic acid before controls and samples.

J. After the last sample peak has been recorded, put all lines into 0.15 N. NaOH for at least 5 minutes. Then transfer all lines to a water wash and wash thoroughly.

Thyroxin (T_4) by Column (Continued):

3. CALCULATIONS:

A. Consider the first elutions of both standards and the samples to contain the full T-4 iodine value of that sample. Plot the curve assigning each standard the full assayed value of that standard. Draw the best straight line through the plotted points. It will approximate a straight line with the highest standard falling off a little from linearity

B. Read the control and patients T-4 iodide concentration in Mg%. from the standard curve.

C. To check the elution ratio, read T-4 iodide values in mg%. from the standard curve for the second elutions of the standards, controls, and patients sera. The elution ratio is the distribution of the T-4 iodide in the first and second eluates compared to the total of the two.

For Example:

First Elution:	3.7 micrograms%
Second Elution:	0.5 micrograms%
	4.2 micrograms%

$$\% \text{ of first elution} = \frac{3.7}{4.2} \times 100 = 91\%$$

The Elution ratio is therefore 91:9

D. With uniform technique, the elution ratio should be consistent from sample to sample. When a slight amount of non-hormonal organic iodide is present, it is eluted from the column with the thyroxin and its presence would be indicated in the second elution. When there is a marked deviation from a previously established elution ratio, a non-hormonal organic iodide contamination is indicated and so noted on the report.

NORMAL VALUES: 3.5 - 6.5 micrograms%

REFERENCES:

1. Bio-Rad Laboratories Instruction Manual for Standard II, 1971.

Thyroxin (T_4) by Column (Continued):

2. Kessler, G. and Pileggi, V. J.: "A Semi-automated Non-incineration Technique for Determining Serum Thyroxin", Clin. Chem. Vol. 16, Nov. 5, 1970.

CLINICAL INTERPRETATION:

At the time of birth the level of T_4 of the infant is the same as the maternal blood. The T_4 increases until the third month of life, after which it begins to drop and reaches adult euthyroid levels by one year of life. A decrease in T_4 occurs late in life and may be related to a hypometabolic state. There is no difference in the normal range of T_4 in males and females. The T_4 increases during the third month of pregnancy. This rise relates to an increase in thyroxin-binding globulin. This same rise occurs in women receiving estrogens or oral contraceptive tablets.

A low T_4 signifies hypometabolism or hypothyroidism relating to thyroidectomy, myxedema, chronic thyroiditis, Riedel's Struma, cretinism, or destruction of the thyroid by cancer.

A high T_4 indicates hyperthyroidism due to diffuse hyperplasia or a hyperactive adenoma of the thyroid or subacute thyroiditis.

Non-thyroid disorders such as nephrosis, cirrhosis or malnutrition may lower the T_4 while acute hepatitis may raise the T_4. Many drugs or x-ray contrast media containing iodine will falsely elevate the T_4. Thus, it would then be necessary to perform a T_4 by competitive protein-binding (The Murphy-Pattee Test) to circumvent this type of spurious elevation.

Technicon ®
AutoAnalyzer ®
Methodology

T-4 Determination

AutoAnalyzer—Automated Bio-Bromine Flow Diagram

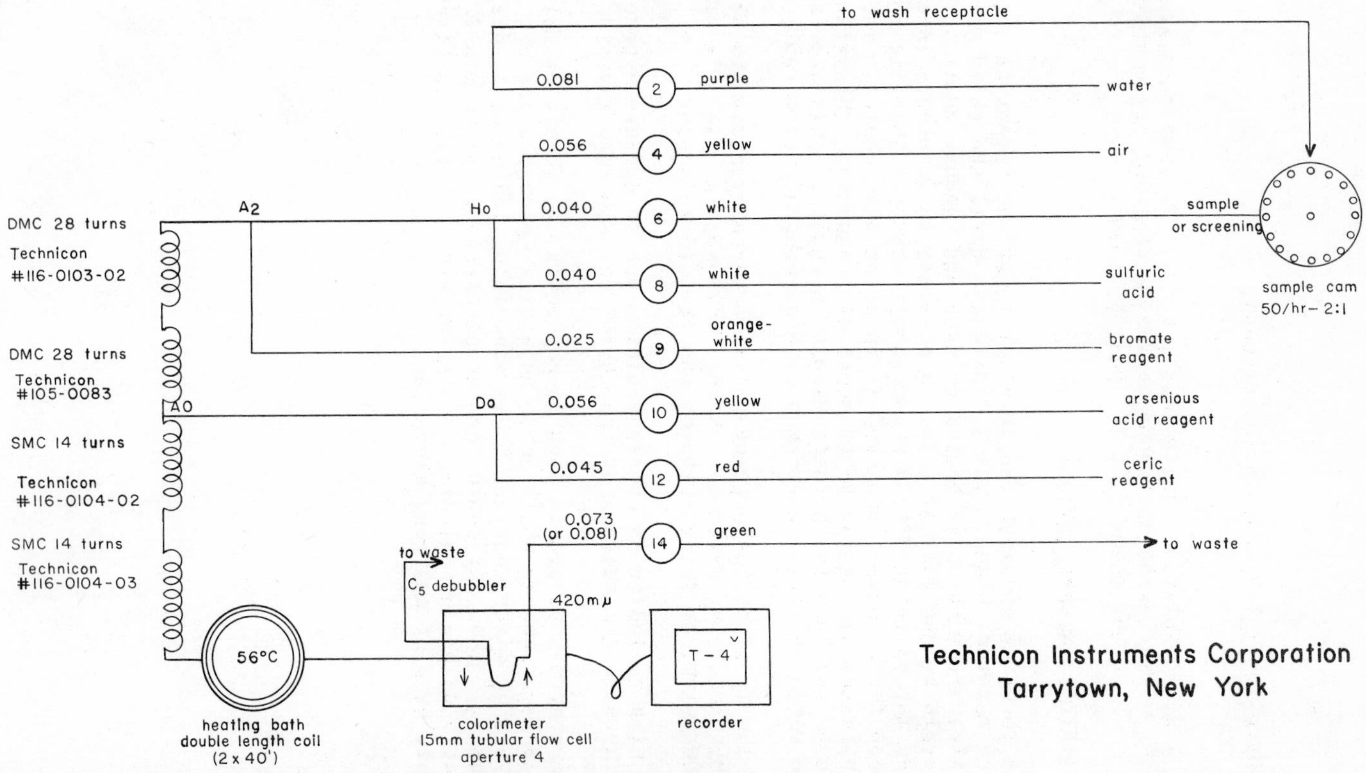

Technicon Instruments Corporation
Tarrytown, New York

GLUTAMATE-OXALOACETATE TRANSAMINASE

PRINCIPLE: Glutamate-oxaloacetate transaminase (GOT) catalyzes the reaction:

1). L-Aspartate + alpha-Oxoglutarate $\rightleftharpoons$ L-Glutamate + Oxaloacetate

2). Oxaloacetate + NADH + $H^+ \xrightleftharpoons{MDH}$ Malate + NAD^+

The activity of the transaminase is measured by the conversion of L-aspartate and alpha-oxoglutarate into glutamate and oxaloacetate. In an indicator reaction with malic dehydrogenase, oxaloacetate and NADH are converted into malate and NAD. Enzyme activity is measured by the rate of decrease in absorbance of NADH.

SPECIMEN: 0.5 ml. of serum. Serum should be separated from cells as soon as possible. Only minimum hemolysis allowed. Specimen stable at 4°C. up to 72 hours.

REAGENTS AND EQUIPMENT:

(Biochemica Test Combination Kit - TC A1 - 15971 TGAE)

1. Stock Buffer + L-Aspartate (Bottle No. 1)
   0.1 M. Phosphate buffer, pH 7.4, and 0.04 M. L-Aspartate.

2. Stock NADH (Bottle No. 2)
   0.012 M. NADH.

3. Malate Dehydrogenase/Lactate Dehydrogenase (Bottle No. 3)
   0.25 mg. each of MDH and LDH.

4. Alpha-Oxoglutarate (Bottle No. 4)
   0.25 M. alpha-oxoglutarate.

5. Working Buffer (Solution 1)
   Dissolve contents of Bottle No. 1 in 100 ml. redistilled water in clean glass container. Dissolve Bottle No. 2 in 1.5 ml. redistilled water. Use Bottle No. 3 as is (liquid). Combine the three bottles to make "Working Buffer". Be sure to rinse contents of bottles.

6. Working Substrate, alpha-Oxoglutarate (Solution 2)
   Dissolve contents of Bottle No. 4 in 3 ml. redistilled water (directly into bottle). Do not add to Working Buffer.

Glutamate-Oxaloacetate Transaminase (Continued):

NOTE: Solutions 1 and 2, while kept separate and at 4°C., are stable for 4 weeks. Do <u>not pipette</u> from either of these. Use Cornwall syringe for dispensing Solution 1 and Eppendorf pipette with clean disposable tip for solution 2. These reagents are sufficient for 25 determinations.

7. <u>Beckman DBG</u>
Instrument with thermostated cuvette well at 37°C.; Beckman 10 inch recorder with expanded scale.

PROCEDURE:

1. To 3.1 ml. of buffered substrate (Solution 1), add 0.5 ml. of serum with 500 lambda Eppendorf pipette.

2. Mix. Allow to come to temperature for 5 minutes in a 37°C. water bath.

3. Add 0.1 ml. of alpha-oxoglutarate (Solution 2). Mix by gentle inversion with parafilm.

4. Place in DBG, and using recorder, scan for 1 minute of linearity.

5. CALCULATION:

ϵ NADH at 340 nm. = 6.22×10^3 Liter/Mole x cm.

$$\frac{\Delta A}{\epsilon x d} \times 10^6 \times \frac{TV}{SV} \times \frac{1}{Time} = \text{IU/Liter or mU/ml.}$$

$$\frac{\Delta A}{1} \times \frac{1}{6.22 \times 10^3 \times 1} \times 10^6 \times \frac{3.7}{0.5} \times \frac{1}{1} = \text{mU/ml.}$$

$$\Delta A \times F = \text{mU/ml.}$$
$$F = 1190.$$

NORMAL VALUES: 5 - 40 mU/ml.

REFERENCES:

1. Karmen, A., et al: <u>J. Clin. Invest</u>., 34:1261, 1955.

2. Bergmeyer, H. U.: METHODS OF ENZYMATIC ANALYSIS, Academic Press, 1968.

Glutamate-Oxaloacetate Transaminase (Continued):

3. Henry, R. J., Chiamari, N., Golub, O. J. and Berkman: Am. J. Clin. Path., 34:381, 1960.

CLINICAL INTERPRETATION:

Increased serum levels of glutamic oxalacetic transaminase and glutamic pyruvic transaminase include enzyme release from injured cells and increased production of the enzyme by various cells are related to various diseases. The greatest amount of GOT and GPT is present in the liver followed by lesser amounts in heart muscle and skeletal muscle. A small amount is present in the kidney, pancreas, red blood cells, and in the lungs. The activity of GOT in the red cells is about ten times the normal serum level. Thus, in hemolytic states there may be an elevation of GOT.

The GPT activity is in greatest amounts in the liver with lesser amounts in various other tissues. The usual reason for requesting a GPT determination is when there is a question of hepatocellular disease.

Elevation of GOT in the serum occurs in the following conditions:

1. Acute hepatitis
2. Acute myocardial infarction
3. Active cirrhosis
4. Infectious mononucleosis with hepatitis
5. Toxic hepatic necrosis due to carbon tetrachloride or other chemicals toxic to liver
6. Carcinoma metastatic to liver or leukemic infiltration of liver
7. Acute pancreatitis
8. Trauma to skeletal muscle, surgical, accidental, crush injury, irradiation of skeletal muscle
9. Pseudohypertrophic muscular dystrophy
10. Dermatomyositis
11. Acute hemolytic anemia
12. Acute renal disease, specifically recent renal infarction
13. Severe burns with extensive epidermal injury
14. Cardiac catheterization and angiography with injury of heart muscle
15. Recent brain trauma with brain necrosis

Serum elevations of GPT occur in the following conditions:

1. Acute liver necrosis

Glutamate-Oxaloacetate Transaminase (Continued):

2. Less marked elevation in necrosis of skeletal muscle, heart, kidney, and pancreas
3. Infectious mononucleosis with hepatitis

Decreased levels of GOT occur in:

1. Beriberi
2. Lactic acidosis
3. Pregnancy
4. Chronic renal hemodialysis
5. Low pyridoxine

Prolonged and severe exercise tends to raise the GOT levels in the blood. Healthy males tend to exhibit slightly higher transaminase levels than females. There is a slight decrease in transaminase levels during normal pregnancy. When the transaminase levels are elevated in the blood this usually results from cellular destruction or a change in the cell membrane permeability with release of the transaminase of the cellular cytoplasm into the blood.

The commonst causes for an elevated transaminase level are heart and liver diseases. GOT activity in the blood rises within the first 18 hours following an acute myocardial infarction. Maximum levels are reached within 48 hours and return to normal within 5 to 7 days. The degree of elevation of GOT is related to the amount of myocardial muscle that has become necrotic. An elevation of serum GOT is particularly helpful to the clinician diagnostically when the electrocardiogram does not indicate the presence of myocardial infarct. Patients who suffer from angina pectoris generally do not have an increase in the GOT. Extensions of the myocardial infarct will cause persistent and subsequent elevations of the GOT levels. Occasionally elevated serum transaminase levels may be seen in patients with severe tachycardia or arrhythmia with a cardiac rate greater than 160 per minute. Active myocarditis secondary to either viral, metabolic, rheumatic, or bacterial cause is associated with elevated transaminase levels. Here again, the severity of the myocarditis influences the rise of the GOT levels. Patients who undergo cardiac catheterization and angiography may have elevations in transaminase levels. Presumably the cardiac muscle is injured during the procedure, with release of GOT into the blood. Utilization of opiates and direct current counter shock will elevate GOT. Patients who suffer from congestive heart failure most likely have GOT and GPT elevations on the basis of hepatic congestion and centrilobular necrosis of the liver. The elevated levels of GOT in the urine in patients who have myocardial infarction are usually correlated with the elevated GOT levels in the blood. In some patients the elevated urinary concentrations of the enzyme remain elevated for a longer period of time than in the serum, simulating the situation in acute pancreatitis where the urinary levels at times remain elevated over those of the blood. In myocardial infarction, the level usually rises over 200 units, and at times there is a great increase in GOT reaching 1000 units especially if there is an associated heart failure with central lobular liver necrosis. When infarction of the myocardium is suspected, daily measurements of GOT activity are more useful than a single determination which may not detect the transient peak elevation. Pericarditis usually does not elevate the GOT levels. If the GOT is elevated in patients who have pericarditis, then most likely there is subepicardial necrosis of the heart or perhaps congestive heart failure secondary to the pericarditis.

Hepatic Disease

The liver is the richest source of both the GOT and GPT transaminases. Thus, any damage to the parenchymal iver cells will result in elevations in both transaminases. An increase in serum transaminase activity thus occurs in neoplastic, inflammatory or degenerative lesions of the liver. The extent of rise of transaminase usually reflects the severity of the hepatic damage. The highest values have been obtained in acute hepatic necrosis which, for example, are associated with chemical poisoning such as carbon tetrachloride. In these patients the rise might be as much as 25,000 units. In acute viral hepatitis, the rise may be upwards to 5,000 units. Increase in serum GOT and GPT may be present for as long as one to four weeks before clinical jaundice in a hepatitis patient is apparent. Thus, they may be used as biochemical indicators or detectors of the carrier state of the anticteric patient. Furthermore, one or more elevated levels may be found in twenty-five per cent of asymptomatic individuals in the first five months of the post convalescent period.

Elevated GOT when present, unaccompanied by symptoms or other significant laboratory findings such as in increased BSP, does not warrant restriction of activity in convalescent phase.

Lower values such as in the range of some 100 to 500 units are usually found in cirrhosis due to alcoholism, biliary cirrhosis, and in toxic hepatitis due to drugs. The increase in GPT usually exceeds that of GOT in viral hepatitis. GPT frequently is elevated at the early phases of viral hepatitis before other liver function tests become abnormal. Thus, this enzyme determination is especially useful in detecting anicteric hepatitis, and hepatitis which may accompany infectious mononucleosis. One must keep in mind that with extremely severe liver damage, the rise in GOT may not be sustained and may fall because of great damage to the cells which interferes with continued and persistent enzyme production. Patient with neoplasms which are metastatic to the liver may have increases in the GOT activity. This rise in the GOT activity is due to the necrosis of liver cells in and surrounding the metastatic cancer.

Serum GOT levels are only elevated in sporadic instances in cases of obstructive liver disease. GOT is excreted in the bile, and although increased biliary pressure may stimulate hepatic cells to release GOT this is not a consistent finding. Furthermore, biliary obstruction tends to elevate GOT usually in post-cholecystectomy patients. This may be in part due to biliary pressure being exerted against the liver cells with abscence of the gallbladder. The trans aminase elevation in both GPT and GOT seen in infectious mononucleos is presumably due to the hepatic damage. Elevations of GPT being more striking than GOT.

Several other important causes for elevations of serum transaminase are recognized. It has been observed that patients with acute infarction of the kidney with necrosis of renal tubular cells will cause an elevated serum transaminase. Renal insufficiency without infarction of the kidney will not cause an elevated serum transaminase.

Transaminase is present in pancreatic acinar cells. Thus, in patien with acute pancreatitis, there may be a moderate rise in the GOT levels in the blood. GOT is present in red cells and in white cells in the peripheral blood. Thus, in patients who have hemolytic anemi and in patients with leukemia, there will be a rise in the GOT level In leukemia, the rise in GOT results from proliferation of the leukemic cells with production of GOT by the leukemic cells and also liberation from leukemic cells which are undergoing lysis. An early rise in GOT following a burn has been observed. Burns which extend over 10% of the body surface may liberate a sufficient amount of GOT from damaged epidermal cells. Blister fluid from burns contains a large amount of GOT. GPT may be elevated in patients who have extensive burns. This elevation may suggest liver damage secondary to the burn. A small amount of GPT may also be liberated from the damaged epidermal cells. It is frequent to observe elevations following surgery. The GOT is probably liberated from the surgical wound especially from skeletal muscle. Patients who are undergoing cardiac surgery invariably have elevations of GOT for a number of days following the surgical procedure.

GOT is also present in the brain. Thus, in patients with cerebral disease, there will be an increase of GOT in the cerebral spinal fluid and in the blood. There is some evidence that the transaminases do not freely traverse the blood brain barrier easily. Thus, there will be a greater elevation in the cerebral spinal fluid than in the blood.

Skeletal muscle is a rich source of transaminases, specifically GOT. After severe muscle trauma, e.g. accidents or surgical operations, elevations of GOT are found. These serum levels bear a direct relationship to the CPK and aldolase levels indicating a skeletal muscle origin. In the muscular dystrophies, marked elevations are found in both GOT and CPK. 80% of a total of approximately 500 patients with progressive muscular dystrophy had elevated transaminase levels. However, it should be noted that the transaminases are less useful than CPK or aldolase in the diagnosis of muscular disease.

Drug Effects

A whole host of drugs can give spurious elevations of SGOT. These include:

Amantadine	Colchicine	Lincomycin
Ampicillin	Cycloserine	Methotrexate
Anabolic Agents	Erythromycin	Methyl dopa
Androgens	Ethionamide	Nafcillin
Cephalothin	Gentamycin	Nalidixic Acid
Chloroquin	N-Hydroxyacetamide	Opiates
Clofibromate	Indomethacin	Oxacillin
Cloxacillin	Isoniazid	Phenothiazines

Some GOT elevation is caused by hepatic parenchymal cell damage due to drugs. Some drugs interfere with the method of analysis of the GOT; specifically with erythromycin and the tranquilizer group of drugs such as the phenothiazines. These two drug groups cause a formation of a hydrazone without the presence of liver damage. The hydrazone method entails usually the transformation of hydrazine to hydrazone in the presence of GOT. However, if there is an erythromycin or phenothiazine drug present in the serum a hydrazone forms from the combination of one drug and hydrazine, thus giving a false color reaction. This effect can be shown to be spurious if the GOT level is assayed by the NAD - NADH method. Discontinuence of the drug leads to normal levels within 24 hours. Erythromycin, polycillin, postaphcillin, coumadin, chlorpromazine, the oral contraceptives, and methyl dopa all have been reported to cause an increased GOT. GOT may be elevated in ketoacidosis associated with starvation or diabetic acidosis. The ketoacids combine with the colorimetric dye utilized in the SMA 12/60 instrument giving a spurious elevation.

Low levels of GOT have been reported in conditions where elevated pyruvate or lactate is present in the blood. These conditions are Beriberi due to thiamine deficiency, uncontrolled diabetes mellitus with acidosis, and at times liver disease. The low levels of GOT arise because of an artefact induced during the pre-incubation period in the chemical reaction for GOT. A consumption of the substrates by the elevated pyruvate in the serum being tested for GOT occurs, and because of this consumption of substrate by lactate or pyruvate, there will be a spurious decrease in GOT below expected levels. Pyridoxine deficiency in chronic renal dialysis patients and in pregnancy will result in low GOT and GPT.

GLUTAMATE-PYRUVATE TRANSAMINASE

PRINCIPLE: GPT converts L-alanine and alpha-oxoglutarate into glutamate and pyruvate. In an indicator system with LDH, pyruvate and NADH are converted into lactate and NAD. GPT is measured by the decrease in absorbance of NADH at 340 nm. (ϵ NADH at 340 nm. is 6.22×10^3 liter/mole x cm.)

SPECIMEN: 0.5 ml. serum. Serum should be separated from cells as soon as possible. Only minimum hemolysis allowed. Specimen stable at 4°C. up to 72 hours.

REAGENTS AND EQUIPMENT:

(Biochemica Test Combination Kit, TC-HI 15978 TGAH)

1. Stock Buffer + DL-Alanine (Bottle No. 1)
   Buffer, pH 7.4, 0.08 M. DL-alanine, 0.1 M. Phosphate

2. Stock NADH (Bottle No. 2)
   0.012 M. NADH

3. LDH (Bottle No. 3)
   0.25 mg. LDH

4. Alpha-Oxoglutarate (Bottle No. 4)
   0.25 M.

5. Working Buffer (Solution 1)
   Dissolve bottle No. 1 in 100 ml. redistilled water (use clean glass container). Dissolve contents of Bottle No. 2 in 1.5 ml. water; add to buffer (rinse bottle); add contents of bottle No. 3 to buffer (rinse bottle). Mix by swirling.

6. Working Substrate, alpha-oxoglutarate (Solution 2)
   Add 3.0 ml. of redistilled water to bottle No. 4. Add directly to bottle; mix by inversion. Do NOT add to the working buffer.
   NOTE: Solutions 1 and 2, while kept separate and at 4°C., are stable for 4 weeks. Do not pipette from either of these.

   Use Cornwall syringe for dispensing Solution 1. Use Eppendorf pipette with disposable tip for Solution 2.

   These reagents sufficient for 25 determinations.

Glutamate-Pyruvate Transaminase (Continued):

7. Beckman DBG
   Instrument with thermostated cuvette well at 37°C. Beckman 10 inch recorder with expanded scale.

PROCEDURE:

1. Into 10 mm. square civette add 3.1 ml. of Working Buffer (Solution 1) + 0.5 ml. of serum with a 500 microliter Eppendorf pipette).

2. Mix. Place in 37°C. water bath for 5 minutes.

3. Add 0.1 ml. of alpha-oxoglutarate (Solution 2). Mix by inverting with parafilm.

4. Place in DBG. Record reaction for 1 minute of linearity using recorder.

5. CALCULATIONS:

$$\frac{\Delta A}{\epsilon \mathrm{x} d} \times 10^6 \times \frac{TV}{SV} \times \frac{1}{Time} = \mathrm{IU/Liter\ or\ mU/ml.}$$

NADH at 340 nm. = 6.22×10^3 Liter/Mole x cm.

$$\frac{\Delta A}{6.22 \times 10^3 \times 1} \times 10^6 \times \frac{TV}{SV} \times \frac{1}{1}$$

$$\frac{\Delta A}{1} \times \frac{1}{6.22 \times 10^3 \times 1} \times 10^6 \times \frac{3.7}{0.5} \times \frac{1}{1}$$

$\Delta A \times F$ = mU/ml.
$F = 1190$

NORMAL VALUES: 5 - 35 mU/ml.

REFERENCES:

1. Wroblewski, F. and LaDue, J.S.: Ann. Intern. Med. 45:80, 1956.

2. Wroblewski, F. and LaDue, J.S.: Proc. Soc. Exp. Biol. Med., 91:569, 1966.

Glutamate-Pyruvate Transaminase (Continued):

CLINICAL INTERPRETATION:

Refer to Interpretation section under the Glutamate-Oxaloacetate Transaminase Procedure on page 377.

TRIGLYCERIDES - BOEHRINGER MANNHEIM DETERMINATION OF GLYCEROL AND TRIGLYCERIDES IN SERUM (U.V. Method with NADH)

PRINCIPLE: Serum lipids consist mainly of triglycerides (neutral fats), cholesterol and phospholipids. In this method serum is saponified with alcoholic potassium hydroxide to hydrolyze triglyceride to fatty acids and glycerol. (Phospholipids are hydrolyzed to alpha and beta-glycerophosphates, stable in alkaline solution. Since the enzyme glycerokinase (GK) is specific for glycerol and does not act upon alpha and beta-glycerophosphates there is no interference by phospholipid.)

The saponification splits triglyceride into fatty acids and free glycerol. Glycerol is phosphorylated with adenosine-5'-triphosphate (ATP) in a reaction catalyzed by the enzyme glycerokinase (GK) to result in glycero-1-phosphate and adenosine-5'-diphosphate (ADP):

$$\text{Glycerol} + \text{ATP} \rightleftharpoons \text{Glycero-1-phosphate} + \text{ADP}$$

In an auxiliary reaction with the enzyme pyruvate kinase (PK), the ADP and phsophoenolpyruvate (PEP) are transferred into ATP and pyruvate.

$$\text{ADP} + \text{PEP} \rightleftharpoons \text{ATP} + \text{Pyruvate}$$

In the indicator reaction the enzyme LDH catalyzes the reaction of pyruvate and NADH to lactate and NAD.

$$\text{Pyruvate} + \text{NADH} + H^+ \rightleftharpoons \text{Lactate} + \text{NAD}^+$$

The amount of NADH consumed during the reaction is equivalent to the amount of glycerol present in the sample. NADH can be measured by its absorbancy at 340 nm. The method described is based on studies of M. Eggstein and F. H. Kreutz [3,4].

SPECIMEN: Use fresh serum free from hemolysis. The determination of triglyceride should always be preceded by a determination of free glycerol. Strongly lipemic sera should be diluted (1 + 9 with physiological saline) prior to the glycerol determination. 200 micr liters of serum is used for the saponification. 0.5 ml. of serum is used for determining free glycerol.

Triglycerides (Continued):

REAGENTS:

Solution 1. Buffer (0.1 M. Triethanolamine, pH 7.6; 0.004 M. $MgSO_4$); dissolve the contents of bottle No. 1 with 150 ml. redistilled water. The solution is stable for approximately 2 months at room temperature.

Solution 2. NADH/ATP/PAP (0.006 M. NADH, 0.33 M. ATP, 0.011 M. PEP); reconstitute 1 bottle of solution No. 2 with 2.0 ml. of redistilled water. The solution is stable for approximately 2 weeks at $+4^oC$. (Stability of this solution may be prolonged by aliquoting and freezing.)

Solution 3. LDH/PK (2.0 mg. LDH/ml.; 1.0 mg. PK/ml.); Use suspension of bottle No. 3 undiluted. Stable for approximately 1 year at $+4^oC$.

Solution 4. GK (2.0 mg. GK/ml.);
Use suspension of bottle No. 4 undiluted. Stable for approximately 1 year at $+4^oC$.

Solution 5. Alcoholic potassium hydroxide (0.5 N.);
Dissolve 3.3 gm. potassium hydroxide pellets, A. R. free from glycerol, in approximately 10 ml. redistilled water. After cooling dilute to 100 ml. with denatured ethyl alcohol.
NOTE: Methanol may not be used.

Solution 6. Magnesium sulfate (0.15 M.);
Dissolve 3.7 gm. $MgSO_4 \cdot 7H_2O$ A. R. in 100 ml. of distilled water.

COMMENTS ON PROCEDURE:

A. If solution No. 2 is improperly stored, the ATP may deteriorate. This would result in a considerable reduction of NADH prior to the glycerol reaction. The glycerol reaction will then be limited to the amount of remaining NADH and lead to false low values. After completion of the reaction, a drop of diluted glycerol added to the cuvette should cause further decrease in optical density.

B. Some sera will result in a continuous linear decrease of optical density after the reaction is completed. In these

Triglycerides (Continued):

cases, after the 10 minutes, take additional readings at minute intervals for 2 or 3 minutes. Multiply the O.D./minute of this "creep" by 10 (the reaction time) and subtract this figure from the ΔO.D. (E_1 - E_2) to obtain the "corrected" ΔO.D.

C. For absolute accuracy a reagent blank should be determined once for each Lot Number. The determination is carried out by using water instead of serum. The optical density difference resulting after the addition of Solution No. 4 is to be deducted from the optical density difference obtained with serum.

D. The three critical areas in pipetting are the adding of serum and alcoholic KOH in the saponification step and the addition of serum or supernatant in the glycerol determination. This laboratory uses Eppendorf pipettes and syringes with Cheney Adaptors.

PROCEDURE:

Set up two sets of centrifuge tubes for each serum, reagent blank, and standards. First set is for the determination of total glycerol which requires saponification. Second set is for the determination of "free" glycerol which needs no saponification.

A. Saponification

1. Pipette into centrifuge tube:

	Blank	Test
Distilled water	0.2 ml.	-
Serum	-	0.2 ml.
Alcoholic KOH	0.5 ml.	0.5 ml.

Mix, close centrifuge tube with clean stopper or parafilm and allow to stand in water bath for 30 minutes at 70°C., then allow to cool to room temperature in cold water. Add 1.0 ml. $MgSO_4$, mix and spin.

2. Into the second set of centrifuge tubes pipette:

	Blank	Test
$MgSO_4$ Solution	1.0 ml.	1.0 ml.
Serum	-	0.2 ml.
Distilled water	0.2 ml.	-

Triglycerides (Continued):

3. Mix well, and then add cold

	Blank	Test
Alcoholic KOH	0.5 ml.	0.5 ml.

to each tube, one at a time, stopping to mix each tube at once. Stopper and centrifuge. Use 0.5 ml. of the clear supernatant fluid respectively for glycerol assay.

B. Determination of Glycerol

1. Pipette into cuvette: Glass cuvette: 1.0 cm. light path.

	Routine Assay
Solution 1.	2.50 ml.
Solution 2.	0.10 ml.
Serum or supernatant fluid	0.50 ml.
Solution 3.	0.02 ml.

2. Cover with parafilm and mix -- allow to stand for 10 minutes at room temperature; measure optical density E_1. Wavelength: 340 nm. Blank machine with air. Read reagent blank and record. Set machine to 1.0 O.D. with R. B. then read tests.

   Solution 4. 0.02 ml.

3. Mix and wait until reaction stops (approximately 10 minutes), or take 3 - 5 readings at 2 minute intervals and extrapolate E_2 to the time of the addition of suspension 4; refer to "Biochemica - Test Combinations, Principles and Practice".

$$E_1 - E_2 = \Delta E$$

4. With optical density differences exceeding 0.800 (340 nm.), mix 0.1 ml. serum or filtrate with 0.9 ml. physiological saline and repeat assay with 0.5 ml. of this dilution. Result must be multiplied by 10.

5. CALCULATION

   The ratio of assay volumes between routine and semi-micro assays is not exactly 2.5:1. The difference, however, is less than 1.0% and can be neglected. For both procedures the same factors can be applied.

Triglycerides (Continued):

Free Glycerol: If a 0.5 ml. serum sample is used directly.

$$\frac{\Delta O.D. \times Va \times {}^{S}M.W.}{\times E. \times Vs} = \text{micrograms/ml.}$$

Va = Total assay volume - 3.14 ml.
Vs = Sample volume - 0.50 ml.
E = Light path - 1.0 cm.
Absorbancy coefficient - 340 nm. = 6.22
M.W. = Molecular Weight glycerol = 92
To convert micrograms/ml. to mg%., a factor of 1/10 (100/1000) must be added.

Resolving the above formula, the factor for free glycerol then is:

340 nm.: O.D. x 9.28 = mg%. free glycerol in serum.

Total Glycerol: Due to saponification and neutralization, the sample is diluted 1 to 8.5.

The factor for total glycerol is then:

$$\frac{\Delta O.D. \times 3.14 \times 92}{6.22 \times 1 \times 0.5 \times 1/8.5} = \text{micrograms/ml.}$$

340 nm.: O.D. x 78.6 = mg%. Total glycerol in serum.

Triglycerides:

Total Glycerol - mg%. (Serum with saponification)
Free Glycerol - mg%. (Serum without saponification)
"Glyceride glycerol" - mg%.

The mg%. of "Glyceride Glycerol" (glycerol derived from triglycerides) is now converted into mg%. of triglycerides. This is accomplished by equating the average molecular weight of triglycerides (885) to the molecular weight of glycerol (92).

$$\frac{885}{92} = X \frac{\text{(mg\%. triglyceride)}}{\text{1.0 mg\%. glycerol}}$$

X = 9.62

Triglycerides (Continued):

Mg%. "Glyceride - glycerol" x 9.62 = mg%. Triglycerides (neutral fat).

Example:

0.5 ml. serum each was determined prior and after saponification. Wavelength: 366 nm. lightpath 1.0 cm.

	Free Glycerol	Total Glycerol After saponification
Prior to addition of GK: E_1	0.748	0.695
10 minutes after addition of GK: E_2	0.665	0.560
11 minutes after addition of GK:	0.663	0.557
12 minutes after addition of GK:	0.661	0.554
13 minutes after addition of GK:	0.659	0.551
Δ O.D./minute of "creep"	0.002	0.003
$E_1 - E_2$	0.083	0.135
10 minute "creep (10 x Δ O.D./minute creep)	0.020	0.030
"Corrected" ΔO.D. for calculation	0.063	0.105

NORMAL VALUES IN SERUM:

Free Glycerol:	0.5 - 1.7 mg%.
Glyceride - Glycerol:	7.7 - 17.9 mg%.
Neutral Fat:	40 - 150 mg%.

REFERENCES:

1. Eggstein, M.: Klin. Waschr., 44:276, 1966.

2. Eggstein, M. and F. H. Dreutz: Klin. Waschr., 44:262, 1966.

3. Schmidt, F. H., et al.: Klin. Chem. and Klin. Biochem., 6:156, 1968.

4. Pinter, J.K., et al.: Arch. Biochem. & Biophys., 121:404, 1967.

Triglycerides (Continued):

CLINICAL INTERPRETATION:

Hyperlipemia is reflected by an opalescent to milky or turbid appearance of serum which is called lactescence. It may indicate the postprandial rise of triglycerides. Triglycerides may be elevated without a lactescence. Phospholipids comprise the largest fraction of serum lipids while cholesterol is the second largest serum fraction. Glycerides usually as triglycerides comprise the third largest serum lipid fraction. Small amounts of mono and diglycerides are present. Glycerides are important in the transport of fatty acids. Exogenous glycerides are from a dietary source absorbed from the small intestine. Endogenous glycerides reflect transport from the liver to adipose tissue and other organs. These endogenous glycerides appear in the liver from:

1). Glucose conversion
2). Free fatty acid conversion
3). Exogenous glycerides

Triglycerides are markedly elevated in Type I hyperlipoproteinemia and moderately elevated in Types III, IV, and V. It is normal or slightly elevated in Type II.

Secondary causes of elevated triglyceride levels are diabetes mellitus, pancreatitis, alcoholism, glycogen storage disease, hypothyroidism, nephrosis, pregnancy, oral contraceptives and gout.

Triglyceride levels are decreased in dieting individuals, hyperthyroidism, lipid lowering drugs, malabsorption syndrome, and hereditary abeta or hypobetalipoproteinemia.

URIC ACID DETERMINATION
(Uricase Method)

PRINCIPLE: The enzyme uricase as a substrate is specific for uric acid, and catalyzes the following reaction:

$$\text{Uric Acid} + O_2 + H_2O \longrightarrow \text{allantoin} + H_2O_2 + CO_2$$

Uric acid has an absorbance peak at 293 nm.; the methodology takes advantage of this fact, measuring the decrease in absorbance due to the oxidation of uric acid in the sample by the uricase added. The absorbance decrease is proportional to the amount of uric acid present in the sample, the reaction being allowed to go to completion. At the dilutions at which one is working, the method is specific for uric acid and is not interfered with by substances in the sample.

SPECIMEN: 50 microliters of serum. Plasma may not be used for this procedure. 50 microliters of centrifuged urine which has been diluted 1:10 (1.0 ml. + 9.0 ml.) with physiological saline. If original urine is cloudy, warm to 60°C. for 10 minutes, mix, then centrifuge. Fresh serum and urine samples are preferred if possible.

REAGENTS:

Boehringer Mannheim "Uric Acid Test Combination", Catalogue #15986 THAA (U.V. Method); for approximately 30 determinations. The test combination includes the following:

1. 0.2 M. Borate Buffer, pH 9.5 (Bottle #1)
   Dissolve the contents of bottle #1 in 100 ml. of redistilled water. Stable for several months at room temperature; protect from evaporation.

2. Uricase, 2.0 mg./ml. (Bottle #2)
   The contents of this bottle are used undiluted. Store at refrigerator temperature; stable for several months.

3. Uric Acid Stock Standard, 1.0 mg./ml.
   Into a 100 ml. volumetric flask, place 100 mg. uric acid, and 60 mg. lithium carbonate (Li_2CO_3). Add approximately 50 ml. redistilled water and dissolve contents at 60°C. Allow flask to cool and bring to volume with re-distilled water. Stable up to one year at refrigerator temperature.

4. Uric Acid Working Standard, 0.1 mg./ml.
   Place 10 ml. of stock standard in a 100 ml. volumetric flask and bring to volume with re-distilled water.

Uric Acid Determination (Continued):

COMMENTS ON PROCEDURE:

1. The ten minute timing need not be exact to the second, but should not be extended beyond thirty seconds.

2. The reference against air should be checked before each reading.

3. The reaction appears to be linear at 15 mg%. uric acid and beyond.

4. Factor Explanation:
With a 1.0 cm. light path, the oxidation of 1 microgram of uric acid per ml. corresponds to an absorbance decrease of 0.075 at 293 nm.

$$\frac{\Delta A}{0.075} \times \frac{TV}{SV} = \text{ug. Uric Acid/ml.}$$

$$\Delta A \times \frac{1}{.075} \times \frac{3.07}{0.05} = \text{ug./ml.}$$

Multiplication by 0.1 = Concentration in mg./100 ml.

$\Delta A \times F$ = Mg./100 ml.

F = 81.8 for serum

F = 818 for 1:10 dilution of urine

5. If one wishes to check the accuracy of the enzymatic system, a series of standards may be pipetted into the buffer and run as specimens.

Working Standard	5 Lambda	25 Lambda	50 Lambda
Buffer	3.0 ml.	3.0 ml.	3.0 ml.
Concentration	1.0 mg%.	5.0 mg%.	10.0 mg%.

PROCEDURE:

A. Determining the absorbance of the Uricase. (A_e)
   1. The following procedure is carried out once with each new "test combination kit" in order to determine the absorbance contributed by that lot of enzyme.

Uric Acid Determination (Continued):

2. Place 3.0 ml. of borate buffer, Bottle No. 1, into two quartz cuvettes. Reserve one cuvette as a reference cell.

3. Pipette 20 microliters of uricase, Bottle No. 2, into the other cuvette and mix thoroughly. Determine the absorbance of this solution against the buffer alone at 293 nm.

B. Analysis of Samples.

1. Pipette 3.0 ml. of borate buffer into a quartz cuvette; prepare one cuvette for each sample to be analyzed.

2. Deliver 50 microliters of serum or diluted urine into a cuvette with a T.C. pipette and mix well by inversion. Determine the initial absorbance of the uric acid against air at 293 nm. (A_1)

3. After taking the initial reading, deliver 20 microliters of uricase into the cuvette, mix again, and allow to stand at room temperature (20° - 25^oC.) for ten minutes.

4. Take the second absorbance reading (A_2) against air at 293 nm.

5. CALCULATION:

$(A_1 + A_e) - A_2 = \Delta A$ for the sample.

for serum: $\Delta A \times 81.8$ = mg%. uric acid
for urine: $\Delta A \times 818$ = mg%. uric acid

NORMAL VALUES:

Serum: 1. Men 2.6 - 6.8 mg%.
2. Women 2.0 - 6.3 mg%.

Urine: 0.25 - 0.75 gm./24 hours.

REFERENCES:

1. Praetorius, Elith, in Bergemeyer: METHODS OF ENZYMATIC ANALYSIS, Academic Press, pg. 500 - 501, 1965.

2. Henry, R. B.: CLINICAL CHEMISTRY: PRINCIPLES AND TECHNIQUES, Hoeber, pg. 280, 283 - 287, 1964.

Uric Acid Determination (Continued):

CLINICAL INTERPRETATION:

Uric acid, an end product of purine metabolism or a catabolite of tissue nucleic acids, may be increased by either overproduction or inability of excretion. The classical cause is gout. However, the commonest cause for an elevated value in hospitalized patients is renal failure. When there is rapid cell proliferation as in polycythemia, leukemia or lymphoma, the uric acid may be elevated, as it may be when there is tissue necrosis from whatever cause. Inflammatory conditions frequently result in hyperuricemia. Psoriasis is associated with hyperuricemia due to rapid cell turnover.

In metabolic acidosis in which keto-acids are at times elevated, there is hyperuricemia because of inhibition of uric acid excretion. Utilization of thiazide diuretics and lead poisoning are less common etiologies for hyperuricemia. Recently minor elevations have been ascribed to achievement oriented individuals, especially those in stressful situations.

Low values are found in individuals taking uricosuric drugs such as cortisone, and those on Xanthine oxidase inhibitors such as Allopurinol. Wilson's disease and the Fanconi syndrome are rare causes for hypouricemia.

URINE UROBILINOGEN
(Rapid Two Hour Test)

PRINCIPLE: Urobilinogen, along with other aldehyde-reacting substances found in urine, reacts with Ehrlich's Aldehyde Reagent to produce a red color, the intensity of which is proportional to the amount of pigment present.

SPECIMEN: There is a greater quantity of urobilinogen excreted in the urine during the afternoon and evening than during the morning. In order to make the test more quantitative, the total urine excreted during a two hour period in the afternoon should be used (preferably 1:00 p.m. to 3:00 p.m.). No preservative is used, and the test should be run as soon as possible, since urobilinogen is unstable.

REAGENTS AND EQUIPMENT:

1. Sodium Acetate, a saturated aqueous solution
   Approximately 1100 gm. of sodium acetate containing 3 molecules of water of crystallization is added to 800 ml. of deionized water (0.8 ml. water dissolves 1.0 gm. of sodium acetate). The mixture is heated to approximately 60°C. When the solution cools, there should be a large excess of crystals. (DO NOT PIPETTE FROM THIS BOTTLE.)

2. Ehrlich's Aldehyde Reagent, Watson's modification
   Dissolve 0.7 gm. of paradimethylaminobenzaldehyde in a mixture of 150 ml. concentrated HCl and 100 ml. deionized water. The reagent is stable if stored in a brown bottle. (DO NOT PIPETTE FROM THIS BOTTLE.)

3. Stock Standard, Phenol Red (LaMotte Chemical Co., Chestertown, Md.)
   Dissolve 22 mg. Phenol red (pH 6.8 - 8.5; Code 2411-C), in 100 ml. of 0.05%. NaOH. Store in a brown polyehtylene bottle. (DO NOT PIPETTE FROM THIS BOTTLE.)

4. Working Standard
   Dilute the stock standard 1:100 with 0.05% NaOH to make a 0.22 mg%. solution, which is equivalent in color to 0.35 mg%. urobilinogen. Store in a brown polyethylene bottle. (DO NOT PIPETTE FROM THIS BOTTLE.)

5. Beckman DU Spectrophotometer
   The working standard must first be read on the Beckman DU before proceeding with the unknowns at a wavelength of 562 nm. against a 0.05% NaOH Blank, and should have an O.D. of .385 $\pm$ .002.

Urine Urobilinogen (Continued):

COMMENTS ON PROCEDURE:

1. The determination of urobilinogen with aldehyde reagent is not specific for urobilinogen. Other aldehyde-reacting chromogens, notably indols, are excreted in the urine. Abnormal porphyrins may also add to the colór development.

2. The greatest source of error is inadequate mixing of the urine and reagent during each step of the procedure.

PROCEDURE:

1. Measure and record volume of urine

2. If a known positive is available it should be used as a control and run like patient specimen.

3. Two tubes should be set up for each unknown, standard, and control when available.

   A. UNKNOWN

      1). <u>Patient Blank</u>:
      Combine 2.5 ml. Ehrlich's and 5.0 ml. sodium acetate and mix well. Add 2.5 ml. of urine and mix.
      NOTE: There should not be any pink to purple color development in the blank.

      2). <u>Patient Test</u>:
      To 2.5 ml. of urine, add 2.5 ml. of Ehrlich's Reagent. Mix well for 15 seconds, then <u>immediately</u> add 5.0 ml. of sodum acetate, and mix thoroughly.
      NOTE: Sodium acetate stops the color reaction of Ehrlich's Reagent and also develops the color complex formed.

      If the urine is cloudy or precipitated upon adding Ehrlich's, repeat the test on a bile-free filtrate (equal amounts of urine and 10% BaCl filtered), and multiply calculations x 2.

   B. STANDARD
      1). <u>Blank</u>:
      Deionized water.

Urine Urobilinogen (Continued):

2). Standard:
0.35 mg%. urobilinogen standard.

4. Read unknowns, standard, and control (when available) against their respective blanks on the Beckman DU at 562 nm.

5. CALCULATIONS:

$$\frac{\text{O.D. Unk.}}{\text{O.D. Std.}} \times 0.35 \text{ (Conc. of Std.)} \times \frac{10 \text{ (Vol. Final Sol.)}}{2.5 \text{ (Vol. Urine used)}}$$

= Ehrlich units/100 ml. Urine

$$\frac{\text{O.D. Unk.}}{\text{O.D. Standard}} \times 1.4 = \text{Ehrlich units/100 ml. Urine}$$

$$\frac{1.4}{100} = .014 \text{ (Factor)} \quad \text{THUS:}$$

$$\frac{\text{O.D. Unk.}}{\text{O.D. Std.}} \times 0.014 \times \text{vol. of Urine} = \text{Ehrlich units/2 hours.}$$

NOTE: Report Ehrlich units/2 hours.

NORMAL VALUES: 0.3 - 1.0 Ehrlich Units/2 Hours

REFERENCES:

1. Davidsohn and Wells: TODD-SANFORD CLINICAL DIAGNOSIS BY LABORATORY METHODS, 13th Ed., Saunders, pg. 548 - 550, 1963.

2. Henry, R. B.: CLINICAL CHEMISTRY: PRINCIPLES AND TECHNIQUES, Hoeber, pg. 611 - 613, 1964.

CLINICAL INTERPRETATION:

The total urinary excretion of urobilinogen is 3 Ehrlich Units. A diurnal excretion of 1.5 Ehrlich Units exists during a 2 hour period from 1:00 to 3:00 p. m. Urobilinogen is colorless while urobilin is amber in color.

An increase in urinary urobilinogen occurs in hemolytic anemia because of increased production of unconjugated and conjugated

Urine Urobilinogen (Continued):

bilirubin. Conjugated bilirubin is converted to stercobilinogen (urobilinogen) in the colon by the action of E. coli. Hepatic disease also results in an increase in urinary urobilinogen because with liver disease, the urobilinogen remains in the blood stream instead of being removed from the enterohepatic circulation by the hepatic cell.

Decreased urinary urobilinogen occurs in intra or extrahepatic obstructive jaundice because conjugated bilirubin does not reach the colon for conversion by E. coli, or when E. coli function is negated by broad spectrum antibiotic usage or severe diarrhea.

VITAMIN A AND "CAROTENE" DETERMINATION

PRINCIPLE: Vitamin A and "carotene" are split from their protein complexes and saponification effected by warming with ethanolic KOH. Vitamin A and "carotene" are then extracted into petroleum ether. The extract is evaporated to dryness, the residue dissolved in chloroform and reacted with glycerol dichlorohydrin. Vitamin A forms a blue color immediately which changes in 2 minutes to a violet color, which is measured at 550 nm.

SPECIMEN: 1.0 ml. of serum specimen.

REAGENTS:

1. Ethanolic KOH, 1.0 N.
   1 volume 1.0 N. KOH plus 10 volumes of absolute ethanol. This mixture must be prepared fresh. The stock solution of KOH (56 gm./L.) can be used for several months. Denatured ethanol, formula 3A (95% ethanol, 5% methanol) is satisfactory.

2. Petroleum Ether, A. R. grade
   B. P. 30 - 60°C.

3. Chloroform

4. Activated Glycerol Dichlorohydrin
   "Glycerol dichlorohydrin, Eastman Kodak 657 (for vitamin A determination)." To 4.5 ml. of this reagent add 1.0 drop concentrated HCl. The color produced by reagent without HCl added and stored at room temperature decreases in intensity about 0.3 - 1.8% per month. The stability of this modified reagent is unknown.

PROCEDURE:

The procedure must be carried out in the absence of direct light, preferably in dim light.

1. Place 1.0 ml. serum into a glass stoppered, 15 ml. centrifuge tube and add 1.0 ml. ethanolic KOH solution. Mix and heat at 60°C. for 20 minutes.

2. Add 4.0 ml. petroleum ether, stopper, and shake for 10 minutes. A shaking machine is convenient.

Vitamin A and "Carotene" Determination (Continued):

3. Centrifuge for 30 seconds. Transfer supernate to a test tube (ca. ½" x 5"). Re-extract with 2.0 ml. petroleum ether and combine extracts.

4. Place in 37°C. water bath and evaporate to dryness with a stream of N_2. Rinse down tube walls with 0.5 ml. $CHCl_3$; re-evaporate.

5. Dissolve residue in 0.2 ml. $CHCl_3$, add 0.8 ml. activated glycerol dichlorohydrin, and mix, being careful not to entrap air bubbles in the solution.

6. Immediately transfer to Lowry-Bessey cuvet (1.0 cm. light path, capacity 1.0 ml. or less) and measure absorbance at 550 nm. at exactly 2.0 minutes after reagent addition, and at 960 nm. at 4 minutes against reagent blank composed of 0.2 ml. $CHCl_3$ plus 0.8 ml. activated glycerol dichlorohydrin. If A_{960} is less than 0.8, read absorbance immediately at 830 nm.

CALCULATION:

$$\text{Micrograms Vitamin A alcohol/100 ml.} = \frac{A_{550,\ corr}}{0.088} \times 100$$

$$\text{where } A_{550,\ corr} = A_{550} - \frac{0.012}{0.300} A_{960} \quad \text{or}$$

$$= A_{550} - \frac{0.012}{0.097} A_{830}$$

When using A_{960}:

$$\text{Micrograms "Carotene"/100 ml.} = \frac{A_{960}}{0.300} \times 100$$

When using A_{830}:

$$\text{Micrograms "Carotene"/100 ml.} = \frac{A_{830}}{0.097} \times 100$$

COMMENTS ON PROCEDURE:

1. Standardization

To standardize the determination of vitamin A, a solution of either the alcohol or the acetate equivalent to 5.0 mg./100 ml.

of vitamin A alcohol is prepared in $CHCl_3$. Any of the available commercial preparations of known assay can be used. To 1.0 ml. of this standard, or any dilution thereof in $CHCl_3$, is added 4.0 ml. glycerol dichlorohydrin. The solution is mixed with care to avoid entrapment of air bubbles and read against a similarly prepared reagent blank (1.0 ml. $CHCl_3$ plus 4.0 ml. reagent) at 550 nm. in a cuvet with a 1.0 cm. light path. The standard containing 5.0 mg./100 ml. corresponds to a serum concentration of 1000 micrograms/100 ml. The $CHCl_3$ standards should be protected from light and used within 2 hours of preparation.

In 1949, the Committee on Biological Standards of W. H. O. redefined the International Unit (I.U.) of vitamin A as the biologic activity of 0.344 micrograms vitamin A acetate or 0.400 micrograms of the free alcohol. The U.S.P. reference standard adopted in 1948 conforms to the same equivalents. Both the free alcohol and acetate are available from several sources (e.g., Nutritional Biochemical; Mann; California Corporation; Eastman Kodak). U.S.P. Vitamin A Reference Solution can be obtained from U.S. Pharmacopeial Convention. This is _trans_-vitamin A acetate in cottonseed oil. Each capsule contains approximately 250 mg. of a solution prepared to contain in each gram 34.4 mg. of _trans_-vitamin A acetate, equivalent to 30 mg. of vitamin A alcohol. To use, snip off the end of the capsule, expel, and weigh the solution. Discard any unused portion. The capsules are stored at room temperature (60 - 80°F.) in a dry place, protected from light.

To standardize the test for carotene, prepare a $CHCl_3$ solution of a good grade of beta-carotene containing 1.0 mg./100 ml. To 1.0 ml. add 4.0 ml. of glycerol dichlorohydrin. Mix with care to avoid entrapment of air bubbles and measure absorbance at 4 minutes at 830 and/or 960 nm. in a cuvet with 1.0 cm. light path, against a similarly prepared reagent blank (1.0 ml. $CHCl_3$ plus 4.0 ml. reagent). This standard corresponds to a serum concentration of 200 micrograms/100 ml.

One unit of beta-carotene is defined as 0.6 micrograms. Beta-carotene is stable for years if stored _in vacuo_ or over inert gas.

Standardization should be checked with each lot of glycerol dichlorohydrin purchased.

Vitamin A and "Carotene" Determination (Continued):

2. Color Stability
   The color produced with vitamin A is stable over the period extending from 2 to 10 minutes after addition of reagent. The color formed with carotene must not be allowed to stand much longer than 4 minutes before reading (significant decrease by 7 minutes).

3. Mutual Interference
   100 micrograms vitamin A per 100 ml. serum does not contribute significantly to the readings used in determining carotene. 100 micrograms beta-carotene per 100 ml. serum gives an absorbance at 550 nm. equivalent to ca. 7.4 micrograms vitamin A per 100 ml. In the procedure as performed in our laboratory using the modified reagent the equivalent was found to be about 13.

4. Stability of Serum
   Both carotene and vitamin A are stable in serum at $4^{o}C$. or in the frozen state for at least 2 days. In 1 month at room temperature the carotene deteriorates by 90% and the vitamin A by 55%.

ACCURACY AND PRECISION:

Regarding "Carotene", the result obtained is actually "total carotenoids, in terms of beta-carotene." The accuracy of the method is impossible to assess with the data available since the contribution of lutein and lycopene, the other major pigments, to results has not been worked out. Nevertheless, "carotene" values obtained from the glycerol dichlorohydrin reaction agree quite well with values calculated from the absorbance at 440 nm. of petroleum ether extracts. Values for "carotene" may be up to 11% lower after saponification. The explanation for this is not certain but presumably it could be the result of destruction. It has also been reported that carotene is to some extent destroyed during or immediately after evaporation of a solution of it to dryness. Presumably this results from oxidation, which is minimized in the procedure presented by carrying out this step in a N_2 atmosphere.

The precision of the method is about $\pm$ 15% at normal levels.

NORMAL VALUES:

Carotene:
The normal adult range for "carotene" in serum is about 50 - 300 micrograms/100 ml. Values for women may be slightly higher

Vitamin A and "Carotene" Determination (Continued):

than those for men. The normals for children over 2 years of age are about the same as for adults. Normal values for infants are quite low, e.g., up to about 50 micrograms/100 ml. in infants less than 1 week old.

Vitamin A:
The normal adult range of serum vitamin A is about 20 - 80 micrograms/100 ml. Values for adult men may be slightly higher than those for women. The normals for children over 2 years of age appear to be the same as for adults. The normals for infants appear to be somewhat lower, about 10 - 60 micrograms/100 ml.

REFERENCE: Henry, J. B.: CLINICAL CHEMISTRY: PRINCIPLES AND TECHNIQUES, Hoeber, pg. 704 - 709, 1964.

CLINICAL INTERPRETATION:

Carotenemia, a condition causing yellow pigmentation of plasma and skin, is usually caused by individuals ingesting excessive yellow vegetables containing carotene. It may cause an individual to appear jaundiced. It also occurs in conditions associated with an increase in blood lipids. It has been found in diabetes mellitus, nephrotic syndrome, hypothyroidism, and hepatic disease.

The major cause for a lower than normal blood carotene is malabsorption due to conditions such as sprue. The determination of serum carotene is one of the important determinations in the evaluation of malabsorption which results in a low serum carotene.

SERUM VITAMIN B_{12} ESTIMATION USING LACTOBACILLUS LECHMANNII

PRINCIPLE: Vitamin B_{12} is determined utilizing the following microbiological procedure. The determination of serum vitamin B_{12} is essential when a macrocytic megaloblastic anemia is present.

SPECIMEN:

1. Serum samples arriving at the laboratory should be labelled and then placed in the deep freeze without delay, that is 30 minutes of receipt. They should not be unfrozen until time of assay.

2. If sample consists of clotted blood, the clot should be "rimmed" with a sterile, acid-washed rod. (A previously unused, sterile wooden applicator stick is also satisfactory.)

3. The tube is then centrifuged for 5 minutes at "3/4 speed", and the supernatant serum aspirated (using previously unused, capillary pipette, or an acid-washed standard pipette).

4. This serum is placed in a sterile, acid-washed glass container (or disposable container), is labelled, and promptly placed in deep freeze.
   NOTE: Cap or plug for container must also be sterile, and should either be acid-washed, or washed twelve times in tap water, and rinsed three times in distilled water.

5. Record if serum sample is hemolyzed.

REAGENTS USED IN ASSAY:

1. <u>0.1% NaCN</u>
   In a 100 ml. volumetric flask, dissolve 0.1 gm. NaCN and dilute to volume with distilled water.

2. <u>Acetate Buffer</u>, pH 4.5
   400 ml. distilled water is placed in a 1 liter volumetric flask, to which is added 22.8 ml. glacial acetic acid and 160 ml. 1.0 N. NaOH. The solutions are mixed and then diluted up to volume with distilled water and mixed again.

3. <u>Cyanocobalamine</u> (Vitamin B_{12})
   Fresh Stock Solution is made up every six months and stored in the refrigerator.

Serum B_{12} Assay (Continued):

4. $\underline{B_{12}}$ Medium (Difco #0457-15)
   Make up desired quantity each time using 8.5 gm. in 120 ml. of distilled water.

   NOTE: All equipment (including stirrers) used in the preparation of these reagents must be acid-washed, or very thoroughly washed (12 times in tap water, and then rinsed three times in distilled water).

GENERAL COMMENTS: This is a microbiological assay and is dependent upon meticulous attention to detail for results to be meaningful.

1. Trace contamination with B_{12} will completely nullify the significance of the assay, and thus ALL equipment must be acid-washed or extremely thoroughly cleaned, washed 12 times in tap water and rinsed 3 times with distilled water.

2. Bacterial contamination is a problem that is dealt with by autoclaving. However, delays between the various steps of the assay will reduce the reliability of this technique. If any delay becomes necessary, it should always be timed to occur after an autoclaving and before the solutions are handled again.

3. Handling of the initial serum specimens should be minimal, and once deep frozen they should not be thawed out until time of assay.

4. Certain phases of the assay procedures are dependent upon very accurate pipette measurement. These are marked with an asterisk, "*".

PREPARATION OF BACTERIA FOR B_{12} ASSAYS:

A. Initial Preparation and Subsequent Maintenance of Culture

   1. The initial basic cultures are prepared by rehydrating the lyophilized organisms, *Lactobacillus leichmannii*.

      These organisms are placed in tubes containing the appropriate Difco Culture Broth and are then incubated at 37°C. for 48 hours.

Serum B_{12} Assay (Continued):

These basic cultures are used for the first subcultures, and then are stored in the refrigerator for possible use if difficulties arise with subsequent cultures.

2. Fresh agar stabs must be prepared every 2 weeks for *Lactobacillus leichmannii*.

   Using overnight broth culture, plunge flamed stab wire into culture and then down into 10 ml. of agar in one stab. Set up 2 agar cultures.

   Incubate at 37°C. for at least 48 hours, and then store in refrigerator.

   NOTE: Retain previously used stab for 2 further weeks.

B. Subcultures

1. Flame wire loop and place it down side of agar culture tube to cool it. Then place wire loop into the middle of the bacterial growth.
2. Put wire loop into 10 ml. of inoculum broth in a 40 ml. test tube and agitate. Set up 2 tubes.
3. Plug the broth tube with cottonwool and incubate at 37°C. overnight.

C. Harvesting Bacteria

1. Centrifuge the broth culture for 15 minutes at 2,000 rpm. in the cold.
2. Pour off the broth leaving a pellet of bacteria.
3. Add sterile, distilled water to the 35 ml. mark and mix the bacterial pellet into this using a sterile, cottonwool plugged pipette.
4. Centrifuge for 15 minutes at 2,000 rpm in the cold and decant the water.
5. Add 5.0 ml. of sterile distilled water using a sterile pipette and resuspend the bacteria.

Serum B_{12} Assay (Continued):

6. Take 0.2 ml. of this suspension and add it to 20 ml. of sterile distilled water. Thoroughly mix this dilute suspension by means of a sterile, cottonwool plugged pipette.

7. Add *one drop of this dilute suspension into each assay tube (except the blank for each specimen) using a sterile cottonwool plugged Pasteur pipette.

D. Media and Equipment

1. Every 2 weeks prepare fresh "Difco" broth and agar for *Lactobacillus leichmannii* (light brown). These should be autoclaved in 40 ml. test tubes (150 x 20 mm.) in 10 ml. amounts using cottonwool plugs.

   Special medium for the actual assay of B_{12} is also provided by Difco, but this should be made up freshly for each assay.

   Instructions on how to make up these preparations will be found on the label of the various media bottles. The particular Difco Code Numbers are as follows:

   Lactobacillus leichmannii

Broth	0320-15
Agar	0319-15
Medium	0457-15

2. All equipment used for bacterial preparation must be sterile, and when broth or agar stabs are inoculated, the mouths of the tubes should be flamed.

3. The wire loops and wire stabs used for the culture should be labelled and kept in a glass jar.

4. Always have a spare vial of lyophilized *Lactobacillus leichmannii* stored in the refrigerator. (Can be obtained from: The American Type Culture Company, 12301 Parklawn Drive, Rockville, Maryland 20852)

5. Checking Bacteria
   Every 4 weeks send spare broth culture tube (from sub-culture routine) to a Bacteriology Laboratory for a check on culture purity.

Serum B_{12} Assay (Continued):

AUTOCLAVE ROUTINE:

1. Check "Shut Out" Valve — open (counter-clockwise).

2. Check "By-Pass" Valve — closed (clockwise).

3. Liquid load selection (Drying time = 0).

4. Time selection adjusted.

5. Set temperature timing device.

6. Set pressure in "jacket" of autoclave (on left hand pressure dial) for the level required for particular run.

A. First Autoclave (5 lbs./square inch) - 108°C./20 Minutes

   1. Set indicator at 220° F. (105°C.) - timing for 20 minutes.

   2. Set pressure in jacket at 10 lbs. pressure (use Pressure-reducing Valve).

   3. Close door - bring pressure inside chamber to 4 lbs./square inch in 1 minute - then adjust to 5 lbs./square inch in both chambers (using Pressure-reducing Valve).

   4. When 20 minutes completed and cooling starts, slowly reduce pressure in chamber by very gradually turning "By Pass" Valve counterclockwise ¼ to ½ turn.

Pressure down	1 minute 5 lbs. - 4 lbs.
in 5 minutes	1 minute 4 lbs. - 3 lbs.
by gradual	1 minute 3 lbs. - 2½ lbs.
reduction	1 minute 2½ lbs.- 2 lbs.
	1 minute 2 lbs. - 1½ lbs.

Machine now turns off.

B. Second Autoclave (10 lbs./square inch) - 115°C./6 Minutes

   1. Set indicator at 230°F. - (110°C.) - timing for 6 minutes.

   2. Set pressure in jacket at 15 lbs. pressure (use Pressure-reducing Valve).

Serum B_{12} Assay (Continued):

3. Close door - bring pressure inside chamber to 9 lbs./square inch in 1 minute - then adjust to 10 lbs./square inch in both chambers (using Pressure-reducing Valve).

4. When 6 minutes are completed and cooling starts, slowly reduce pressure in chamber by very gradually turning "By Pass" Valve <u>counterclockwise ¼ to ½ turn</u>.

Pressure down	1 minute	10 lbs.	-	7 lbs.
in 5 minutes	1 minute	7 lbs.	-	5 lbs.
by gradual	1 minute	5 lbs.	-	3½ lbs.
reduction	1 minute	3½ lbs.	-	2½ lbs.
	1 minute	2½ lbs.	-	1½ lbs.

Machine now turns off.

PROCEDURE:

1. Prepare growth of <u>Lactobacillus</u> <u>leichmannii</u> in broth the night before using (See "Bacterial Preparation").

2. Using 40 ml. graduated centrifuge tubes and take,

	2.0 ml.* serum (bulb pipette)
Total = 20 ml.	0.4 ml. of 0.1% NaCN (1.0 ml. pipette)
	1.0 ml. Na Acetate buffer (10 ml. pipette)
	16.6 ml. Distilled water

Plug tubes with cottonwool.

3. Autoclave at 5 lbs. pressure for 20 minutes at 220°F. (See Autoclave Protocol).

4. Cool tubes.

5. While tubes are cooling, prepare B_{12} Standard Curve. B_{12} Standard Stock Solution consists of:

a). 1 ampule Vitamin B_{12} (cyanocobalamine) 1,000 ugm./ml.
b). Dilute this in 200 ml. distilled water (5 ugm./ml.)
c). Take 1.0 ml. of this and add to 100 ml. water in volumetric flask. This gives a stock solution of 50 mugm./ml.

It should be stored in the refrigerator (4°C.) for weekly use.

Serum B_{12} Assay (Continued):

B_{12} Standard Curve

Solution "A"	0.5 ml. Stock Solution + 9.5 ml. distilled water (2.5 mugm./ml.)
Solution "B"	0.5 ml. of Solution "A" + 9.5 ml. distilled water (0.125 mugm./ml.)
Solution "C"	2.0 ml. of Solution "A" + 8.0 ml. distilled water (0.5 mugm./ml.)

6. Make 1:25 dilution of 0.1% NaCN by putting 2.0 ml. of 0.1% NaCN into 48 ml. of distilled water.

7.

Row No.	1	2	3	4	5	6	7
1:25 Dil. NaCN	1.0	1.0	1.0	1.0	1.0	1.0	1.0
*Solution "B"	0	0.1	0.2	0.4	0.8	0	0
*Solution "C"	0	0	0	0	0	0.4	0.8
Distilled water	1.0	0.9	0.8	0.6	0.2	0.6	0.2
Final Amount							
B_{12}							
mugm.	0	.0125	.025	.05	0.1	0.2	0.4
(or, in uugm.	0	12.5	25	50	100	200	400

This gives a total volume of 2.0 ml. in each tube. The tubes must be set up in quadruplicate. Cover temporarily with parafilm

8. Now take autoclaved (and cooled) tubes from Step No. 4 and centrifuge at 2,000 rpm for 10 minutes.
   NOTE: Cover with parafilm having removed cotton plugs first. Pour off the supernatant fluid into clean tubes.

9. Using supernatant fluid, prepare unknowns for assay:

   1st Row: 1.0 ml.* supernatant + 1.0 ml. of 1:50 dil. 0.1% NaCN. (1.0 ml. of 0.1% NaCN + 49 ml. water)

   2nd Row: 2.0 ml.* supernatant

   Both rows should be done in quadruplicate.

10. Add 3.0 ml. of Difco B_{12} Medium to each assay tube - both Standard and unknowns. Cap and autoclave at 10 lbs. for 6 minutes. (Using caps that have been rendered B_{12}-free by thorough washing; twelve times in tap water and three times in distilled water.)

Serum B_{12} Assay (Continued):

11. After tubes have cooled, place one* drop of Lactobacillus leichmannii inoculum in 3 of the 4 tubes of each serum dilution, the fourth tube being set aside as a blank. (This must be done with a sterile Pasteur Pipette, taking care to avoid contamination and only inoculating with one drop).

12. Cap all tubes and incubate for 18 hours at 37°C. (including the blanks).

13. If growth is excessive, it may be necessary to add 2.0 ml. of distilled water to each tube with an automatic syringe before reading.

14. Shake all tubes well before transferring fluid to cuvettes. Read at 600 nm. in a spectrophotometer.

    Read each dilution against its own blank which must be used to zero the instrument.

15. Plot Standard Curve on arithmetic graph paper, using spectrophotometric growth densities against the known amount of B_{12} per tube. (See Example of Standard Curve).

16. CALCULATIONS:

    Calculate the amount of B_{12} in the unknown serum samples by reading directly from the Standard Curve (utilizing the mean value of the 3 Spectrophotometric readings on the 3 tubes of each dilution). Each serum sample will have 3 readings at dilutions of 1 in 10 (Row No. 1) and dilutions of 1 in 5 (Row No. 2).

EXAMPLE

	Mean of 3 Photometer Readings	B_{12} (uugm./ml.)	Dilution	Actual B_{12} Value (uugm./ml.)	Final Average
Row "1"	46	24.5	1:10	245	250 uugm. per ml.
Row "2"	88	51.1	1:5	255.5	

NORMAL VALUES: 150 - 1,500 uugm./ml. (i.e. pg./ml.)

REFERENCES:

1. Spray, G. H.: "An Improved Method for the Rapid Estimation of Vitamin B_{12} In Serum", Clin. Sci., 14:661, 1955.

Serum B_{12} Assay (Continued):

2. Spray, G. H.: "The Estimation and Significance of the Level of Vitamin B_{12} in Serum", J. Postgrad. Med., 38:35, 1962.

3. Spray, G. H. and Witts, L. J.: "Results of Three Years Experience with Microbiological Assay of Vitamin B_{12} in Serum", Brit. Med. Journal, 1:295, 1958.

CLINICAL INTERPRETATION:

Vitamin B_{12} is extrinsic factor. It is only absorbed in the presence of a specific protein substance, intrinsic factor. The site of absorption of vitamin B_{12} is the small intestine, the terminal ileum. The manner of absorption of vitamin B_{12} is uncertain, but is seems probable that the vitamin is bound by intrinsic factor and that the intrinsic factor is bound by receptors in the intestine in the presence of calcium.

Pernicious anemia patients in relapse have low serum levels, below 100 micromicrograms/ml. Patients who have megaloblastic anemia secondary to pregnancy, partial or total gastrectomy, blind loops following surgery, malabsorption syndrome due to various causes, and utilization of anticonvulsants or chemotherapy, also have low serum vitamin B_{12} levels.

Serum vitamin B_{12} is elevated in acute and chronic myelogenous leukemi di Guglielmo syndrome, and liver disease.

Assays of vitamin B_{12} and folic acid blood levels in various disorders are characterized by megaloblastic anemia show that in pernicious anemia in relapse, the serum vitamin B_{12} is low and the serum folic acid is variable. When vitamin B_{12} is injected, a rapid fall of folic acid occurs within two hours. When folic acid is given to pernicious anemia patients in relapse, a decrease in serum vitamin B_{12} occurs.

In nutritional megaloblastic anemia, folic acid serum levels are low. Folic acid levels are low in renal patients on chronic hemodialysis, patients taking anti-folic acid antagonistic chemotherapeutic drugs for cancer, patients on anti-convulsants and oral contraceptives, and hemolytic anemia patients.

The metabolic roles of vitamin B_{12} and folic acid are closely related. As a deficiency of vitamin B_{12} becomes pronounced, the requirement for folic acid may increase.

D-XYLOSE ABSORPTION TEST

PRINCIPLE: Xylose absorption, as judged by blood levels or 5 hour excretion is a test primarily for upper small bowel absorption. This pentose is not digested in the bowels. About 60% of absorbed xylose is metabolized via fructose-6-phsophate and through the Kreb's Cycle; the remainder is excreted within 5 hours in the urine.

The determination of pentose is based upon the formation of furfural and its reaction with aniline. Specificity for pentoses is achieved by using conditions sufficiently mild so that other potential furfural precursors (glucose, glucuronic acid, ascorbic acid) do not react.

SPECIMEN: A urine specimen is collected for 5 hours after an oral dose of D-xylose has been administered. The usual dosage is 5.0 gm. The urine should be preserved in the freezer. Benzoic acid may be added to urine collection to make a saturated solution (3.4 gm./L.) should a freezer not be available.

REAGENTS:

1. p-Bromoaniline Reagent, 2% (Eastman Kodak)
   Thiourea is added to glacial acetic acid in excess of the amount that will dissolve. Approximately 4.0 gm. of thiourea per 100 ml. of acetic acid are used. Decant 100 ml. of acetic acid saturated with thiourea and dissolve 2.0 gm. of pure p-bromoaniline in it. Keep the p-bromoaniline reagent in a dark glass bottle. The reagent is stable for seven days.

2. Stock Standard, 500 mg%.
   A Stock Solution of D-xylose is made up by dissolving 500 mg. or reagent grade D-xylose in 100 ml. of saturated benzoic acid solution.

3. Working Standard I, 10 mg%.
   Prepare a 1:50 dilution from the Stock Standard using deionized water. Prepare fresh daily.

4. Working Standard II, 20 mg%.
   Prepare a 1:25 dilution from the Stock Standard using deionized water. Prepare fresh daily.

COMMENTS ON PROCEDURE:

1. The color formed is proportional to the pentose concentration and reaches a maximum in about 60 minutes, and remains constant

D-Xylose Absorption Test (Continued):

for a further 30 minutes at a temperature of 20 - 25°C. and then slowly fades.

2. Since the intensity and stability of this colored product is affected by light and temperature, a Standard Solution of pentose should be set up with each run.

3. When an unknown reads higher than the optical density of the highest standard, dilute with the blank after setting the spectrophotometer with the blank. Multiply the answer with appropriate dilution factor to obtain milligrams per ml. of D-xylose before calculating Gms. excreted in 5 hours.

4. With normal renal function, normal values are 16 - 33% of oral dose.

PROCEDURE:

1. Dilute urine as follows with deionized water:

Less than 200 ml. total volume	- 1:200 dilution
200 to 500 ml. total volume	- 1:50 dilution
Over 500 ml. total volume	- 1:25 dilution

2. Set up 2 16 x 100 mm. pyrex tubes for each standard and specimen. Label one tube "Test" and the other "Blank". Pipette 1.0 ml. of the dilution and standards into appropriate tubes. Add 5.0 ml. of p-Bromoaniline Reagent to all tubes.

3. Incubate the standard and specimen "Test" tubes in a 70°C. water bath for 10 minutes. Cool to room temperature as soon as possible. The "Blank" standard and specimen tubes are <u>not</u> heated and serve as individual blanks.

4. Set tubes in the dark for 70 minutes.

5. Using individual "Blanks" to zero spectrophotometer, read corresponding "Tests" at 520 nm.

6. CALCULATION:
Use the O.D. of the Standard closest to the O.D. of the unknown.

D-Xylose Absorption Test (Continued):

Standard I $\frac{\text{O.D. Unknown}}{\text{O.D. Standard}}$ x 0.1 x dilution = mgs. Xylose/ ml.

Standard II $\frac{\text{O.D. Unknown}}{\text{O.D. Standard}}$ x 0.2 x dilution = mgs. Xylose/ml.

Mgs./ml. x $\frac{\text{5 hour volume}}{1000}$ = Gms. D-Xylose excreted in 5 hours.

Then calculate percentage of dose excreted in 5 hours.

NORMAL VALUES: 16 - 33% of Oral Dose

REFERENCE: Roe, J. H. and Rice, E. W.: "Photometric Method for the Determination of Free Pentoses in Animal Tissues", J. Biol. Chem., 173:507, 1948.

CLINICAL INTERPRETATION:

A clinically useful procedure for determination of carbohydrate malabsorption is the D-Xylose Test. The urinary excretion of D-xylose after oral administration of a 5 gm. test dose is determined. In normal individuals, the serum xylose concentration rises to an average of 10 mg%. in two hours falling to resting levels in 5 hours. Normal individuals excrete 25% of the test dose in their urine in 5 hours. Subnormal urinary excretion suggests the presence of the malabsorption syndrome due to such causes as sprue, amyloidosis, lymphoma, or Whipple's disease. Kidney function must be normal, otherwise D-xylose excretion into the urine will be abnormal. D-xylose absorption and excretion is one of the most useful screening procedures for demonstration of intestinal malabsorption.